Lecture Notes in Physics

Volume 918

The Lecture Notes in Physics

The series Lecture Notes in Physics (LNP), founded in 1969, reports new developments in physics research and teaching-quickly and informally, but with a high quality and the explicit aim to summarize and communicate current knowledge in an accessible way. Books published in this series are conceived as bridging material between advanced graduate textbooks and the forefront of research and to serve three purposes:

- to be a compact and modern up-to-date source of reference on a well-defined topic
- to serve as an accessible introduction to the field to postgraduate students and nonspecialist researchers from related areas
- to be a source of advanced teaching material for specialized seminars, courses and schools

Both monographs and multi-author volumes will be considered for publication. Edited volumes should, however, consist of a very limited number of contributions only. Proceedings will not be considered for LNP.

Volumes published in LNP are disseminated both in print and in electronic formats, the electronic archive being available at springerlink.com. The series content is indexed, abstracted and referenced by many abstracting and information services, bibliographic networks, subscription agencies, library networks, and consortia.

Proposals should be sent to a member of the Editorial Board, or directly to the managing editor at Springer:

Christian Caron
Springer Heidelberg
Physics Editorial Department I
Tiergartenstrasse 17
69121 Heidelberg/Germany
christian.caron@springer.com

More information about this series at http://www.springer.com/series/5304

Alexander Dosch • Gary P. Zank

Transport Processes in Space Physics and Astrophysics

Problems and Solutions

Springer

Alexander Dosch
CSPAR
University of Alabama in Huntsville
Huntsville
Alabama, USA

Gary P. Zank
CSPAR
University of Alabama in Huntsville
Huntsville
Alabama, USA

ISSN 0075-8450 ISSN 1616-6361 (electronic)
Lecture Notes in Physics
ISBN 978-3-319-24878-3 ISBN 978-3-319-24880-6 (eBook)
DOI 10.1007/978-3-319-24880-6

Library of Congress Control Number: 2015955637

Springer Cham Heidelberg New York Dordrecht London

Printed on acid-free paper

Springer International Publishing AG Switzerland is part of Springer Science+Business Media (www.springer.com)

Preface

This book was written during my stay at the Center for Space Plasma and Aeronomic Research (CSPAR) at the University of Alabama in Huntsville, between the years 2010 and 2014. During that time, Prof. Dr. Gary Zank was giving the lecture on *Transport Processes in Astrophysics*, which included many problems from the fields of statistics, transport theory (particle transport and turbulence transport), diffusion theory, and many more. I had the great pleasure to teach some of the classes and to help him grade the classwork. Since many of the problems were quite complex and intricate at times, I found it conducive to collect and solve those problems in a comprehensive way and to organize them in this book.

Therefore, this book is what it is: a solution manual providing detailed and extensive descriptions of the solutions to nearly all problems given in the lecture notes *Transport Processes in Space Physics and Astrophysics (Lecture Notes in Physics)* by Gary P. Zank, see [5], and on multiple occasions in this book we refer to the lecture notes for further reading.

At the beginning of each section, we give a brief introduction and repetition of all information necessary to understand the problems and solutions. However, it has to be clear that this book cannot substitute the lecture notes and, therefore, should be understood as a supplement to it. The reader is also referred to any textbooks in physics and math related to this topic.

Note that some problems and equations have been altered from the lecture notes to be more consistent with the terminology in this book or to clarify the concept of the problem itself. For example, Problem 3.3 has been extended by some subquestions to provide a better idea of the steps necessary to solve the problem. Some equations have been corrected, for example Eq. (2.112), compare with Eq. (3.33) in the lecture notes, and for the sake of brevity (and to avoid too many repetitions) some problems have been omitted altogether. For example, the problems in Sect. 5.2 from the lecture notes (Transport Equation for Relativistic Charged Particles) have not been included in this book, since they are similar to the preceding Sect. 5.1. (5.1.1 The Focussed Transport equation).

I want to thank Prof. Dr. Gary Zank for this wonderful opportunity to write this book and the freedom I had in writing it. The many conversations and discussions

I had with him about this solution book and his lecture notes gave me a deep insight into the field of *Transport Processes in Astrophysics* and certainly a better understanding for many of the problems in this book. I want to thank him especially for reading the manuscript and his numerous explanations, which made the solutions much more readable and understandable. However, despite careful readings of the manuscript, I cannot rule out any errors, typos, or mislabeling. Needless to say that all remaining errors are my own.

I also want to thank my colleagues at CSPAR, for their valuable discussions and comments regarding some specific problems.

Lastly, I want to thank all the students who so tenaciously and thoroughly worked through all the problems. Their comments and discussions helped tremendously to improve the readability of this solution book. With their valuable comments and discussions, I was able to go into more detail where it was necessary and leave things out that were only of minor importance.

I hope that this book might be helpful not only to students but also to researchers and anyone who is interested in the exciting field of astrophysics.

Kaiserslautern, Germany
July 2015

Alexander Dosch

Contents

Chapter 1
Statistical Background

1.1 Probability Set Function

Suppose we perform n independent experiments under identical conditions. If an outcome A results n_A times, then the probability that A occurs is

$$P(A) = \lim_{n\to\infty} \frac{n_A}{n}. \tag{1.1}$$

Let $\mathcal{C}$ be the set of all possible outcomes of a random experiment, then $\mathcal{C}$ is called the *sample space*. An outcome is a point or element in the sample space. The sample space can be finite or infinite.

Definition 1.1 If $P(C)$ is defined for a subset C of the sample space $\mathcal{C}$, and $C_1, C_2, C_3, \ldots$ are disjoint subsets of the sample space $\mathcal{C}$, then $P(C)$ is called the *probability set function* of the outcome of the random experiment if

- $P(C) \geq 0$
- $P(C_1 \cup C_2 \cup C_3 \ldots) = P(C_1) + P(C_2) + P(C_3) + \ldots$
- $P(\mathcal{C}) = 1$.

Theorem 1.1 *For each $C \subset \mathcal{C}$, $P(C) = 1 - P(C^*)$, where C^* denotes the complement of C.*

Theorem 1.2 $P(\emptyset) = 0$.

Theorem 1.3 *If C_1 and C_2 are subsets of $\mathcal{C}$ such that $C_1 \subset C_2$, then $P(C_1) \leq P(C_2)$.*

(continued)

A. Dosch, G.P. Zank, *Transport Processes in Space Physics and Astrophysics*, Lecture Notes in Physics 918, DOI 10.1007/978-3-319-24880-6_1

Theorem 1.4 *For each* $C \subset \mathcal{C}$, $0 \leq P(C) \leq 1$.

Theorem 1.5 *If* $C_1 \subset \mathcal{C}$ *and* $C_2 \subset \mathcal{C}$, *then*

$$P(C_1 \cup C_2) = P(C_1) + P(C_2) - P(C_1 \cap C_2). \tag{1.2}$$

Problem 1.1 A positive integer from 1 to 6 is randomly chosen by casting a die. Thus the sample space is $\mathcal{C} = \{c : c = 1, 2, 3, 4, 5, 6\}$. Let $C_1 = \{c : c = 1, 2, 3, 4\}$ and $C_2 = \{c : c = 3, 4, 5, 6\}$ be subsets of the sample space $\mathcal{C}$. If the probability set function P assigns a probability of $1/6$ to each $c \in \mathcal{C}$, compute $P(C_1)$, $P(C_2)$, $P(C_1 \cap C_2)$, and $P(C_1 \cup C_2)$!

Solution The probability of each outcome $c \in \mathcal{C}$ is given by $P(c) = 1/6$. The probability to chose a number from the subsets C_1 and C_2 is then given by

$$P(C_1) = P(1) + P(2) + P(3) + P(4) = \frac{4}{6} = \frac{2}{3}$$

$$P(C_2) = P(3) + P(4) + P(5) + P(6) = \frac{4}{6} = \frac{2}{3}.$$

The probability for the intersection and union of the subsets is given by

$$P(C_1 \cap C_2) = P(3) + P(4) = \frac{2}{6} = \frac{1}{3}$$

$$P(C_1 \cup C_2) = P(C_1) + P(C_2) - P(C_1 \cap C_2) = \frac{2}{3} + \frac{2}{3} - \frac{1}{3} = 1.$$

Problem 1.2 Draw a number without replacement from the set $\{1, 2, 3, 4, 5\}$, i.e., choose a number, and then a second from the remaining numbers. The sample space is then given by

$$\begin{aligned} \mathcal{C} = \{c : c = \ & (1,2)(1,3)(1,4)(1,5)(2,1)(2,3)(2,4)(2,5)(3,1)(3,2) \\ & (3,4)(3,5)(4,1)(4,2)(4,3)(4,5)(5,1)(5,2)(5,3)(5,4)\}. \end{aligned}$$

Assume, that all 20 possible results have the same probability $P(c) = 1/20$. Find the probability that an odd digit will be selected (A) the first time (B) the second time and (C) both times.

Solution

A. The probability of finding an odd digit the first time is given by

$$P(\text{odd first}) = \frac{12}{20} = \frac{3}{5}.$$

B. The probability of finding an odd digit the second time is given by

$$P(\text{odd second}) = P(\text{odd \& odd}) + P(\text{even \& odd})$$
$$= \frac{3}{5} \cdot \frac{2}{4} + \frac{2}{5} \cdot \frac{3}{4} = \frac{3}{5}.$$

C. The probability of finding an odd digit both times (first and second) is given by

$$P(\text{odd \& odd}) = \frac{3}{5} \cdot \frac{2}{4} = \frac{3}{10}.$$

Problem 1.3 Draw one card from an ordinary deck of 52 cards and suppose that the probability set function assigns a probability of $1/52$ to each of the possible outcomes c. Let C_1 denote the collection of 13 hearts and C_2 the collection of 4 kings. Compute $P(C_1)$, $P(C_2)$, $P(C_1 \cap C_2)$, and $P(C_1 \cup C_2)$!

Solution The collection of 13 hearts is described by the subset

$$C_1 = \{c : c = \heartsuit 2, \heartsuit 3, \heartsuit 4, \ldots, \heartsuit \text{Ace}\}.$$

Since there are only 13 cards of hearts in the deck, the probability of drawing one card of hearts is given by

$$P(C_1) = \frac{13}{52} = \frac{1}{4}.$$

Similarly, the collection of 4 kings is described by the subset

$$C_2 = \{c : c = \diamondsuit \text{King}, \heartsuit \text{King}, \spadesuit \text{King}, \clubsuit \text{King}\}$$

and the probability that one card drawn from the deck is a king is given by

$$P(C_2) = \frac{4}{52} = \frac{1}{13}.$$

Since there is only one king of hearts in the deck we find for the intersection

$$P(C_1 \cap C_2) = \frac{1}{52}.$$

The union of both subsets (i.e., collecting 13 cards of hearts and the remaining 3 kings) is then given by

$$P(C_1 \cup C_2) = P(C_1) + P(C_2) - P(C_1 \cap C_2) = \frac{13}{52} + \frac{4}{52} - \frac{1}{52} = \frac{4}{13}.$$

Problem 1.4 A coin is tossed as many times as necessary to give one head. The sample space is therefore $\mathcal{C} = \{c : c = H, TH, TTH, \ldots\}$. The probability set function assigns probabilities $1/2, 1/4, 1/8, \ldots$, respectively. Show that $P(\mathcal{C}) = 1$. Suppose the subsets $C_1 = \{c : c = H, TH, TTH, TTTH, TTTTH\}$ and $C_2 = \{c : c = TTTTH, TTTTTH\}$. Compute $P(C_1)$, $P(C_2)$, $P(C_1 \cap C_2)$, and $P(C_1 \cup C_2)$!

Solution The probability of the sample space can be calculated by using the convergence of the geometric series $\sum_{x=0}^{\infty} q^x = 1/(1-q)$ for $q < 1$. For $q = 1/2$ we find

$$\begin{aligned} P(\mathcal{C}) &= \frac{1}{2} + \frac{1}{4} + \frac{1}{8} + \frac{1}{16} + \ldots \\ &= \sum_{n=1}^{\infty} \left(\frac{1}{2}\right)^n = \sum_{n=0}^{\infty} \left(\frac{1}{2}\right)^n - 1 = \frac{1}{1-1/2} - 1 = 1. \end{aligned}$$

The probabilities of the subsets are

$$\begin{aligned} P(C_1) &= \frac{1}{2} + \frac{1}{4} + \frac{1}{8} + \frac{1}{16} + \frac{1}{32} = \frac{31}{32} \\ P(C_2) &= \frac{1}{32} + \frac{1}{64} = \frac{3}{64}. \end{aligned}$$

The probability of the intersection $P(C_1 \cap C_2)$ is the probability of the outcome $c = TTTTH$, so that

$$\begin{aligned} P(C_1 \cap C_2) &= \frac{1}{32} \\ P(C_1 \cup C_2) &= \frac{31}{32} + \frac{3}{64} - \frac{1}{32} = \frac{63}{64}. \end{aligned}$$

Problem 1.5 A coin is tossed until for the first time the same result appears twice in succession. Let the probability for each outcome requiring $n \geq 2$ tosses be $1/2^{n-1}$. Describe the sample space, and find the probability of the events (A) the tosses end before the sixth toss, (B) an even number of tosses is required.

Solution The sample space is given by

$$\mathcal{C} = \left\{ c : c = \begin{cases} (TT),\ (THH),\ (THTT),\ (THTHH),\ (THTHTT),\ \ldots \\ (HH),\ (HTT),\ (HTHH),\ (HTHTT),\ (HTHTHH),\ \ldots \end{cases} \right\}$$

with the probability $P(n) = 1/2^{n-1}$ for $n \geq 2$. As an example, for $n = 2$ the possible outcomes are HH, HT, TH, TT with a probability of $1/4$ for each outcome. The event that the result appears twice in succession (that is either TT or HH) has the probability $P(TT \text{ or } HH) = P(TT) + P(HH) = 1/2$. Similarly, for $n = 3$ there are 8 possible outcomes $(HHH, HHT, HTH, HTT, \ldots)$ with a probability of $1/8$ for each outcome. Obviously, the outcomes TTT and HHH can never occur, since the event of two identical results in succession already would have been fulfilled after two tosses. However, since we are interested in the probability of an outcome it is more important that an outcome *can* occur rather than it *will*. Therefore, the probability for $n = 3$ is $P(THH \text{ or } HTT) = P(THH) + P(HTT) = 1/4$.

A. The event that the tosses end before the sixth toss ($n \leq 5$) can be described by the subset

$$A_1 = \left\{ c : c = \begin{cases} (TT), \ (THH), (THTT), \ (THTHH), (THTHTT) \\ (HH), (HTT), \ (HTHH), (HTHTT), \ (HTHTHH) \end{cases} \right\}.$$

The probability for this event is then given by

$$P(A_1) = \sum_{n=2}^{5} \frac{1}{2^{n-1}} = \frac{1}{2} + \frac{1}{2^2} + \frac{1}{2^3} + \frac{1}{2^4} = \frac{15}{16}.$$

B. For an even number of tosses ($n = 2, 4, 6, 8, \ldots$) we introduce for simplicity $n = 2i$ with $i = 1, 2, 3, 4, \ldots$, so that the probability is given by

$$P(n = \text{even}) = \sum_{i=1}^{\infty} \frac{1}{2^{2i-1}} = 2\sum_{i=1}^{\infty} \frac{1}{2^{2i}} = 2\sum_{i=1}^{\infty} \left(\frac{1}{4}\right)^i = \frac{2}{1-\frac{1}{4}} - 2 = \frac{2}{3},$$

where we used the convergence of the geometric series (see also the previous problem).

Problem 1.6 Find $P(C_1 \cap C_2)$ if the sample space is $\mathcal{C} = C_1 \cup C_2$, $P(C_1) = 0.8$ and $P(C_2) = 0.5$!

Solution Since the sample space is given by $\mathcal{C} = C_1 \cup C_2$ we find immediately $P(\mathcal{C}) = P(C_1 \cup C_2) = 1$. From $P(C_1 \cup C_2) = P(C_1) + P(C_2) - P(C_1 \cap C_2)$ it follows that

$$P(C_1 \cap C_2) = P(C_1) + P(C_2) - P(C_1 \cup C_2) = 0.8 + 0.5 - 1 = 0.3.$$

Problem 1.7 Suppose $C \subset \mathcal{C} = \{c : 0 < c < \infty\}$ with $C = \{c : 4 < c < \infty\}$ and $P(C) = \int_C e^{-x} dx$. Determine $P(C)$, $P(C^*)$, and $P(C \cup C^*)$, where C^* denotes the complement of C.

Solution The probability of subset C is given by

$$P(C) = \int_4^{\infty} e^{-x} dx = -e^{-x}|_4^{\infty} = e^{-4}$$

and the probability of the complementary set $C^* = \{c : 0 < c \leq 4\}$ is calculated by

$$P(C^*) = P(\mathcal{C}) - P(C) = 1 - e^{-4}.$$

Since the subsets C and C^* are disjoint and the sample space is defined by $\mathcal{C} = C + C^*$, we find easily

$$P(\mathcal{C}) = P(C \cup C^*) = P(C) + P(C^*) = \int_0^\infty e^{-x} dx = -e^{-x}\big|_0^\infty = 1.$$

Problem 1.8 If $C \subset \mathcal{C}$ is a subset for which $\int_C e^{-|x|} dx$ exists, where the sample space is given by $\mathcal{C} = \{c : -\infty < x < \infty\}$, then show that this set function is *not* a probability set function. What constant should the integral be multiplied by to make it a probability set function?

Solution The probability of the sample space is per definition $P(\mathcal{C}) = 1$, thus,

$$1 \stackrel{!}{=} P(\mathcal{C}) = \int_{-\infty}^{\infty} e^{-|x|} dx = \int_{-\infty}^{0} e^{x} dx + \int_0^\infty e^{-x} dx.$$

Substituting now $y = -x$ in the first integral and using the results from the previous Problem 1.7 we find

$$P(\mathcal{C}) = \int_0^\infty e^{-y} dy + \int_0^\infty e^{-x} dx = 2 \int_0^\infty e^{-x} dx = 2 \neq 1.$$

Hence the set function is *not* a probability set function. The set function has to be multiplied by $1/2$ to make it a probability set function.

Problem 1.9 If C_1 and C_2 are two arbitrary subsets of the sample space $\mathcal{C}$, show that

$$P(C_1 \cap C_2) \leq P(C_1) \leq P(C_1 \cup C_2) \leq P(C_1) + P(C_2).$$

Solution

A. First, we show that $P(C_1 \cup C_2) \leq P(C_1) + P(C_2)$. From Theorem 1.5 we find

$$P(C_1 \cup C_2) = P(C_1) + P(C_2) - P(C_1 \cap C_2).$$

Since $P(C) \geq 0$, i.e., any probability has to be larger or equal to zero, it follows immediately that

$$P(C_1 \cup C_2) \leq P(C_1) + P(C_2),$$

where we used $P(C_1 \cap C_2) \geq 0$.

B. In the next step we show that $P(C_1) \leq P(C_1 \cup C_2)$. Here we use the fact that the union of both subsets can be described by $C_1 \cup C_2 = C_1 \cup (C_1^* \cap C_2)$, where C_1 and $(C_1^* \cap C_2)$ are disjoint. Hence we have

$$P(C_1 \cup C_2) = P(C_1) + P(C_1^* \cap C_2)$$

and it follows that

$$P(C_1 \cup C_2) \geq P(C_1),$$

since $P(C_1^* \cap C_2) \geq 0$.

C. In the last step we show $P(C_1 \cap C_2) \leq P(C_1)$. Here we use the relation $C_1 = (C_1 \cap C_2) \cup (C_1 \cap C_2^*)$, where $(C_1 \cap C_2)$ and $(C_1 \cap C_2^*)$ are disjoint subsets. We find

$$P(C_1) = P(C_1 \cap C_2) + P(C_1 \cap C_2^*)$$

and it follows immediately that

$$P(C_1) \geq P(C_1 \cap C_2),$$

since $P(C_1 \cap C_2^*) \geq 0$.

1.2 Random or Stochastic Variables

Definition 1.2 Consider a random experiment with sample space $\mathcal{C}$. A function X, that assigns to each outcome $c \in \mathcal{C}$ one and only one real number $x = X(c)$, is a *random variable*, and the space of X is the set of real numbers $\mathcal{A} = \{x : x = X(c), c \in \mathcal{C}\}$.

Random variables can be *discrete* or *continuous*. *Discrete random variables* are those that take on a finite or denumerably infinite number of distinct values. *Continuous random variables* are those that take on a continuum of values within a given range.

Problem 1.10 Select a card from an ordinary deck of 52 playing cards with outcome c. Let $X(c) = 4$ if c is an ace, $X(c) = 3$ for a king, $X(c) = 2$ for a queen, $X(c) = 1$ for a jack, and $X(c) = 0$ otherwise. Suppose $P(C)$ assigns a probability $1/52$ to each outcome c. Describe the probability $P(A)$ on the space $\mathcal{A} = \{x : x = 0, 1, 2, 3, 4\}$ of the random variable X.

Solution Since there are only 4 aces, 4 kings, 4 queens and 4 jacks in the game, we find

$$P(X = 4) = P(X = 3) = P(X = 2) = P(X = 1) = \frac{4}{52} = \frac{1}{13}$$

and for the remaining cards

$$P(0) = \frac{36}{52} = \frac{9}{13}.$$

Problem 1.11 Suppose the probability set function $P(A)$ of the random variable X is $P(A) = \int_A f(x)dx$, where $f(x) = 2x/9$, $x \in \mathcal{A} = \{x : 0 < x < 3\}$. For $A_1 = \{x : 0 < x < 1\}$ and $A_2 = \{x : 2 < x < 3\}$, compute $P(A_1)$, $P(A_2)$, and $P(A_1 \cup A_2)$.

Solution The probabilities for the subsets A_1 and A_2 are given by

$$P(A_1) = \int_0^1 \frac{2x}{9}dx = \left.\frac{x^2}{9}\right|_0^1 = \frac{1}{9}$$

$$P(A_2) = \int_2^3 \frac{2x}{9}dx = \left.\frac{x^2}{9}\right|_2^3 = 1 - \frac{4}{9} = \frac{5}{9}.$$

Since A_1 and A_2 are disjoint we find

$$P(A_1 \cup A_2) = P(A_1) + P(A_2) = \frac{1}{9} + \frac{5}{9} = \frac{2}{3}.$$

Problem 1.12 Suppose that the sample space of a random variable X is given by $\mathcal{A} = \{x : 0 < x < 1\}$. If the subsets $A_1 = \{x : 0 < x < 1/2\}$ and $A_2 = \{x : 1/2 \le x < 1\}$, find $P(A_2)$, if $P(A_1) = 1/4$.

Solution Since the sample space is identical to the union of both subsets, i.e., $A_1 \cup A_2 = \mathcal{A}$, we find with $P(\mathcal{A}) = 1$,

$$P(A_2) = 1 - P(A_1) = 1 - \frac{1}{4} = \frac{3}{4}.$$

1.3 The Probability Density Function

Whenever a probability set function $P(A)$, with $A \subset \mathcal{A}$ and sample space $\mathcal{A}$, can be expressed as

$$P(A) = P(X \in A) = \sum_{A} f(x) \tag{1.3}$$

$$P(A) = P(X \in A) = \int_{A} f(x)\,dx, \tag{1.4}$$

then X is a random variable of *discrete* or *continuous* type, and X has a *discrete* or *continuous* distribution.

The probability $P(A)$ is determined completely by the *probability density function (pdf)*, $f(x)$, whether or not X is a discrete or continuous random variable.

Problem 1.13 Find the constant a that ensures that $f(x)$ is a pdf of the random variable X for

A.

$$f(x) = \begin{cases} a\left(\frac{2}{3}\right)^x & \text{for } x = 1, 2, 3, \ldots \\ 0 & \text{elsewhere} \end{cases}$$

B.

$$f(x) = \begin{cases} f(x) = axe^{-x} & \text{for } 0 < x < \infty \\ 0 & \text{elsewhere.} \end{cases}$$

Solution

A. The sample space is given by $\mathcal{A} = \{x : x = 1, 2, 3, 4, \ldots\}$. By using the convergence of the geometric series $\sum_{x=0}^{\infty} q^x = 1/(1-q)$ for $q < 1$ the probability set function is

$$1 \stackrel{!}{=} P(\mathcal{A}) = \sum_{x=1}^{\infty} a\left(\frac{2}{3}\right)^x = a\left[\sum_{x=0}^{\infty}\left(\frac{2}{3}\right)^x - 1\right] = 2a.$$

Therefore, the constant is $a = 1/2$.

B. The sample space is given by $\mathcal{A} = \{x : 0 < x < \infty\}$. The probability set function is

$$1 \stackrel{!}{=} P(\mathcal{A}) = \int_0^\infty axe^{-x}dx \stackrel{p.I.}{=} a\,[-xe^{-x}]_0^\infty + a\int_0^\infty e^{-x} = -ae^{-x}|_0^\infty = a,$$

where we used integration by parts. The constant is $a = 1$.

Problem 1.14 Consider a function of the random variable X such that

$$f(x) = \begin{cases} ax & 0 \le x < 10 \\ a(20-x) & 10 \le x < 20 \\ 0 & \text{elsewhere.} \end{cases} \tag{1.5}$$

Find a so that $f(x)$ is a probability density function and sketch the graph of the pdf. Compute $P(X \ge 10)$ and $P(15 \le X \le 20)$.

Solution The sample space is given by $\mathcal{C} = \{x : 0 \le x < 20\}$.

$$1 \stackrel{!}{=} \int_0^{10} dx\; ax + \int_{10}^{20} dx\; a(20-x) = 100a.$$

The parameter has to be $a = 1/100$ to obtain a pdf. The graph of the pdf is shown in Fig. 1.1.

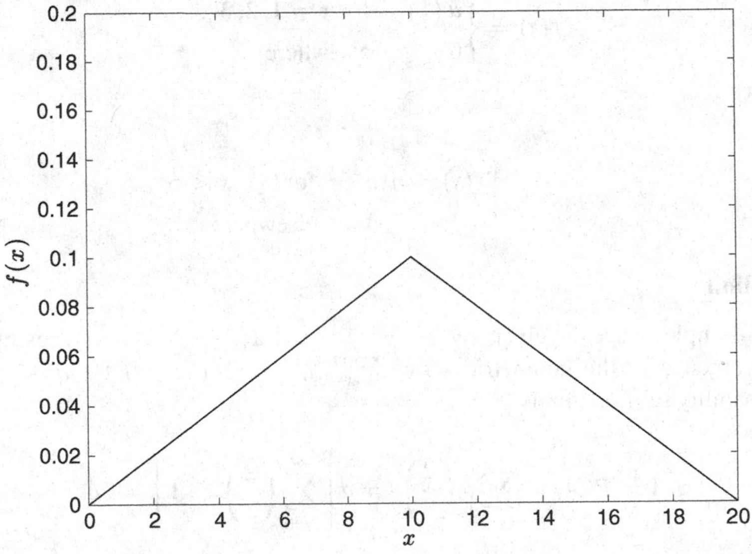

Fig. 1.1 Shown is the pdf from Eq. (1.5) for the parameter $a = 100$

The probabilities are then calculated by

$$P(X \geq 10) = \frac{1}{100}\int_{10}^{20} dx\,(20 - x) = 0.5$$

and

$$P(15 \leq X \leq 20) = \frac{1}{100}\int_{15}^{20} dx\,(20 - x) = 0.125.$$

Problem 1.15 Let $f(x) = x/15, x = 1, 2, 3, 4, 5, 0$ elsewhere, be the pdf of X. Find $P(X = 1 \text{ or } X = 2)$, $P(1/2 < X < 5/2)$, and $P(1 \leq X \leq 2)$.

Solution First we note that this probability density function is discrete and we find

$$P(X = 1 \text{ or } X = 2) = \sum_{x=1}^{2} f(x) = \sum_{x=1}^{2} \frac{x}{15} = \frac{1}{15} + \frac{2}{15} = \frac{1}{5}$$

$$P(\frac{1}{2} < X < \frac{5}{2}) = \sum_{x=1}^{2} \frac{x}{15} = \frac{1}{5}$$

$$P(1 \leq X \leq 2) = \sum_{x=1}^{2} \frac{x}{15} = \frac{1}{5}.$$

Problem 1.16 Compute the probability set functions $P(|X| < 1)$ and $P(X^2 < 9)$ for the following pdfs of X: (A) $f(x) = x^2/18, -3 < x < 3, 0$ elsewhere, and (B) $f(x) = (x + 2)/18, -2 < x < 4, 0$ elsewhere.

Solution

A. The probability set functions for the first pdf are given by

$$P(|X| < 1) = P(-1 < X < 1) = \int_{-1}^{1} \frac{x^2}{18} dx = \left.\frac{x^3}{54}\right|_{-1}^{1} = \frac{1}{27}$$

and

$$P(X^2 < 9) = P(-3 < X < 3) = \int_{-3}^{3} \frac{x^2}{18} dx = \left.\frac{x^3}{54}\right|_{-3}^{3} = 1.$$

B. The probability set functions for the second pdf are given by

$$P(|X| < 1) = P(-1 < X < 1) = \int_{-1}^{1} \frac{x+2}{18} dx = \left.\frac{\frac{x^2}{2} + 2x}{18}\right|_{-1}^{1} = \frac{2}{9}$$

and

$$P(X^2 < 9) = P(-3 < X < 3) = \int_{-2}^{3} \frac{x+2}{18} dx = \left. \frac{\frac{x^2}{2} + 2x}{18} \right|_{-2}^{3} = \frac{25}{36}.$$

Problem 1.17 Given $P(X > a) = e^{-\lambda a}(\lambda a + 1)$, with $\lambda > 0, a \geq 0$, find the pdf of X and $P(X > \lambda^{-1})$.

Solution Let $f(x)$ be a pdf so that

$$P(X > a) = \int_{a}^{\infty} f(x)\, dx = e^{-\lambda a}(\lambda a + 1).$$

Now, we take the derivative of the integral (on the left side) with respect to a and obtain an expression for the pdf at point a. According to the Fundamental Theorem of calculus we find

$$\frac{d}{da} \int_{a}^{\infty} f(x)\, dx = -f(a).$$

The derivative of the term on the right side is

$$\frac{d}{da} \left[(\lambda a + 1) e^{-\lambda a} \right] = -\lambda^2 a e^{-\lambda a}.$$

Since left and right hand side have to be equal we find (after substituting $a \to x$)

$$f(x) = \lambda^2 x e^{-\lambda x}$$

and, thus,

$$P(X > \lambda^{-1}) = \int_{\lambda^{-1}}^{\infty} f(x)\, dx = \frac{2}{e}.$$

Problem 1.18 Let $f(x) = x^{-2}$, with $1 < x < \infty$, 0 elsewhere, be the pdf of X. If $A_1 = \{x : 1 < x < 2\}$ and $A_2 = \{x : 4 < x < 5\}$, find $P(A_1 \cup A_2)$ and $P(A_1 \cap A_2)$.

Solution Since A_1 and A_2 are disjoint subsets we find

$$P(A_1 \cup A_2) = P(A_1) + P(A_2) = \int_{1}^{2} \frac{1}{x^2} dx + \int_{4}^{5} \frac{1}{x^2} dx$$
$$= \left[-\frac{1}{x} \right]_{1}^{2} + \left[-\frac{1}{x} \right]_{4}^{5} = \frac{11}{20}$$

and

$$P(A_1 \cap A_2) = 0.$$

Problem 1.19 Let $f(x, y) = 4xy, 0 < x < 1$ and $0 < y < 1$, 0 elsewhere, be the pdf of X and Y. Find $P(0 < X < \frac{1}{2}, \frac{1}{4} < Y < 1)$, $P(X = Y)$, $P(X < Y)$, and $P(X \le Y)$.

Solution The probability is given by

$$P(0 < X < \frac{1}{2}, \frac{1}{4} < Y < 1) = \int_0^{1/2} dx \int_{1/4}^1 dy 4xy = x^2\Big|_0^{1/2} \cdot y^2\Big|_{1/4}^1 = \frac{15}{64}$$

and

$$P(X < Y) = \int_0^1 \int_0^y 4xy \, dx \, dy = \int_0^1 2y^3 \, dy = \frac{1}{2}.$$

Similarly we find

$$P(X \le Y) = \frac{1}{2}.$$

Since $P(X = k) = 0$, where k is a constant, we find

$$P(X = Y) = 0.$$

Problem 1.20 Given that the random variable X has the pdf

$$f(x) = \begin{cases} \frac{5}{a} & -0.1a < x < 0.1a \\ 0 & \text{elsewhere} \end{cases}$$

and $P(|X| < 2) = 2P(|X| > 2)$, find the value of a.

Solution As a first condition we find

$$P(|X| < 2) = P(-2 < X < 2) = \int_{-2}^2 f(x) \, dx = \int_{-2}^2 \frac{5}{a} \, dx = \frac{5}{a}x\Big|_{-2}^2 = \frac{20}{a}.$$

Obviously, we find the restriction $a \ge 20$, since any probability has to be $P \le 1$. The second condition is derived by

$$\begin{aligned} P(|X| > 2) &= P\left(-\frac{a}{10} < X < -2\right) + P\left(2 < X < \frac{a}{10}\right) \\ &= \int_{-a/10}^{-2} f(x) \, dx + \int_2^{a/10} f(x) \, dx = \int_{-a/10}^{-2} \frac{5}{a} \, dx + \int_2^{a/10} \frac{5}{a} \, dx \\ &= \frac{5}{a}\left[x\Big|_{-a/10}^{-2} + x\Big|_2^{a/10}\right] = 1 - \frac{20}{a}. \end{aligned}$$

Since $P(|X| < 2) = 2P(|X| > 2)$ we find that

$$\frac{20}{a} = 2 - \frac{40}{a},$$

and, by solving this equation for a, that $a = 30$.

1.4 The Distribution Function

Definition 1.3 Suppose a random variable X has the probability set function $P(A)$, and is a 1D set. For a real number x, let $A = \{y : -\infty, y \leq x\}$, so that $P(A) = P(X \in A) = P(X \leq x)$. The probability is thus a function of x, say $F(x) = P(X \leq x)$. The function $F(x)$ is called the *distribution function* of the random variable X. Hence, if $f(x)$ is the pdf of X, we have

$$F(x) = \sum_{y \leq x} f(y) \qquad \text{and} \qquad F(x) = \int_{y \leq x} f(y)\,dy \tag{1.6}$$

for a discrete and a continuous random variable X.

Problem 1.21 Let $f(x)$ be the pdf of a random variable X. Find the distribution function $F(x)$ of X and sketch the graph for

A.

$$f(x) = \begin{cases} 1 & \text{for } x = 0 \\ 0 & \text{elsewhere} \end{cases}$$

B.

$$f(x) = \begin{cases} \frac{1}{3} & \text{for } x = -1, 0, 1 \\ 0 & \text{elsewhere} \end{cases}$$

C.

$$f(x) = \begin{cases} \frac{x}{15} & \text{for } x = 1, 2, 3, 4, 5 \\ 0 & \text{elsewhere} \end{cases}$$

D.

$$f(x) = \begin{cases} 3(1-x)^2 & \text{for } 0 < x < 1 \\ 0 & \text{elsewhere} \end{cases}$$

E.

$$f(x) = \begin{cases} x^{-2} & \text{for } 1 < x < \infty \\ 0 & \text{elsewhere} \end{cases}$$

F.

$$f(x) = \begin{cases} \frac{1}{3} & \text{for } 0 < x < 1 \text{ and } 2 < x < 4 \\ 0 & \text{elsewhere.} \end{cases}$$

Solution

A. The distribution function is given by

$$F(x) = \begin{cases} 0 & \text{for } x < 0 \\ 1 & \text{for } 0 \leq x \end{cases} \tag{1.7}$$

and the graph is shown in Fig. 1.2.

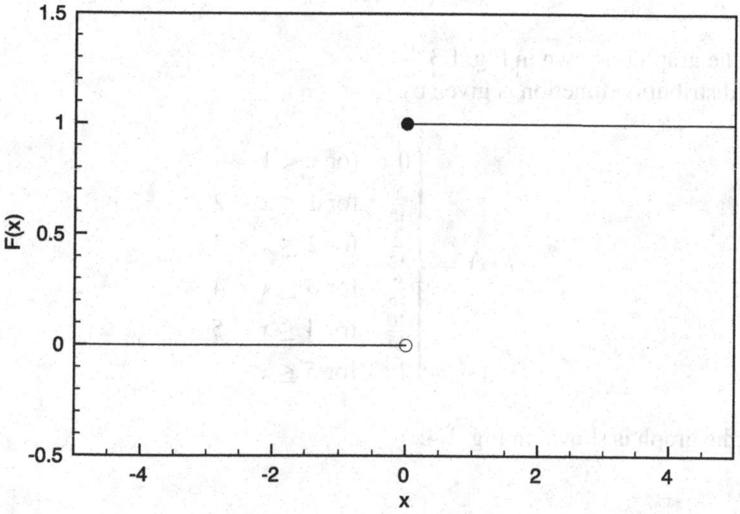

Fig. 1.2 Shown is the probability distribution function Eq. (1.7)

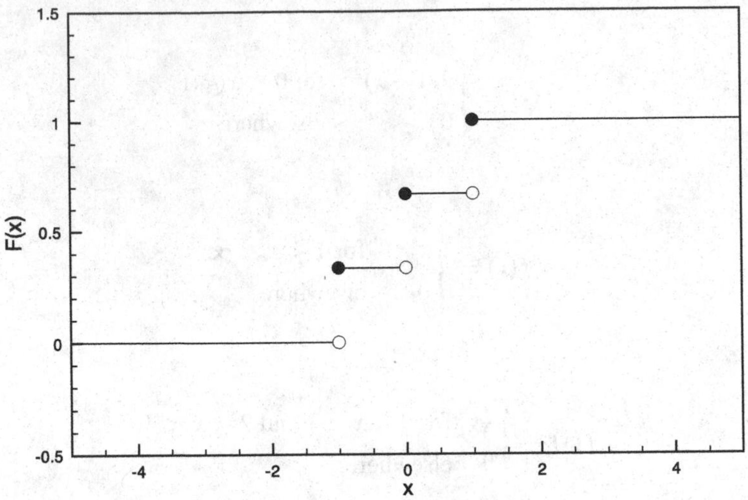

Fig. 1.3 Shown is the probability distribution function Eq. (1.8)

B. The distribution function is given by

$$F(x) = \begin{cases} 0 & \text{for } x < -1 \\ \frac{1}{3} & \text{for } -1 \leq x < 0 \\ \frac{2}{3} & \text{for } 0 \leq x < 1 \\ 1 & \text{for } 1 \leq x \end{cases} \tag{1.8}$$

and the graph is shown in Fig. 1.3.

C. The distribution function is given by

$$F(x) = \begin{cases} 0 & \text{for } x < 1 \\ \frac{1}{15} & \text{for } 1 \leq x < 2 \\ \frac{3}{15} & \text{for } 2 \leq x < 3 \\ \frac{6}{15} & \text{for } 3 \leq x < 4 \\ \frac{10}{15} & \text{for } 4 \leq x < 5 \\ 1 & \text{for } 5 \leq x \end{cases} \tag{1.9}$$

and the graph is shown in Fig. 1.4.

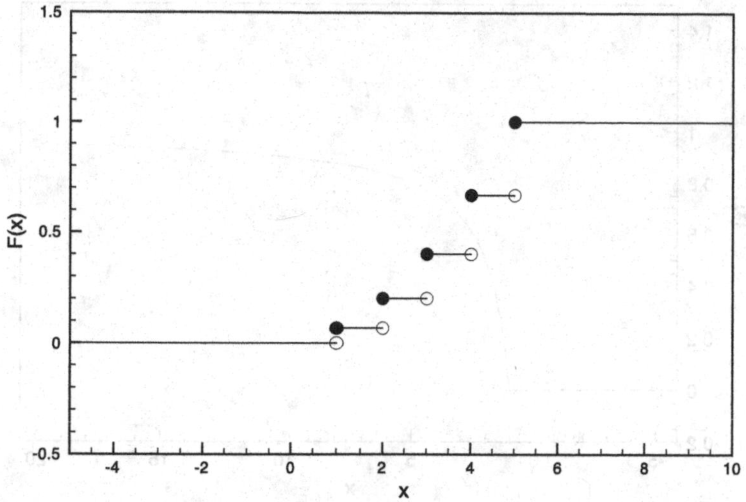

Fig. 1.4 Shown is the probability distribution function Eq. (1.9)

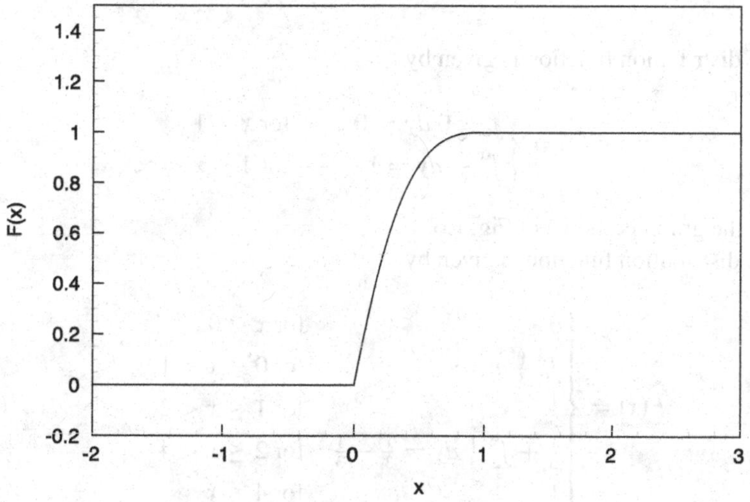

Fig. 1.5 Shown is the probability distribution function Eq. (1.10)

D. The distribution function is given by

$$F(x) = \begin{cases} \int_{-\infty}^{x} 0\, dy = 0 & \text{for } x < 0 \\ \int_{0}^{x} 3(1-y)^2\, dy = -(1-x)^3 + 1 & \text{for } 0 \leq x < 1 \\ 1 & \text{for } 1 \leq x \end{cases} \qquad (1.10)$$

and the graph is shown in Fig. 1.5.

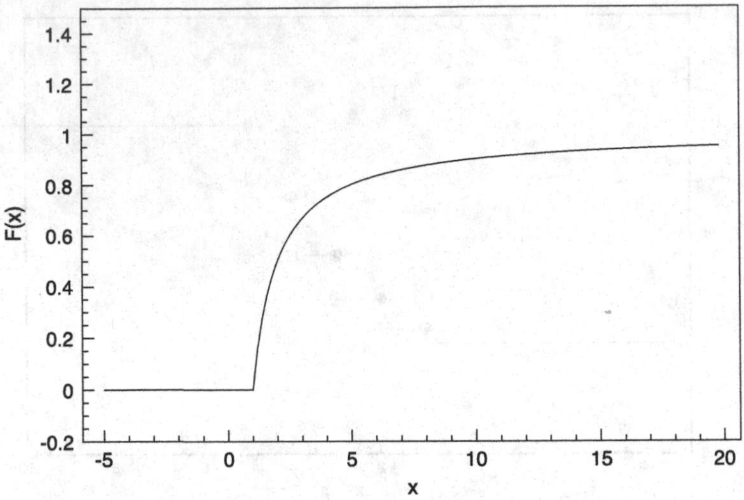

Fig. 1.6 Shown is the probability distribution function Eq. (1.11)

E. The distribution function is given by

$$F(x) = \begin{cases} \int_{-\infty}^{x} 0\, dy = 0 & \text{for } x < 1 \\ \int_{1}^{x} \frac{1}{y^2}\, dy = 1 - \frac{1}{x} & \text{for } 1 \leq x < \infty \end{cases} \tag{1.11}$$

and the graph is shown in Fig. 1.6.

F. The distribution function is given by

$$F(x) = \begin{cases} 0 & \text{for } x < 0 \\ \int_{0}^{x} \frac{1}{3}\, dy = \frac{x}{3} & \text{for } 0 \leq x < 1 \\ \frac{1}{3} & \text{for } 1 \leq x < 2 \\ \frac{1}{3} + \int_{2}^{x} \frac{1}{3}\, dy = \frac{x}{3} - \frac{1}{3} & \text{for } 2 \leq x < 4 \\ 1 & \text{for } 4 \leq x \end{cases} \tag{1.12}$$

and the graph is shown in Fig. 1.7.

Problem 1.22 Given the distribution function

$$F(x) = \begin{cases} 0, & \text{for } x < -1 \\ \frac{x+2}{4}, & \text{for } -1 \leq x < 1 \\ 1 & \text{for } 1 \leq x. \end{cases} \tag{1.13}$$

Sketch $F(x)$ and compute $P(-1/2 < X < 1/2)$, $P(X = 0)$, $P(X = 1)$, and $P(2 < X \leq 3)$!

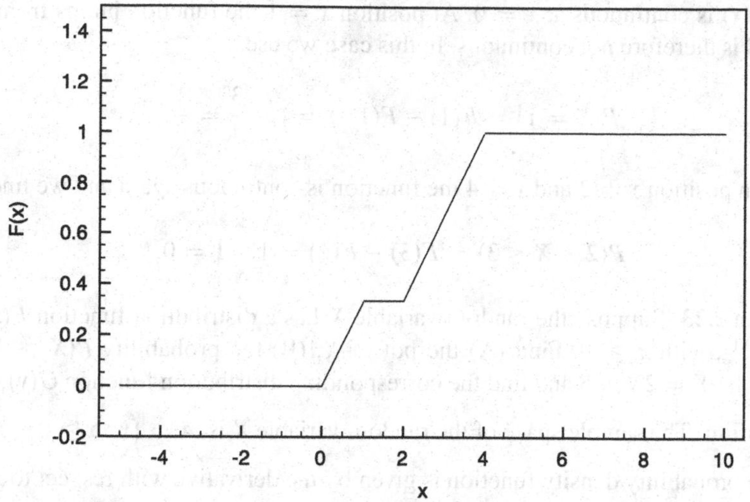

Fig. 1.7 Shown is the probability distribution function Eq. (1.12)

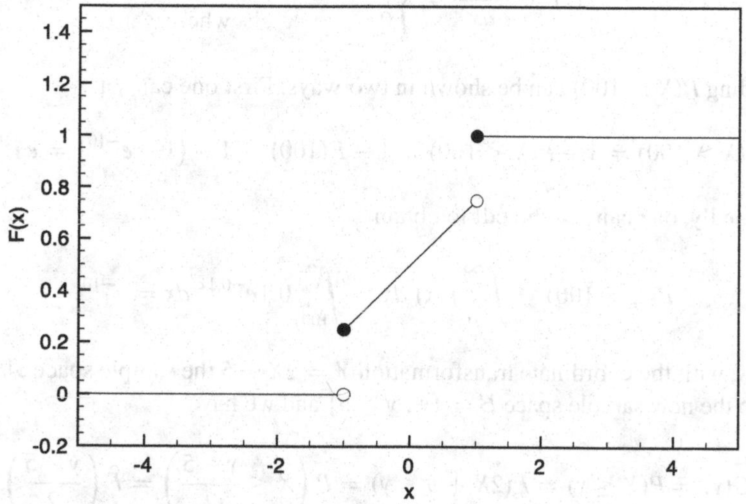

Fig. 1.8 Shown is the probability distribution function given by Eq. (1.13)

Solution The distribution function is shown in Fig. 1.8. For the first two probabilities we find

$$P\left(-\frac{1}{2} < X < \frac{1}{2}\right) = F\left(\frac{1}{2}\right) - F\left(-\frac{1}{2}\right) = \frac{5}{8} - \frac{3}{8} = \frac{2}{8} = \frac{1}{4}$$

$$P(X = 0) = 0,$$

since $F(x)$ is continuous at $x = 0$. At position $x = 1$ the function jumps from 3/4 to 1 and is therefore not continuous. In this case we use

$$P(X = 1) = F(1) - F(1-) = 1 - \frac{3}{4} = \frac{1}{4}.$$

Between position $x = 2$ and $x = 4$ the function is continuous again and we find

$$P(2 < X \leq 3) = F(3) - F(2) = 1 - 1 = 0.$$

Problem 1.23 Suppose the random variable X has a distribution function $F(x) = 1 - e^{-0.1x}$, with $x \geq 0$. Find (A) the pdf of X, (B) the probability $P(X > 100)$, and (C) let $Y = 2X + 5$ and find the corresponding distribution function $G(y)$.

Solution The sample space of the random variable X is $\mathcal{A} = \{x : x \geq 0\}$.

A. The probability density function is given by the derivative with respect to x and we obtain

$$f(x) = \frac{dF(x)}{dx} = \begin{cases} 0.1e^{-0.1x} & x \geq 0 \\ 0 & \text{elsewhere.} \end{cases}$$

B. Finding $P(X > 100)$ can be shown in two ways: First one can set

$$P(X > 100) = 1 - P(X < 100) = 1 - F(100) = 1 - \left(1 - e^{-10}\right) = e^{-10}.$$

Secondly, one can use the pdf to obtain

$$P(X > 100) = \int_{100}^{\infty} f(x)\, dx = \int_{100}^{\infty} 0.1e^{-0.1x}\, dx = e^{-10}.$$

C. Now, with the coordinate transformation $Y = 2X + 5$ the sample space $\mathcal{A}$ maps onto the new sample space $\mathcal{B} = \{y : y \geq 5\}$ and we have

$$G(y) = P(Y \leq y) = P(2X + 5 \leq y) = P\left(X \leq \frac{y-5}{2}\right) = F\left(\frac{y-5}{2}\right).$$

Obviously, we find for the distribution function of the random variable Y,

$$G(y) = \begin{cases} 1 - e^{-0.05(y-5)} & y \geq 5 \\ 0 & \text{elsewhere.} \end{cases}$$

Problem 1.24 Let $f(x) = 1$, $0 < x < 1$, 0 elsewhere, be the pdf of X. Find the distribution function and pdf of $Y = \sqrt{X}$.

Solution The new variable Y has its own sample space, given by

$$\mathcal{B} = \left\{y : y = \sqrt{x} \Rightarrow 0 < y < 1\right\}.$$

The probability set function of the random variable Y is then given by

$$G(y) = P(Y \leq y) = P(\sqrt{X} \leq y) = P(X \leq y^2),$$

where y obeys $0 < y < 1$. It follows immediately that

$$G(y) = \int_{-\infty}^{y^2} f(x)\,dx = \begin{cases} 0 & \text{for } y < 0 \\ \int_0^{y^2} dx = y^2 & \text{for } 0 \leq y < 1 \\ 1 & \text{for } 1 \leq y. \end{cases}$$

The probability density function is defined by $g(y) = G'(y)$ and we find

$$g(y) = \begin{cases} 2y & \text{for } 0 \leq y < 1 \\ 0 & \text{elsewhere.} \end{cases}$$

Problem 1.25 Let $f(x) = x/6$, $x = 1, 2, 3$, 0 elsewhere, be the pdf of X. Find the distribution function and pdf for $Y = X^2$.

Solution Note that the random variable is discrete. The sample space of the random variable X is given by $\mathcal{A} = \{x : x = 1, 2, 3\}$. The new random variable is defined by $Y = X^2$. Thus, the transformation $y = u(x) = x^2$ maps the sample space $\mathcal{A}$ onto $\mathcal{B} = \{y : y = 1, 4, 9\}$. This transformation is, in general, not *injective* and the inverse transformation is given by $x = u^{-1}(y) = \pm\sqrt{y}$. Here, however, the sample space $\mathcal{A}$ has no negative values, therefore, we choose the single-valued inverse function $x = \sqrt{y}$.

The probability density function of the random variable $Y = X^2$ is then given by

$$g(y) = \begin{cases} \frac{\sqrt{y}}{6} & \text{for } y = 1, 4, 9 \\ 0 & \text{else.} \end{cases}$$

The probability distribution function is then given by

$$G(y) = \sum_{x \leq y} g(x) = \begin{cases} 0 & \text{for } y < 1 \\ \frac{1}{6} & \text{for } 1 \leq y < 4 \\ \frac{3}{6} = \frac{1}{2} & \text{for } 4 \leq y < 9 \\ 1 & \text{for } 9 \leq y. \end{cases}$$

Problem 1.26 Suppose that a random variable X has the pdf

$$f(x) = \begin{cases} x + \frac{1}{2} & \text{for } 0 \leq x \leq 1 \\ 0 & \text{elsewhere.} \end{cases}$$

Determine (A) the distribution function $F(x)$, (B) the probability density function of the random variable $Y = X^2$, and (C) the probability $P(Y > 0.36)$.

Solution

A. The distribution function is then given by

$$F(x) = \int_{-\infty}^{x} f(w)\, dw = \begin{cases} 0 & \text{for } x < 0 \\ \int_0^x \left[w + \frac{1}{2}\right] dw = \frac{x^2}{2} + \frac{x}{2} & \text{for } 0 \leq x < 1 \\ 1 & \text{for } 1 \leq x. \end{cases}$$

B. The transformation is given by $u(x) = y = x^2$. The inverse transformation is then given by $u^{-1} = x = \pm\sqrt{y}$. Again, the sample space of X has no negative values, therefore, we choose the single-valued inverse function $u^{-1} = x = \sqrt{y}$. That is

$$G(y) = P(Y \leq y) = P(X^2 \leq y) = P(X \leq \sqrt{y})$$

with the sample space for the random variable Y given by $0 \leq y \leq 1$. The probability distribution function is then

$$G(y) = \begin{cases} \int_{-\infty}^{x=\sqrt{y}} 0\, dx = 0 & \text{for } y < 0 \\ \int_0^{x=\sqrt{y}} \left[x + \frac{1}{2}\right] dx = \frac{y+\sqrt{y}}{2} & \text{for } 0 \leq y < 1 \\ 1 & \text{for } 1 \leq y. \end{cases}$$

The probability density function is then given by $g(y) = G'(y)$,

$$g(y) = \begin{cases} \frac{1}{2} + \frac{1}{4\sqrt{y}} & \text{for } 0 \leq y \leq 1 \\ 0 & \text{elsewhere.} \end{cases}$$

C. The probability $P(Y > 0.36)$ is equivalent to

$$\begin{aligned} P(Y > 0.36) &= 1 - P(Y \leq 0.36) = 1 - G(y = 0.36) \\ &= 1 - \frac{0.36 + \sqrt{0.36}}{2} = 0.52. \end{aligned}$$

Problem 1.27 Suppose that a random variable X has the pdf

$$f(x) = \begin{cases} \frac{1}{2} & \text{for } -1 \le x \le 1 \\ 0 & \text{elsewhere.} \end{cases}$$

Determine (A) the distribution function $F(x)$, (B) the probability density function of the random variable $Y = X^2$, and (C) the probability $P(Y > 0.36)$.

Solution

A. The distribution function is then given by

$$F(x) = \int_{-\infty}^{x} f(w)\, dw = \begin{cases} 0 & \text{for } x < -1 \\ \int_{-1}^{x} \frac{1}{2}\, dw = \frac{x}{2} + \frac{1}{2} & \text{for } -1 \le x \le 1 \\ 1 & \text{for } 1 < x. \end{cases}$$

B. The transformation is given by $u(x) = y = x^2$. The inverse transformation is then given by $u^{-1} = x = \pm\sqrt{y}$. Here, the sample space of X has negative values, therefore,

$$G(y) = P(Y \le y) = P(X^2 \le y) = P(-\sqrt{y} \le X \le \sqrt{y})$$

with the new sample space $\mathcal{B} = \{y : 0 \le y \le 1\}$ for the random variable Y. The probability distribution function is then

$$G(y) = \begin{cases} 0 & \text{for } y < 0 \\ \int_{-\sqrt{y}}^{\sqrt{y}} \frac{1}{2}\, dx = \sqrt{y} & \text{for } 0 \le y \le 1 \\ 1 & \text{for } 1 < y. \end{cases}$$

The probability density function is then given by $g(y) = G'(y)$,

$$g(y) = \begin{cases} \frac{1}{2\sqrt{y}} & \text{for } 0 \le y \le 1 \\ 0 & \text{elsewhere.} \end{cases}$$

C. The probability $P(Y > 0.36)$ is equivalent to

$$P(Y > 0.36) = 1 - P(Y \le 0.36) = 1 - G(y = 0.36) = 1 - \sqrt{0.36} = 0.4.$$

1.5 Expectations and Moments

Definition 1.4 Suppose X is a continuous or discrete random variable with probability density function $f(x)$ and let $u(x)$ be a function of X so that

$$E[u(x)] = \int_{-\infty}^{\infty} u(x)f(x)\,dx; \qquad E[u(x)] = \sum_x u(x)f(x) \tag{1.14}$$

exists, then $E[u(x)]$ is called the *mathematical expectation* or *expected value* of $u(x)$. An important expectation is the *moment-generating* function of a random variable X. Suppose there exists a finite real number t for which the expectation

$$E\left(e^{tX}\right) = \int_{-\infty}^{\infty} e^{tx} f(x)\,dx; \qquad E\left(e^{tX}\right) = \sum_x e^{tx} f(x)\,dx \tag{1.15}$$

(continuous or discrete) exists. Then $M(t) = E\left(e^{tX}\right)$ is the *moment-generating* function, with $M(t = 0) = 1$. In general, for $m > 0$ an integer, the m-th derivative of the moment-generating function generates the m-th moment of the distribution, $M^m(0) = E(X^m)$, so that

$$E(X^m) = \int_{-\infty}^{\infty} x^m f(x)\,dx; \qquad E(X^m) = \sum_x x^m f(x). \tag{1.16}$$

Problem 1.28 Suppose X has the pdf

$$f(x) = \begin{cases} \frac{x+2}{18} & \text{for } -2 < x < 4 \\ 0 & \text{elsewhere.} \end{cases}$$

Find $E(X)$, $E\left[(X+2)^3\right]$, and $E\left[6X - 2(X+2)^3\right]$.

Solution From the definition of the expected value (1.14) we find

$$E(X) = \int_{-2}^{4} \frac{x^2 + 2x}{18}\,dx = \left[\frac{x^3}{54} + \frac{x^2}{18}\right]_{-2}^{4} = 2$$

$$E\left[(X+2)^3\right] = \int_{-2}^{4} \frac{(x+2)^4}{18}\,dx = \left[\frac{(x+2)^5}{90}\right]_{-2}^{4} = \frac{432}{5}$$

$$E\left[6X - 2(X+2)^3\right] = 6E(X) - 2E\left[(X+2)^3\right] = -\frac{804}{5}.$$

Problem 1.29 The *median* of a random variable X is the value x such that the distribution function $F(x) = 1/2$. Compute the median of the random variable X for the pdf

$$f(x) = \begin{cases} 2x & \text{for } 0 < x < 1 \\ 0 & \text{elsewhere.} \end{cases}$$

Solution From the definition of the distribution function we find

$$F(x) = \int_{-\infty}^{x} f(y)\, dy = \int_{0}^{x} 2y\, dy = x^2 \equiv \frac{1}{2} \quad \longrightarrow \quad x = \frac{1}{\sqrt{2}}.$$

The median is $x = 1/\sqrt{2}$.

Problem 1.30 The *mode* of a random variable X is the value that occurs most frequently—sometimes called the *most probable value*. The value a is the mode of the random variable X if

$$f(a) = \max f(x),$$

(for a continuous pdf). The mode is not necessarily unique. Compute the mode and median of a random variable X with pdf

$$f(x) = \begin{cases} \frac{2x}{3} & \text{for } 0 \leq x < 1 \\ \frac{1}{3} & \text{for } 1 \leq x < 3 \\ 0 & \text{elsewhere.} \end{cases}$$

Solution The distribution function is given by

$$F(x) = \int_{-\infty}^{x} f(y)\, dy = \begin{cases} 0 \\ \int_0^x \frac{2y}{3}\, dy \\ \frac{1}{3} + \int_1^x \frac{1}{3}\, dy \\ 1 \end{cases} = \begin{cases} 0 & \text{for } x < 0 \\ \frac{x^2}{3} & \text{for } 0 \leq x < 1 \\ \frac{x}{3} & \text{for } 1 \leq x < 3 \\ 1 & \text{for } 3 \leq x. \end{cases}$$

The median is defined by $F(x) = 1/2$. This is possible only for $1 \leq x < 3$, where the distribution function is

$$\frac{x}{3} = \frac{1}{2} \quad \longrightarrow \quad x = \frac{3}{2}.$$

The mode is the maximum value of the pdf (not the distribution function!). Analyzing the pdf we find, that for $0 \leq x < 1$ the pdf increases linearly, reaching its maximum value at $x = 1$ with $f(x = 1) = 2/3$, then jumps to $1/3$ for $1 \leq x < 3$.

Thus, the mode of the pdf is

$$a = 1 \qquad \longrightarrow \qquad f(a) = \frac{2}{3} = \max f(x).$$

Problem 1.31 Suppose X and Y have the joint pdf

$$f(x, y) = \begin{cases} e^{-x-y} & \text{for } 0 < x < \infty, \quad 0 < y < \infty \\ 0 & \text{elsewhere} \end{cases}$$

and that $u(X, Y) = X$, $v(X, Y) = Y$, and $w(X, Y) = XY$. Show that $E[u(X, Y)] \cdot E[v(X, Y)] = E[w(X, Y)]$.

Solution We calculate

$$E(XY) = \int_0^\infty dy \int_0^\infty dx\, xy\, e^{-x-y} = 1$$
$$E(X) = \int_0^\infty dy \int_0^\infty dx\, x\, e^{-x-y} = 1$$
$$E(Y) = \int_0^\infty dy \int_0^\infty dx\, y\, e^{-x-y} = 1$$

and hence

$$E(X) \cdot E(Y) = 1 = E(XY).$$

Alternatively, we find for $E[w(X, Y)]$

$$E(XY) = \int_0^\infty dy \int_0^\infty dx\, xy\, e^{-x-y}$$

and for $E[u(X, Y)] \cdot E[v(X, Y)]$

$$E(X) \cdot E(Y) = \left[\int_0^\infty dy \int_0^\infty dx\, x\, e^{-x-y}\right] \cdot \left[\int_0^\infty dy \int_0^\infty dx\, y\, e^{-x-y}\right]$$
$$= \left[\int_0^\infty dy\, e^{-y} \int_0^\infty dx\, x\, e^{-x}\right] \cdot \left[\int_0^\infty dy\, y\, e^{-y} \int_0^\infty dx\, e^{-x}\right].$$

Since $\int_0^\infty dx\, e^{-x} = 1$ it follows immediately that

$$E(X) \cdot E(Y) = \int_0^\infty dx\, x\, e^{-x} \cdot \int_0^\infty dy\, y\, e^{-y}$$
$$= \int_0^\infty dy \int_0^\infty dx\, xy\, e^{-x-y} = E(XY) = E[w(X, Y)].$$

Problem 1.32 If X and Y are two exponentially distributed random variables with pdfs

$$f(x) = 2e^{-2x}, \qquad x \geq 0$$
$$f(y) = 4e^{-4y}, \qquad y \geq 0,$$

calculate $E(X+Y)$.

Solution The expected value for $X+Y$ is given by

$$E(X+Y) = E(X) + E(Y) = \int_0^\infty dx\, xf(x) + \int_0^\infty dy\, yf(y)$$
$$= 2\int_0^\infty dx\, x\, e^{-2x} + 4\int_0^\infty dy\, y\, e^{-4y} = \frac{3}{4}.$$

Problem 1.33 Suppose X and Y have the joint pdf

$$f(x,y) = \begin{cases} 2 & \text{for } 0 < x < y, \quad 0 < y < 1 \\ 0 & \text{elsewhere} \end{cases}$$

and that $u(X,Y) = X$, $v(X,Y) = Y$, and $w(X,Y) = XY$. Show that $E[u(X,Y)] \cdot E[v(X,Y)] \neq E[w(X,Y)]$.

Solution For $E[w(X,Y)]$ we find

$$E(XY) = \int_0^1 dy \int_0^y dx\, 2xy = \int_0^1 dy\, y^3 = \frac{1}{4}.$$

For $E[u(X,Y)]$and $E[v(X,Y)]$ we find

$$E(X) = \int_0^1 dy \int_0^y dx\, 2x = \int_0^1 dy\, y^2 = \frac{1}{3}$$

and

$$E(Y) = \int_0^1 dy \int_0^y dx\, 2y = \int_0^1 dy\, 2y^2 = \frac{2}{3}.$$

Obviously, one finds

$$E(X) \cdot E(Y) = \frac{2}{9} \neq \frac{1}{4} = E(XY).$$

Problem 1.34 Let X have a pdf $f(x)$ that is positive at $x = -1, 0, 1$ and zero elsewhere. (A) If $f(0) = 1/2$, find $E(X^2)$. (B) If $f(0) = 1/2$, and if $E(X) = 1/6$, determine $f(-1)$ and $f(1)$.

Solution We know, that the probability of the full sample space has to obey

$$1 \stackrel{!}{=} \sum_{-\infty}^{\infty} f(x) = \sum_{x=-1}^{1} f(x) = f(-1) + f(0) + f(1) = \frac{1}{2} + f(-1) + f(1).$$

Thus

$$f(-1) + f(1) = \frac{1}{2}.$$

A. The expected value of X^2 is then given by

$$E(X^2) = \sum_{x=-1}^{1} x^2 f(x) = (-1)^2 f(-1) + (0)^2 f(0) + (1)^2 f(1)$$
$$= f(-1) + f(1) = \frac{1}{2}. \tag{1.17}$$

B. The expected value of X is given by

$$E(X) = \frac{1}{6} = \sum_{-\infty}^{\infty} x f(x) = \sum_{x=-1}^{1} x f(x) = -f(-1) + f(1). \tag{1.18}$$

Note the minus sign in front of $f(-1)$! From Eqs. (1.17) and (1.18) it follows immediately that

$$f(1) = \frac{1}{3}, \qquad f(-1) = \frac{1}{6}. \tag{1.19}$$

Problem 1.35 A random variable X with an unknown probability distribution has a mean $\mu = 12$ and a variance $\sigma^2 = 9$. Use Chebyshev's inequality to bound $P(6 < X < 18)$ and $P(3 < X < 21)$.

Solution Chebyshev's inequality is given by

$$P(\mu - n\sigma < X < \mu + n\sigma) \geq 1 - \frac{1}{n^2} \tag{1.20}$$

or, alternatively,

$$P(|X - \mu| \geq n) \leq \frac{\sigma^2}{n^2} \qquad \text{and} \qquad P(|X - \mu| < n) \geq 1 - \frac{\sigma^2}{n^2}.$$

From the first probability, $P(6 < X < 18)$, we find with $\mu = 12$ and $\sigma = 3$,

$$\mu - n\sigma = 12 - 3n \stackrel{!}{=} 6 \quad \longrightarrow \quad n = 2$$

$$\mu + n\sigma = 12 + 3n \stackrel{!}{=} 18 \quad \longrightarrow \quad n = 2.$$

Thus,

$$P(6 < X < 18) = P(12 - 3 \cdot 2 < X < 12 + 3 \cdot 2) \geq 1 - \frac{1}{2^2} = \frac{3}{4}.$$

For the second probability, $P(3 < X < 21)$, we find with $\mu = 12$ and $\sigma = 3$,

$$\mu - n\sigma = 12 - 3n \stackrel{!}{=} 3 \quad \longrightarrow \quad n = 3$$

$$\mu + n\sigma = 12 + 3n \stackrel{!}{=} 21 \quad \longrightarrow \quad n = 3.$$

Thus,

$$P(6 < X < 18) = P(12 - 3 \cdot 3 < X < 12 + 3 \cdot 3) \geq 1 - \frac{1}{3^2} = \frac{8}{9}.$$

Problem 1.36 Two distinct integers are chosen randomly without replacement from the first six positive integers. What is the expected value of the absolute value of the difference of these two numbers?

Solution We choose two numbers, X_1 and X_2, from the set $1, 2, 3, 4, 5, 6$ without replacement. Since the new variable Y is the absolute value of the difference, $Y = |X_1 - X_2|$, the order of both numbers is unimportant. In this case the number of all possible outcomes is given by the number of combinations,

$$C_2^6 = \binom{6}{2} = \frac{6!}{2!(6-2)!} = \frac{6!}{2!4!} = 15,$$

where the brackets denote the binomial coefficient (see Eq. (1.29) in Sect. 1.8.1). The sample space is then given by

$$\begin{aligned} \mathcal{C} = \{c : c = {} & (1,2)(1,3)(1,4)(1,5)(1,6)(2,3)(2,4)(2,5) \\ & (2,6)(3,4)(3,5)(3,6)(4,5)(4,6)(5,6)\}. \end{aligned}$$

The sample space of the new variable $Y = |X_1 - X_2|$ is given by

$$\mathcal{A} = \{y : y = 1, 2, 3, 4, 5\},$$

where the probability of getting the result 1 is given by $P(Y = 1) = 5/15$. Similarly, one finds $P(Y = 2) = 4/15$, $P(Y = 3) = 3/15$, $P(Y = 4) = 2/15$ and $P(Y = 5) = 1/15$. The pdf can then be found as

$$f(y) = \begin{cases} \frac{6-y}{15} & \text{for } y = 1, 2, 3, 4, 5 \\ 0 & \text{elsewhere.} \end{cases}$$

The expected value is then simply calculated by

$$E(Y) = \sum_{-\infty}^{\infty} y f(y) = \sum_{y=1}^{5} y f(y)$$
$$= \frac{5}{15} + \frac{8}{15} + \frac{9}{15} + \frac{8}{15} + \frac{5}{15} = \frac{35}{15} = \frac{7}{3} \approx 2.3.$$

In general we can write for the pdf

$$f(y) = \begin{cases} \frac{n-y}{C_2^n} & \text{for } y = 1, 2, \ldots, n-1 \\ 0 & \text{elsewhere} \end{cases}$$

Note that C_2^n can be written as $C_2^n = n(n-1)/2$ and for $n = 6$ we obtain the correct result 15 (see above). For the expected value we can write

$$E(Y) = \sum_{y=1}^{n-1} y f(y) = \frac{2}{n(n-1)} \sum_{y=1}^{n-1} yn - y^2 = n - \frac{2n-1}{3}.$$

In the last step we used the theorem of finite series,

$$\sum_{i=1}^{n} i = \frac{n(n+1)}{2}, \qquad \sum_{i=1}^{n} i^2 = \frac{n(n+1)(2n+1)}{6}.$$

Problem 1.37 Assume that the random variable X has mean μ , standard deviation σ, and moment generating function $M(t)$. Show that

$$E\left(\frac{X-\mu}{\sigma}\right) = 0 \qquad E\left[\left(\frac{X-\mu}{\sigma}\right)^2\right] = 1$$

and

$$E\left\{\exp\left[t\left(\frac{X-\mu}{\sigma}\right)\right]\right\} = e^{-\mu t/\sigma} M\left(\frac{t}{\sigma}\right).$$

Solution Since $E(X) = \mu$ with μ and σ being constants, we find

$$E\left(\frac{X-\mu}{\sigma}\right) = \frac{[E(X) - E(\mu)]}{\sigma} = \frac{[\mu - \mu]}{\sigma} = 0$$

and with $\sigma^2 = E\left[(X-\mu)^2\right]$

$$E\left[\left(\frac{X-\mu}{\sigma}\right)^2\right] = \frac{E\left[(X-\mu)^2\right]}{\sigma^2} = \frac{\sigma^2}{\sigma^2} = 1.$$

With the MacLaurin's series for $M(t/\sigma) = \sum_{n=0}^{\infty} \frac{E(X^n)}{n!}\left(\frac{t}{\sigma}\right)^n$ we find

$$\begin{aligned}
E\left\{\exp\left[t\left(\frac{X-\mu}{\sigma}\right)\right]\right\} &= E\left[e^{Xt/\sigma}e^{-\mu t/\sigma}\right] = E\left[e^{Xt/\sigma}\right]E\left[e^{-\mu t/\sigma}\right] \\
&= e^{-\mu t/\sigma}E\left[\sum_{n=0}^{\infty}\frac{1}{n!}\left(\frac{Xt}{\sigma}\right)^n\right] = e^{-\mu t/\sigma}\sum_{n=0}^{\infty}\frac{E(X^n)}{n!}\left(\frac{t}{\sigma}\right)^n \\
&= e^{-\mu t/\sigma}M\left(\frac{t}{\sigma}\right).
\end{aligned}$$

Problem 1.38 Suppose that $E\left[(X-b)^2\right]$ exists for a random variable X for all real b. Show that $E\left[(X-b)^2\right]$ is a minimum when $b = E[X]$.

Solution The expected value of $(X-b)^2$ is given by

$$\begin{aligned}
E\left[(X-b)^2\right] &= E\left[X^2 - 2bX + b^2\right] = E\left[X^2\right] - E\left[2bX\right] + E\left[b^2\right] \\
&= E\left[X^2\right] - 2bE\left[X\right] + b^2.
\end{aligned}$$

Showing that $E\left[(X-b)^2\right]$ is a minimum when $b = E[X]$ requires $dE/db = 0$ and $d^2E/db^2 > 0$. For the first step we find

$$\frac{d}{db}E\left[(X-b)^2\right] = -2E(X) + 2b \stackrel{!}{=} 0 \quad \Longrightarrow \quad b = E[X]$$

and

$$\frac{d^2}{db^2}E\left[(X-b)^2\right] = 2 > 0$$

shows that $b = E[X]$ is in fact a minimum.

Problem 1.39 Suppose that $R(t) = E\left[e^{(X-b)t}\right]$ exists for a random variable X. Show that $R^m(0)$ is the m-th moment of the distribution about the point b, where m is a positive integer.

Solution If $R(t) = E\left[e^{t(X-b)}\right]$ exists for a random variable X, then $R(t)$ is a moment-generating function with

$$R(t) = E\left[e^{t(X-b)}\right] = \int_{-\infty}^{\infty} dx\, e^{(x-b)t} f(x).$$

The m-th derivation with respect to t is given by

$$\frac{d^m R(t)}{dt^m} = R^m(t) = \int_{-\infty}^{\infty} dx\ (x-b)^m\, e^{(x-b)t} f(x).$$

For $t = 0$ it follows

$$R^m(0) = \int_{-\infty}^{\infty} dx\ (x-b)^m f(x),$$

which is the m-th moment of the distribution with $R^m(0) = E\left[(X-b)^m\right]$.

Problem 1.40 Let $\Psi(t) = \ln M(t)$, where $M(t)$ is the moment-generating function of a distribution. Show that $\Psi'(0) = \mu$ and $\Psi''(0) = \sigma^2$.

Solution We summarize that for a moment-generating function

$$M(0) = 1 \qquad M'(0) = \mu \qquad M''(0) - M'(0)^2 = \sigma^2.$$

The first derivation is then

$$\Psi'(t) = \frac{M'(t)}{M(t)} \quad \Rightarrow \quad \Psi'(0) = \frac{M'(0)}{M(0)} = \mu.$$

The second derivation is then

$$\Psi''(t) = \frac{M''(t)M(t) - M'(t)^2}{M(t)^2} \quad \Rightarrow \quad \Psi''(0) = \frac{M''(0)M(0) - M'(0)^2}{M(0)^2} = \sigma^2.$$

Problem 1.41 Suppose X is a random variable with mean μ and variance σ^2, and assume that the third moment $E\left[(X-\mu)^3\right]$ exists. The ratio $E\left[(X-\mu)^3\right]/\sigma^3$ is a measure of the *skewness* of the distribution. Graph the following pdfs and show that the skewness is negative, zero, and positive respectively:

A. $f(x) = (x+1)/2$, for $-1 < x < 1$ and 0 elsewhere
B. $f(x) = 1/2$, for $-1 < x < 1$ and 0 elsewhere
C. $f(x) = (1-x)/2$, for $-1 < x < 1$ and 0 elsewhere.

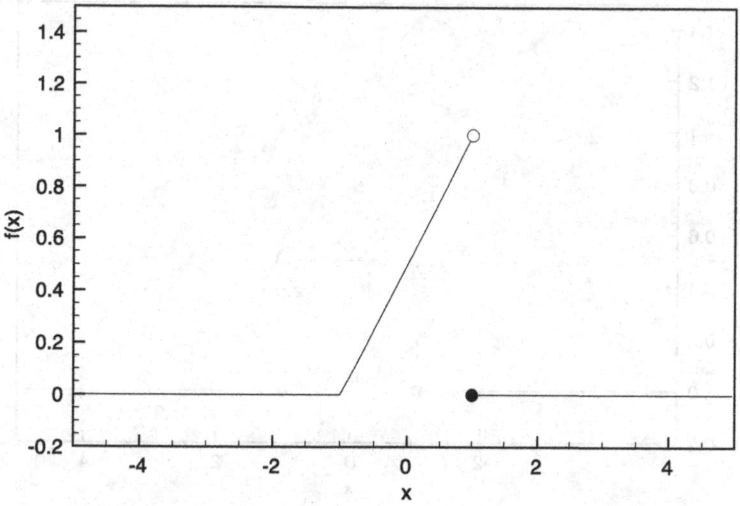

Fig. 1.9 Shown is the pdf $f(x) = (x+1)/2$. For the interval $-1 < x < 1$ the pdf is increasing. For all other values of x the pdf is zero

Solution For the third moment we find the general expression

$$\begin{aligned} E\left[(X-\mu)^3\right] &= E(X^3) - 3\mu E(X^2) + 3\mu^2 E(X) - \mu^3 \\ &= E(X^3) - 3\mu E(X^2) + 2\mu^2 E(X). \end{aligned} \tag{1.21}$$

Therefore, we need to calculate the first, the second and the third moment of each distribution.

A. The graph of the pdf, $f(x) = (x+1)/2$, for $-1 < x < 1$ and 0 elsewhere, is shown in Fig. 1.9. The first, second, and third moment is given by

$$E(X) = \int_{-1}^{1} dx\, x \frac{x+1}{2} = \frac{x^3}{6} + \frac{x^2}{4}\bigg|_{-1}^{1} = \frac{1}{3}$$

$$E(X^2) = \int_{-1}^{1} dx\, x^2 \frac{x+1}{2} = \frac{x^4}{8} + \frac{x^3}{6}\bigg|_{-1}^{1} = \frac{1}{3}$$

$$E(X^3) = \int_{-1}^{1} dx\, x^3 \frac{x+1}{2} = \frac{x^5}{10} + \frac{x^4}{8}\bigg|_{-1}^{1} = \frac{1}{5}.$$

For the variance we find $\sigma^2 = E(X^2) - E(X)^2 = 2/9$. The skewness is therefore negative with

$$\frac{E\left[(X-\mu)^3\right]}{\sigma^3} = \frac{\frac{1}{5} - 3\frac{1}{3}\frac{1}{3} + 2\frac{1}{9}\frac{1}{3}}{\left(\frac{2}{9}\right)^{3/2}} \approx -0.57.$$

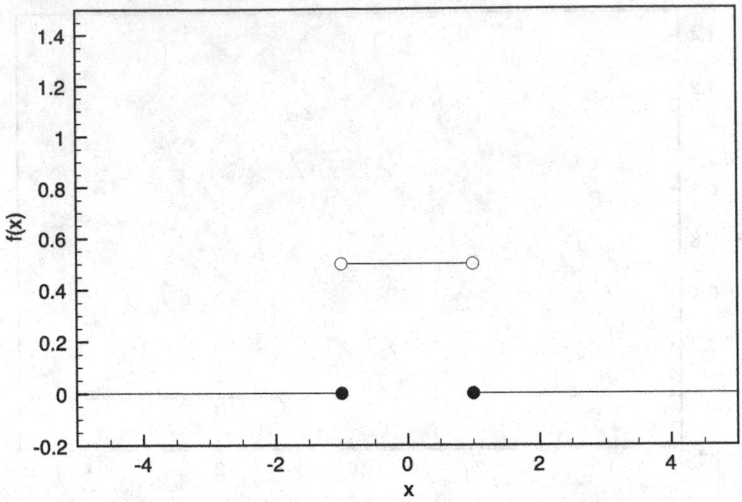

Fig. 1.10 Shown is pdf $f(x) = 1/2$. For the interval $-1 < x < 1$ the curve is constant. For all other values of x the pdf is zero

B. The graph of the pdf, $f(x) = 1/2$, for $-1 < x < 1$ and 0 elsewhere, is shown in Fig. 1.10. The moments are

$$E(X) = \int_{-1}^{1} dx\,\frac{x}{2} = \frac{x^2}{4}\Big|_{-1}^{1} = 0$$

$$E(X^2) = \int_{-1}^{1} dx\,\frac{x^2}{2} = \frac{x^3}{6}\Big|_{-1}^{1} = \frac{1}{3}$$

$$E(X^3) = \int_{-1}^{1} dx\,\frac{x^3}{2} = \frac{x^4}{8}\Big|_{-1}^{1} = 0.$$

For the variance we find $\sigma^2 = E(X^2) - E(X)^2 = 1/3$. According to Eq. (1.21) the skewness is zero with

$$\frac{E\left[(X-\mu)^3\right]}{\sigma^3} = 0.$$

C. The graph is shown in Fig. 1.11. The moments are given by

$$E(X) = \int_{-1}^{1} dx\,x\frac{1-x}{2} = \frac{x^2}{4} - \frac{x^3}{6}\Big|_{-1}^{1} = -\frac{1}{3}$$

$$E(X^2) = \int_{-1}^{1} dx\,x^2\frac{1-x}{2} = \frac{x^3}{6} - \frac{x^4}{8}\Big|_{-1}^{1} = \frac{1}{3}$$

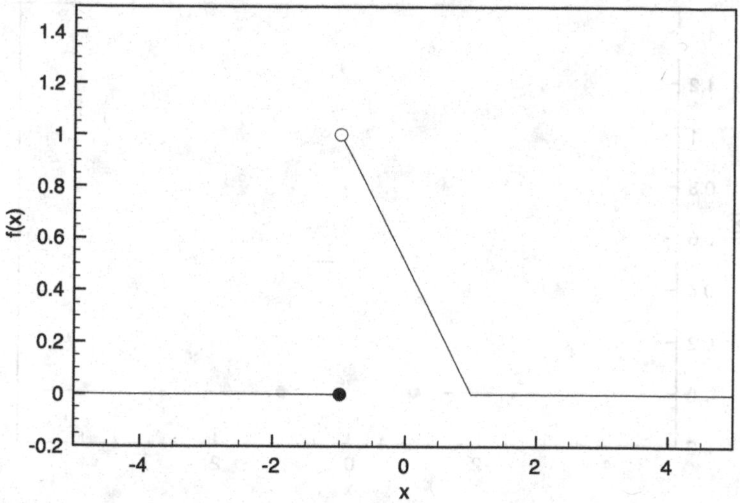

Fig. 1.11 Shown is the pdf $f(x) = (1-x)/2$. For the interval $-1 < x < 1$ the curve is decreasing. For all other values of x the pdf is zero

$$E(X^3) = \int_{-1}^{1} dx\, x^3 \frac{1-x}{2} = \left. \frac{x^4}{8} - \frac{x^5}{10} \right|_{-1}^{1} = -\frac{1}{5}.$$

For the variance we find $\sigma^2 = E(X^2) - E(X)^2 = 2/9$. The skewness is positive with

$$\frac{E\left[(X-\mu)^3\right]}{\sigma^3} = \frac{-\frac{1}{5} + 3\frac{1}{3}\frac{1}{3} - 2\frac{1}{9}\frac{1}{3}}{\left(\frac{2}{9}\right)^{3/2}} \approx +0.57.$$

Problem 1.42 Suppose X is a random variable with mean μ and variance σ^2, and assume that the fourth moment $E\left[(X-\mu)^4\right]$ exists. The ratio $E\left[(X-\mu)^4\right]/\sigma^4$ is a measure of the *kurtosis* of the distribution. Graph the following pdfs and show that the kurtosis is smaller for the first distribution.

A. $f(x) = 1/2$, for $-1 < x < 1$ and 0 elsewhere
B. $f(x) = 3(1-x^2)/4$, for $-1 < x < 1$ and 0 elsewhere

Solution For the fourth moment we find

$$E\left[(X-\mu)^4\right] = E(X^4) - 4\mu E(X^3) + 6\mu^2 E(X^2) - 3\mu^4 \tag{1.22}$$

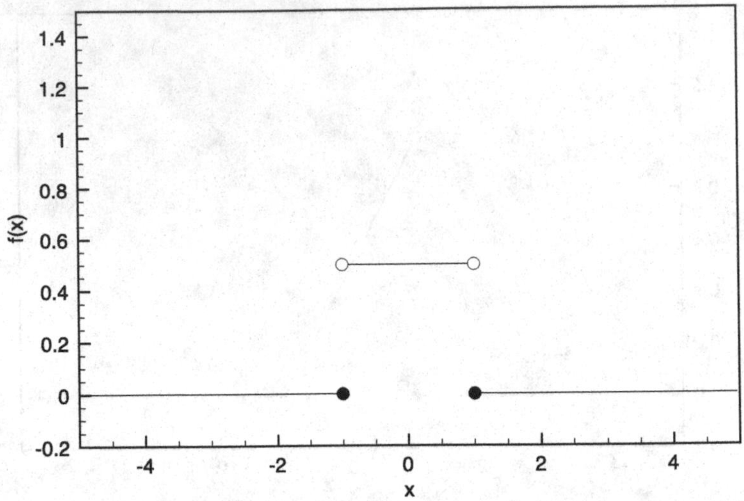

Fig. 1.12 Shown is the pdf $f(x) = 1/2$, for $-1 < x < 1$ and 0 elsewhere

A. The graph for the pdf, $f(x) = 1/2$, for $-1 < x < 1$ and 0 elsewhere, is shown in Fig. 1.12. The first four moments are given by

$$E(X) = \int_{-1}^{1} dx \, \frac{x}{2} = \left. \frac{x^2}{4} \right|_{-1}^{1} = 0$$

$$E(X^2) = \int_{-1}^{1} dx \, \frac{x^2}{2} = \left. \frac{x^3}{6} \right|_{-1}^{1} = \frac{1}{3}$$

$$E(X^3) = \int_{-1}^{1} dx \, \frac{x^3}{2} = \left. \frac{x^4}{8} \right|_{-1}^{1} = 0$$

$$E(X^4) = \int_{-1}^{1} dx \, \frac{x^4}{2} = \left. \frac{x^5}{10} \right|_{-1}^{1} = \frac{1}{5}.$$

For the variance we find $\sigma^2 = E(X^2) - E(X)^2 = 1/3$. The kurtosis is then given by

$$\mathcal{K} = \frac{E\left[(X-\mu)^4\right]}{\sigma^4} = \frac{\frac{1}{5}}{\frac{1}{9}} = \frac{9}{5} = 1.8.$$

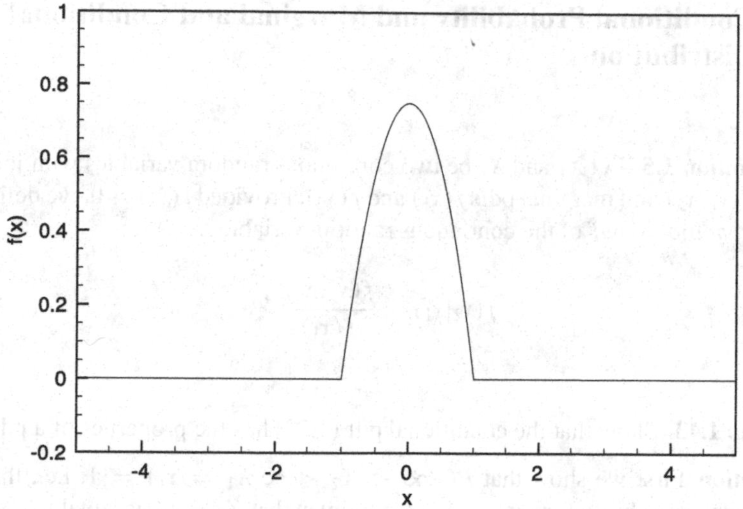

Fig. 1.13 Shown is the pdf $f(x) = 3(1-x^2)/4$, for $-1 < x < 1$ and 0 elsewhere

B. The graph of the pdf, $f(x) = 3(1-x^2)/4$, for $-1 < x < 1$ and 0 elsewhere, is shown in Fig. 1.13. The first four moments are then given by

$$E(X) = \int_{-1}^{1} dx \frac{3}{4}(x - x^3) = \frac{3}{4}\left[\frac{x^2}{2} - \frac{x^4}{4}\right]_{-1}^{1} = 0$$

$$E(X^2) = \int_{-1}^{1} dx \frac{3}{4}(x^2 - x^4) = \frac{3}{4}\left[\frac{x^3}{3} - \frac{x^5}{5}\right]_{-1}^{1} = \frac{1}{5}$$

$$E(X^3) = \int_{-1}^{1} dx \frac{3}{4}(x^3 - x^5) = \frac{3}{4}\left[\frac{x^4}{4} - \frac{x^6}{6}\right]_{-1}^{1} = 0$$

$$E(X^4) = \int_{-1}^{1} dx \frac{3}{4}(x^4 - x^6) = \frac{3}{4}\left[\frac{x^5}{5} - \frac{x^7}{7}\right]_{-1}^{1} = \frac{3}{35}.$$

For the variance we find $\sigma^2 = E(X^2) - E(X)^2 = 1/5$. The kurtosis is then given by

$$\mathcal{K} = \frac{E\left[(X-\mu)^4\right]}{\sigma^4} = \frac{\frac{3}{35}}{\frac{1}{25}} = \frac{15}{7} = 2.14.$$

Obviously, the kurtosis of the first probability distribution ($\mathcal{K} = 1.8$) is smaller than the kurtosis of the second probability distribution ($\mathcal{K} = 2.14$).

1.6 Conditional Probability and Marginal and Conditional Distribution

Definition 1.5 Let X_1 and X_2 be two continuous random variables with joint pdf $f(x_1, x_2)$ and marginal pdfs $f(x_1)$ and $f(x_2)$. Provided $f(x_1) > 0$, we define the *conditional pdf* of the continuous random variable X_2 as

$$f(x_2|x_1) = \frac{f(x_1, x_2)}{f(x_1)}. \tag{1.23}$$

Problem 1.43 Show that the conditional pdf (1.23) has the properties of a pdf!

Solution First we show that $P(-\infty < X_2 < \infty | X_1 = x_1) = 1$, i.e., that the conditional probability for $-\infty < X_2 < \infty$, given that $X_1 = x_1$, is equal to one. We find

$$P(-\infty < X_2 < \infty | X_1 = x_1) = \int_{-\infty}^{\infty} dx_2 f(x_2|x_1) = \int_{-\infty}^{\infty} dx_2 \frac{f(x_1, x_2)}{f(x_1)}$$
$$= \frac{1}{f(x_1)} \int_{-\infty}^{\infty} dx_2 f(x_1, x_2) = \frac{1}{f(x_1)} f(x_1) = 1,$$

where $f(x_1)$ is the marginal pdf of $f(x_1, x_2)$. Secondly, since $f(x_1) \geq 0$ and $f(x_1, x_2) \geq 0$, it follows that $f(x_2|x_1) \geq 0$.

Problem 1.44 Consider the joint pdf

$$f(x_1, x_2) = \begin{cases} \frac{1}{4} x_1 \left(1 + 3x_2^2\right) & \text{for } 0 < x_1 < 2, \quad 0 < x_2 < 1 \\ 0 & \text{elsewhere.} \end{cases}$$

Show that $\int f(x_1, x_2)\, dx_1 dx_2 = 1$. Find $P[(X_1, X_2) \in A]$, where

$$A = \left\{ f(x_1, x_2) | 0 < x_1 < 1, \frac{1}{4} < x_2 < \frac{1}{2} \right\}.$$

Determine also $f_1(x_1), f_2(x_2), f(x_1|x_2)$, and $P(1/4 < X_1 < 1/2 | X_2 = 1/3)$.

Solution First we show, that the above pdf is indeed a pdf,

$$\int_0^2 dx_1 \int_0^1 dx_2\, f(x_1, x_2) = \frac{1}{4} \int_0^2 dx_1 \int_0^1 dx_2\, x_1 \left(1 + 3x_2^2\right)$$
$$= \frac{1}{4} \int_0^2 dx_1\, x_1 \int_0^1 dx_2 + \frac{3}{4} \int_0^2 dx_1\, x_1 \int_0^1 dx_2\, x_2^2$$

$$= \frac{1}{4}\left[\frac{x_1^2}{2}\bigg|_0^2 x_2\bigg|_0^1\right] + \frac{3}{4}\left[\frac{x_1^2}{2}\bigg|_0^2 \frac{x_2^3}{3}\bigg|_0^1\right] = \frac{1}{4}[2\cdot 1] + \frac{3}{4}\left[2\cdot\frac{1}{3}\right] = 1.$$

Next we calculate $P[(X_1, X_2) \in A]$, which can also be written as

$$P\left(0 < X_1 < 1, \frac{1}{4} < X_2 < \frac{1}{2}\right) = \frac{1}{4}\int_0^1 dx_1 \int_{1/4}^{1/2} dx_2\, x_1\left(1+3x_2^2\right) = \frac{23}{512}.$$

The marginal pdfs for $f_1(x_1)$, and $f_2(x_2)$ are calculated by

$$f_1(x_1) = \int_0^1 dx_2\, f(x_1,x_2) = \frac{1}{4}\int_0^1 dx_2\, x_1\left(1+3x_2^2\right) = \frac{x_1}{2}$$

$$f_2(x_2) = \int_0^2 dx_1\, f(x_1,x_2) = \frac{1}{4}\int_0^2 dx_1\, x_1\left(1+3x_2^2\right) = \frac{1}{2}\left(1+3x_2^2\right).$$

The conditional pdfs $f(x_1|x_2)$ and $f(x_2|x_1)$ are calculated by

$$f(x_1|x_2) = \frac{f(x_1,x_2)}{f_2(x_2)} = \frac{\frac{1}{4}x_1\left(1+3x_2^2\right)}{\frac{1}{2}\left(1+3x_2^2\right)} = \frac{x_1}{2}$$

$$f(x_2|x_1) = \frac{f(x_1,x_2)}{f_1(x_1)} = \frac{\frac{1}{4}x_1\left(1+3x_2^2\right)}{\frac{x_1}{2}} = \frac{1}{2}\left(1+3x_2^2\right).$$

Note that the conditional and marginal pdfs are identical, which means that the stochastic variables X_1 and X_2 are stochastically independent (see Sect. 1.7) Lastly, we calculate the conditional probability $P(1/4 < X_1 < 1/2|X_2 = 1/3)$, which can be written as

$$P(1/4 < X_1 < 1/2|X_2 = 1/3) = \int_{1/4}^{1/2} dx_1\, f(x_1|x_2) = \int_{1/4}^{1/2} dx_1\, \frac{x_1}{2} = \frac{3}{64}.$$

Problem 1.45 Two random variables X_1 and X_2 have the joint pdf

$$f(x_1,x_2) = \begin{cases} x_1 + x_2 & \text{for } 0 < x_1 < 1 \text{ and } 0 < x_2 < 1 \\ 0 & \text{elsewhere.} \end{cases}$$

Find the conditional mean and variance of X_2 given $X_1 = x_1$, and $0 < x_1 < 1$.

Solution The idea is as follows: First we calculate the marginal pdf $f(x_1)$ in order to calculate the conditional pdf of X_2, which is given by Eq. (1.23). With the conditional pdf of X_2 we can then easily derive the conditional mean $E[X_2|x_1]$.

- The marginal pdf of X_1 is given by $f(x_1) = \int_{-\infty}^{\infty} f(x_1, x_2)\, dx_2$, so that

$$f(x_1) = \int_0^1 dx_2\ [x_1 + x_2] = \begin{cases} x_1 + \frac{1}{2} & \text{for } 0 < x_1 < 1 \\ 0 & \text{elsewhere.} \end{cases}$$

- According to Eq. (1.23) the conditional pdf of X_2 is then

$$f(x_2|x_1) = \begin{cases} \frac{x_1+x_2}{x_1+\frac{1}{2}} & \text{for } 0 < x_1 < 1 \text{ and } 0 < x_2 < 1 \\ 0 & \text{elsewhere.} \end{cases}$$

- The conditional mean of X_2 given that $X_1 = x_1$ with $0 < x_1 < 0$ is then

$$E(X_2|x_1) = \int_{-\infty}^{\infty} dx_2\, x_2 f(x_2|x_1) = \int_0^1 dx_2\, x_2 \frac{x_1 + x_2}{x_1 + \frac{1}{2}}$$
$$= \frac{1}{x_1 + \frac{1}{2}} \int_0^1 dx_2\ \left[x_1 x_2 + x_2^2\right] = \frac{1}{3} \frac{3x_1 + 2}{2x_1 + 1}.$$

Similarly, we find

$$E(X_2^2|x_1) = \int_{-\infty}^{\infty} dx_2\, x_2^2 f(x_2|x_1) = \int_0^1 dx_2\, x_2^2 \frac{x_1 + x_2}{x_1 + \frac{1}{2}}$$
$$= \frac{1}{x_1 + \frac{1}{2}} \int_0^1 dx_2\ \left[x_1 x_2^2 + x_2^3\right] = \frac{1}{6} \frac{4x_1 + 3}{2x_1 + 1}.$$

- The conditional variance is simply given by

$$\sigma_{X_2} = E(X_2^2|x_1) - E(X_2|x_1)^2 = \frac{1}{18} \frac{6x_1^2 + 6x_1 + 1}{(2x_1 + 1)^2}.$$

Problem 1.46 Suppose the conditional pdf of X_1 given $X_2 = x_2$ is

$$f(x_1|x_2) = \begin{cases} c_1 \frac{x_1}{x_2^2} & \text{for } 0 < x_1 < x_2 \text{ and } 0 < x_2 < 1 \\ 0 & \text{elsewhere} \end{cases}$$

and the marginal pdf of X_2 is

$$f(x_2) = \begin{cases} c_2 x_2^4 & \text{for } 0 < x_2 < 1 \\ 0 & \text{elsewhere.} \end{cases}$$

Compute (A) the constants c_1 and c_2, (B) the joint pdf of X_1 and X_2, (C) the probability $P(1/4 < X_1 < 1/2 \mid X_2 = 5/8)$ and (D) $P(1/4 < X_1 < 1/2)$.

Solution

A. The constant of the marginal pdf of the random variable X_2 can be calculated by

$$1 \stackrel{!}{=} \int_{-\infty}^{\infty} dx_2 f(x_2) = c_2 \int_0^1 dx_2\, x_2^4 = \frac{c_2}{5} \quad \Longrightarrow \quad c_2 = 5.$$

The constant c_1 is then calculated by (see also Problem 1.43, properties of a conditional pdf)

$$1 \stackrel{!}{=} \int_{-\infty}^{\infty} dx_1 f(x_1|x_2) = \int_0^{x_2} dx_1\, c_1 \frac{x_1}{x_2^2} = \frac{1}{2} c_1 \quad \Longrightarrow \quad c_1 = 2.$$

B. The joint probability is then

$$f(x_1, x_2) = f(x_2) f(x_1|x_2) = \begin{cases} 10 x_1 x_2^2 & \text{for } 0 < x_1 < x_2 < 1 \\ 0 & \text{elsewhere.} \end{cases}$$

C. For the probabilities we find

$$P(1/4 < X_1 < 1/2\,|X_2 = 5/8) = \int_{\frac{1}{4}}^{\frac{1}{2}} dx_1 f(x_1|5/8)$$

$$= \frac{64}{25} \int_{\frac{1}{4}}^{\frac{1}{2}} dx_1\, 2x_1 = \frac{12}{25}$$

D. and

$$P(1/4 < X_1 < 1/2) = \int_{\frac{1}{4}}^{\frac{1}{2}} dx_1 f(x_1) = \int_{\frac{1}{4}}^{\frac{1}{2}} dx_1 \int_{x_1}^{1} dx_2 f(x_1, x_2)$$

$$= \int_{\frac{1}{4}}^{\frac{1}{2}} dx_1 \int_{x_1}^{1} dx_2\, 10 x_1 x_2^2 = \frac{449}{1536}.$$

Problem 1.47 Suppose that the joint pdf of X_1 and X_2 is

$$f(x_1, x_2) = \begin{cases} c x_1^2 x_2 & \text{for } x_1^2 \le x_2 < 1 \\ 0 & \text{elsewhere.} \end{cases}$$

(A) Determine the value of the constant c and (B) evaluate the marginal pdfs $f_1(x_1)$ and $f_2(x_2)$. (C) Then calculate $P(X_1 \geq X_2)$ (Hint: sketch the region where $f(x_1, x_2) \geq 0$.).

Solution First we note that the maximum value of the variable x_1^2 is x_2, i.e., $\max\left(x_1^2\right) = x_2$. Since x_2 has to be smaller than 1 we find that $x_1^2 < 1$, which limits the variable x_1 to the range $-1 < x_1 < 1$. It follows that $\min\left(x_1^2\right) = 0$, so that the above pdf is defined in the (more precise) range $0 \leq x_1^2 \leq x_2 < 1$. We could also write

$$f(x_1, x_2) = \begin{cases} cx_1^2 x_2 & \text{for } 0 \leq x_1^2 \leq x_2 < 1 \\ 0 & \text{elsewhere.} \end{cases}$$

A closer inspection of the limits reveals the following two relations:

- For any given x_2 with $0 \leq x_2 < 1$ we find that the variable x_1 has to comply with the limits $0 \leq x_1^2 \leq x_2$ and, thus, $-\sqrt{x_2} \leq x_1 \leq \sqrt{x_2}$.
- For any given x_1 with $-1 < x_1 < 1$ we find that the variable x_2 has to comply with the limits $x_1^2 \leq x_2 < 1$.

A. First we calculate the constant c.

(i) Using the fact that the probability over the entire sample space has to be equal to 1, we can write

$$1 \stackrel{!}{=} \int_{-\infty}^{\infty} dx_1 \int_{-\infty}^{\infty} dx_2\, f(x_1, x_2) = \int_{-1}^{1} dx_1 \int_{x_1^2}^{1} dx_2\, f(x_1, x_2)$$
$$= c \int_{-1}^{1} dx_1\, x_1^2 \int_{x_1^2}^{1} dx_2\, x_2 = \frac{c}{2} \int_{-1}^{1} dx_1\, x_1^2 \left(1 - x_1^4\right) = c\frac{4}{21}.$$

We find for the constant $c = 21/4$.

(ii) Alternatively, we could also have written

$$1 \stackrel{!}{=} \int_{-\infty}^{\infty} dx_1 \int_{-\infty}^{\infty} dx_2\, f(x_1, x_2) = \int_{0}^{1} dx_2 \int_{-\sqrt{x_2}}^{\sqrt{x_2}} dx_1\, f(x_1, x_2)$$
$$= c \int_{0}^{1} dx_2\, x_2 \int_{-\sqrt{x_2}}^{\sqrt{x_2}} dx_1\, x_1^2 = \frac{c}{3} \int_{0}^{1} dx_2\, x_2 \left[x_1^3\right]_{-\sqrt{x_2}}^{\sqrt{x_2}}$$
$$= \frac{2}{3} c \int_{0}^{1} dx_2\, x_2^{5/2} = \frac{2}{3} c \frac{2}{7} \left[x_2^{7/2}\right]_0^1 = c\frac{4}{21}.$$

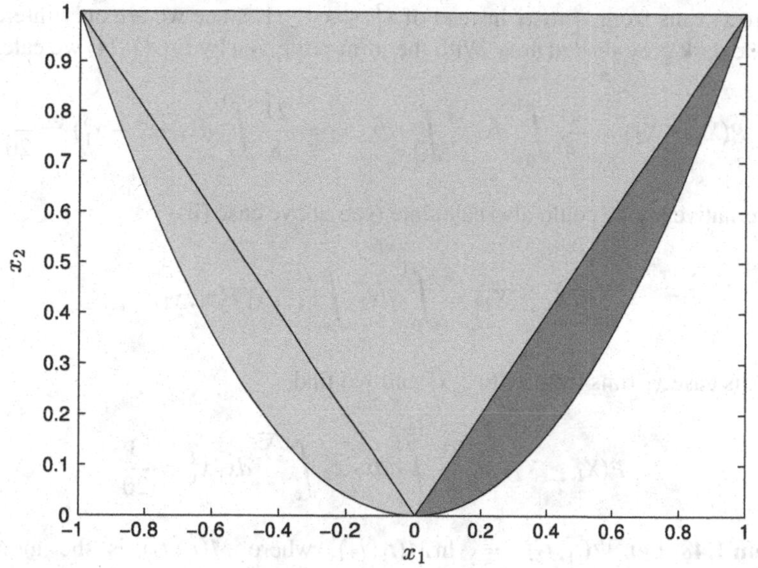

Fig. 1.14 Shown is the area for which the pdf, given by Eq. (1.24), is defined (*light and dark grey*). The *dark grey* area is the area for which $P(X_1 \geq X_2)$

The joint pdf is given by

$$f(x_1, x_2) = \begin{cases} \frac{21}{4} x_1^2 x_2 & \text{for } 0 \leq x_1^2 \leq x_2 < 1 \\ 0 & \text{elsewhere.} \end{cases} \tag{1.24}$$

B. The marginal pdfs are then given by

$$f_1(x_1) = \begin{cases} \frac{21}{4} x_1^2 \int_{x_1^2}^{1} dx_2\, x_2 = \frac{21}{8}\left(x_1^2 - x_1^6\right) & \text{for } -1 < x_1 < 1 \\ 0 & \text{elsewhere} \end{cases}$$

$$f_2(x_2) = \begin{cases} \frac{21}{4} x_2 \int_{-\sqrt{x_2}}^{\sqrt{x_2}} dx_1\, x_1^2 = \frac{21}{6} x_2^{5/2} = \frac{7}{2} x_2^{5/2} & \text{for } 0 \leq x_2 < 1 \\ 0 & \text{elsewhere.} \end{cases}$$

It can easily be shown, that $\int f(x_1)dx_1 = 1 = \int f(x_2)dx_2$.

C. Lastly, we calculate $P(X_1 \geq X_2)$. Since $0 \leq x_1^2 \leq x_2 < 1$, we need to determine where $x_1 \geq x_2$. Figure 1.14 shows for which values the pdf is defined (light and dark grey area). For $P(X_1 \geq X_2)$ the pdf is defined by the dark grey area only. We calculate (see above case (i))

$$P(X_1 \geq X_2) = \int_0^1 dx_1 \int_{x_1^2}^{x_1} dx_2 f(x_1, x_2).$$

Here x_2 runs from x_1^2 to x_1 instead of $x_1^2 \leq x_2 \leq 1$, since we are only interested in the dark grey shaded area. With the joint pdf given by Eq. (1.24) we calculate

$$P(X_1 \geq X_2) = \frac{21}{4} \int_0^1 dx_1\, x_1^2 \int_{x_1^2}^{x_1} dx_2\, x_2 = \frac{21}{8} \int_0^1 dx_1\, \left(x_1^4 - x_1^6\right) = \frac{3}{20}.$$

Alternatively, one could also calculate (see above case (ii))

$$P(X_1 \geq X_2) = \int_0^1 dx_2 \int_{x_2}^{\sqrt{x_2}} dx_1 f(x_1, x_2).$$

In this case x_1 runs from x_2 to $\sqrt{x_2}$ and we find

$$P(X_1 \geq X_2) = \frac{21}{4} \int_0^1 dx_2\, x_2 \int_{x_2}^{\sqrt{x_2}} dx_1\, x_1^2 = \frac{3}{20}.$$

Problem 1.48 Let $\Psi(t_1, t_2) = \ln M(t_1, t_2)$, where $M(t_1, t_2)$ is the moment-generating function of X and Y. Show that $(k = 1, 2)$

$$\frac{\partial \Psi(0,0)}{\partial t_k} = \mu_k, \qquad \frac{\partial^2 \Psi(0,0)}{\partial t_k^2} = \sigma_k^2, \qquad \frac{\partial^2 \Psi(0,0)}{\partial t_1 \partial t_2} = \mathrm{Cov}$$

yields the means, the variances, and the covariance of the two random variables.

Solution We repeat briefly that

$$M(0,0) = 1, \qquad \frac{\partial M(0,0)}{\partial t_k} = \mu_k, \qquad \frac{\partial^2 M(0,0)}{\partial t_k^2} - \left[\frac{\partial M(0,0)}{\partial t_k}\right]^2 = \sigma_k^2,$$

and that the covariance is defined as

$$\mathrm{Cov} = \frac{\partial^2 M(0,0)}{\partial t_1 \partial_2} - \mu_1 \mu_2.$$

With that we can now easily derive

$$\frac{\partial \Psi(t_1, t_2)}{\partial t_k} = \frac{1}{M(t_1, t_2)} \frac{\partial M(t_1, t_2)}{\partial t_k},$$

so that

$$\frac{\partial \Psi(0,0)}{\partial t_k} = \frac{1}{M(0,0)} \frac{\partial M(0,0)}{\partial t_k} = \mu_k.$$

Secondly, we have

$$\frac{\partial^2 \Psi(t_1,t_2)}{\partial t_k^2} = \frac{1}{M(t_1,t_2)^2}\left[\frac{\partial^2 M(t_1,t_2)}{\partial t_k^2}M(t_1,t_2) - \left(\frac{\partial M(t_1,t_2)}{\partial t_k}\right)^2\right],$$

so that

$$\frac{\partial^2 \Psi(0,0)}{\partial t_k^2} = \frac{1}{M(0,0)^2}\left[\frac{\partial^2 M(0,0)}{\partial t_k^2}M(0,0) - \left(\frac{\partial M(0,0)}{\partial t_k}\right)^2\right] = \sigma_k^2 .$$

Lastly, we have

$$\frac{\partial^2 \Psi(t_1,t_2)}{\partial t_1 \partial t_2} = \frac{\partial^2 \ln M(t_1,t_2)}{\partial t_1 \partial t_2} = \frac{\partial}{\partial t_1}\left[\frac{\partial \ln M(t_1,t_2)}{\partial t_2}\right] = \frac{\partial}{\partial t_1}\left[\frac{1}{M(t_1,t_2)}\frac{\partial M(t_1,t_2)}{\partial t_2}\right]$$
$$= \frac{1}{M(t_1,t_2)}\frac{\partial^2 M(t_1,t_2)}{\partial t_1 \partial t_2} - \frac{1}{M(t_1,t_2)^2}\frac{\partial M(t_1,t_2)}{\partial t_1}\frac{\partial M(t_1,t_2)}{\partial t_2}.$$

From that it follows immediately that

$$\frac{1}{M(0,0)}\frac{\partial^2 M(0,0)}{\partial t_1 \partial t_2} - \frac{1}{M(0,0)^2}\frac{\partial M(0,0)}{\partial t_1}\frac{\partial M(0,0)}{\partial t_2} = \frac{\partial^2 M(0,0)}{\partial t_1 \partial t_2} - \mu_1\mu_2 = \mathrm{Cov}\,.$$

Problem 1.49 Given the joint pdf of X_1 and X_2,

$$f(x_1,x_2) = \begin{cases} 21x_1^2x_2^3 & \text{for } 0 < x_1 < x_2 < 1 \\ 0 & \text{elsewhere,} \end{cases}$$

find the conditional mean and variance of X_1 given $X_2 = x_2$ with $0 < x_2 < 1$.

Solution In order to calculate the conditional pdf $f(x_1|x_2) = f(x_1,x_2)/f(x_2)$, which is needed to calculate the conditional mean and variance, we calculate first the marginal pdf $f(x_2)$,

$$f(x_2) = \int_{-\infty}^{\infty} dx_1\, f(x_1,x_2) = 21x_2^3\int_0^{x_2} dx_1\, x_1^2 = \begin{cases} 7x_2^6 & \text{for } 0 < x_2 < 1 \\ 0 & \text{elsewhere.} \end{cases}$$

The conditional pdf is then simply

$$f(x_1|x_2) = \frac{f(x_1,x_2)}{f(x_2)} = \begin{cases} 3\frac{x_1^2}{x_2^3} & \text{for } 0 < x_1 < x_2 < 1 \\ 0 & \text{elsewhere.} \end{cases}$$

The conditional mean is then calculated by

$$E(X_1|x_2) = \int_{-\infty}^{\infty} dx_1\, x_1 f(x_1|x_2) = \frac{3}{x_2^3}\int_0^{x_2} dx_1\, x_1^3 = \frac{3}{4}x_2$$

for $0 < x_2 < 1$. The second moment of X_1 given $X_2 = x_2$ is then

$$E(X_1^2|x_2) = \int_{-\infty}^{\infty} dx_1\, x_1^2 f(x_1|x_2) = \frac{3}{x_2^3}\int_0^{x_2} dx_1\, x_1^4 = \frac{3}{5}x_2^2$$

for $0 < x_2 < 1$. The conditional variance is then

$$\begin{aligned} E\Big[[X_1 - E(X_1|x_2)]^2\Big| x_2\Big] &= E(X_1^2|x_2) - E(X_1|x_2)^2 \\ &= \frac{3}{5}x_2^2 - \frac{9}{16}x_2^2 = \frac{3}{80}x_2^2. \end{aligned}$$

Problem 1.50 Five cards are drawn at random without replacement from a deck of 52 cards. The random variables X_1, X_2, and X_3 denote the number of spades, the number of hearts, and the number of diamonds that appear among the 5 cards respectively. Determine the joint pdf of X_1, X_2, and X_3. Find the marginal pdfs of X_1, X_2, and X_3. What is the joint conditional pdf of X_2 and X_3 given that $X_1 = 3$?

Solution The joint pdf is given by

$$f(x_1, x_2, x_3) = \begin{cases} \dfrac{C_{x_1}^{13} C_{x_2}^{13} C_{x_3}^{13} C_{5-x_1-x_2-x_3}^{13}}{C_5^{52}} & \text{for } x_{1,2,3} = 0, 1, \ldots, 5 \\ 0 & \text{elsewhere,} \end{cases}$$

where we used the binomial coefficient (1.29), see also Problem 1.36. The marginal pdfs are then given by

$$f(x_1) = \frac{C_{x_1}^{13}}{C_5^{52}} \sum_{x_2=0}^{5-x_1} \sum_{x_3=0}^{5-x_1-x_2} C_{x_2}^{13} C_{x_3}^{13} C_{5-x_1-x_2-x_3}^{13}$$

$$f(x_2) = \frac{C_{x_2}^{13}}{C_5^{52}} \sum_{x_1=0}^{5-x_2} \sum_{x_3=0}^{5-x_1-x_2} C_{x_1}^{13} C_{x_3}^{13} C_{5-x_1-x_2-x_3}^{13}$$

and

$$f(x_3) = \frac{C_{x_3}^{13}}{C_5^{52}} \sum_{x_1=0}^{5-x_3} \sum_{x_2=0}^{5-x_3-x_1} C_{x_1}^{13} C_{x_2}^{13} C_{5-x_1-x_2-x_3}^{13}.$$

For the joint conditional pdf of X_2 and X_3 given that $X_1 = 3$ we find

$$f(x_2, x_3|x_1 = 3) = \frac{f(x_1 = 3, x_2, x_3)}{f(x_1 = 3)}$$
$$= \frac{C^{13}_{x_2} C^{13}_{x_3} C^{13}_{2-x_2-x_3}}{\sum_{x_2=0}^{2} \sum_{x_3=0}^{2-x_2} C^{13}_{x_2} C^{13}_{x_3} C^{13}_{2-x_2-x_3}}.$$

Problem 1.51 Suppose that the joint pdf of X and Y is given by

$$f(x, y) = \begin{cases} 2 & \text{for } 0 < x < y, \quad 0 < y < 1 \\ 0 & \text{elsewhere.} \end{cases}$$

Show that the conditional means are $(1+x)/2$ for $0 < x < 1$ and $y/2$ for $0 < y < 1$, and the correlation function of X and Y is $\rho = 1/2$. Show also that the variance of the conditional distribution of Y given $X = x$ is $(1-x)^2/12$ for $0 < x < 1$, and that the variance of the conditional distribution of X given $Y = y$ is $y^2/12$ for $0 < y < 1$.

Solution A closer inspection reveals:

A. For any given $X = x$ the variable Y complies with $x < y < 1$.
B. For any given $Y = y$ the variable X complies with $0 < x < y$.

- First we calculate the marginal pdfs:

$$f(x) = \int_{-\infty}^{\infty} dy f(x, y) = \int_x^1 dy\, 2 = 2y|_x^1 = 2(1-x) \qquad \text{for } 0 < x < 1$$
$$f(y) = \int_{-\infty}^{\infty} dx f(x, y) = \int_0^y dx\, 2 = 2x|_0^y = 2y \qquad \text{for } 0 < y < 1.$$

It can easily be shown, that $\int f(x)dx = 1 = \int f(y)dy$. Also, the random variables X_1 and X_2 are *not* stochastically independent, since $f(x_1, x_2) \neq f(x_1)f(x_2)$.

- Next we calculate the conditional pdfs:

$$f(x|y) = \frac{f(x, y)}{f(y)} = \frac{2}{2y} = \frac{1}{y} \qquad \text{for } x < y < 1 \text{ and } 0 < x < 1$$
$$f(y|x) = \frac{f(x, y)}{f(x)} = \frac{2}{2(1-x)} = \frac{1}{(1-x)} \qquad \text{for } 0 < x < y \text{ and } 0 < y < 1.$$

- The conditional mean of X given $Y = y$ is

$$E(X|y) = \int_{-\infty}^{\infty} dx\, x f(x|y) = \int_0^y dx \frac{x}{y} = \frac{1}{2y} x^2 \Big|_0^y = \frac{y}{2}$$

for $0 < y < 1$. The conditional mean of Y given $X = x$ is

$$E(Y|x) = \int_{-\infty}^{\infty} dx\, y f(y|x) = \int_x^1 dy \frac{y}{1-x}$$
$$= \frac{1}{1-x}\frac{y^2}{2}\bigg|_x^1 = \frac{1-x^2}{2(1-x)} = \frac{1+x}{2}$$

for $0 < x < 1$.

- The conditional variances can be calculated by

$$\sigma_{x|y} = E\left([X - E(X|y)]^2\,|y\right) = E(X^2|y) - E(X|y)^2, \tag{1.25}$$

so that

$$\sigma_{x|y} = E(X^2|y) - E(X|y)^2 = \int_0^y dx \frac{x^2}{y} - \frac{y^2}{4} = \frac{x^3}{3y}\bigg|_0^y - \frac{y^2}{4} = \frac{y^2}{3} - \frac{y^2}{4} = \frac{y^2}{12}$$

for $0 < y < 1$ and

$$\sigma_{y|x} = E(Y^2|x) - E(Y|x)^2 = \int_x^1 dy \frac{y^2}{1-x} - \frac{(1+x)^2}{4}$$
$$= \frac{y^3}{3(1-x)}\bigg|_x^1 - \frac{(1+x)^2}{4} = \frac{1-x^3}{3(1-x)} - \frac{(1+x)^2}{4}$$
$$= \frac{1+x+x^2}{3} - \frac{1+2x+x^2}{4} = \frac{4+4x+4x^2-3-6x-3x^2}{12}$$
$$= \frac{1-2x+x^2}{12} = \frac{(1-x)^2}{12},$$

where we used $1 - x^3 = (1-x)(1+x+x^2)$.

- To calculate the correlation coefficient (function), ρ, which is defined as

$$\rho = \frac{E\left[(X-\mu_x)(Y-\mu_y)\right]}{\sigma_x\sigma_y} = \frac{E(XY) - \mu_x\mu_y}{\sigma_x\sigma_y}, \tag{1.26}$$

we need to determine μ_x, μ_y, and $E(XY)$, as well as $E(X^2)$ and $E(Y^2)$. We start with

$$\mu_x = E(X) = \int_{-\infty}^{\infty} dx \int_{-\infty}^{\infty} dy\, x f(x,y)$$
$$= \int_0^1 dy \int_0^y dx\, 2x = \int_0^1 dy\, y^2 = \frac{1}{3}$$

and

$$\mu_y = E(Y) = \int_{-\infty}^{\infty} dx \int_{-\infty}^{\infty} dy\, y f(x,y)$$
$$= \int_0^1 dx \int_x^1 dy\, 2y = \int_0^1 dx\,(1-x^2) = \left[x - \frac{x^3}{3}\right]_0^1 = \frac{2}{3}.$$

The expected value of XY is

$$E(XY) = \int_0^1 dy \int_0^y 2xy = \int_0^1 dy\, y^3 = \frac{1}{4}.$$

For the variances we need to calculate

$$E(X^2) = \int_0^1 dy \int_0^y dx\, 2x^2 = \frac{2}{3}\int_0^1 dy\, y^3 = \frac{1}{6}$$
$$E(Y^2) = \int_0^1 dx \int_x^1 dy\, 2y^2 = \frac{2}{3}\int_0^1 dy\,(1-x^3) = \frac{1}{2}.$$

The variances are then calculated by

$$\sigma_x^2 = E(X^2) - E(X)^2 = \frac{1}{6} - \frac{1}{9} = \frac{1}{18}$$
$$\sigma_y^2 = E(Y^2) - E(Y)^2 = \frac{1}{2} - \frac{4}{9} = \frac{1}{18}.$$

Finally, we find for the correlation coefficient (function),

$$\rho = \frac{E(XY) - \mu_x\mu_y}{\sigma_x\sigma_y} = 18\left[\frac{1}{4} - \frac{1}{3}\cdot\frac{2}{3}\right] = \frac{18}{36} = \frac{1}{2}.$$

Problem 1.52 Let $f(t)$ and $F(t)$ be the pdf and the distribution function of the random variable T. The conditional pdf of T given $T > t_0$, t_0 a fixed time, is defined by $f(t|T > t_0) = f(t)/[1 - F(t_0)]$, $t > t_0$, 0 elsewhere. This kind of pdf is used in survival analysis, i.e., problems of time until death, given survival until time t_0. Show that $f(t|T > t_0)$ is a pdf. Let $f(t) = e^{-t}$, $0 < t < \infty$, 0 elsewhere, and compute $P(T > 2|T > 1)$.

Solution Since there are no negative times ($t > 0$), we find for the distribution function $F(t_0) = \int_0^{t_0} dt f(t)$ and for the pdf $f(t)$

$$1 = \int_0^{\infty} dt f(t) = \int_0^{t_0} dt f(t) + \int_{t_0}^{\infty} dt f(t) = F(t_0) + \int_{t_0}^{\infty} dt f(t),$$

from which it follows that

$$\int_{t_0}^{\infty} dt f(t) = 1 - F(t_0). \tag{1.27}$$

For the conditional pdf we find

$$\int_0^{\infty} dt f(t|T > t_0) = \frac{1}{1 - F(t_0)} \int_{t_0}^{\infty} dt f(t) = 1,$$

where we used Eq. (1.27) in the last step. With $f(t) = e^{-t}, 0 < t < \infty, 0$ elsewhere, we find from the definition of the distribution function that $F(t_0 = 1) = 1 - e^{-1}$. The probability of $P(T > 2|T > 1)$ is then given by

$$P(T > 2|T > 1) = \int_2^{\infty} dt f(t|T > 1) = \frac{1}{1 - F(1)} \int_2^{\infty} dt f(t)$$
$$= e \int_2^{\infty} dt\, e^{-t} = e^{-1}.$$

1.7 Stochastic Independence

Definition 1.6 Let the random variables X_1 and X_2 have the joint pdf $f(x_1, x_2)$ and marginal pdfs $f(x_1)$ and $f(x_2)$. The random variables X_1 and X_2 are *stochastically independent* if and only if $f(x_1, x_2) = f(x_1)f(x_2)$. Otherwise they are *stochastically dependent*.

Problem 1.53 Let the joint pdf of X_1 and X_2 be

$$f(x_1, x_2) = \begin{cases} x_1 + x_2 & \text{for } 0 < x_1 < 1, \text{ and } 0 < x_2 < 1 \\ 0 & \text{elsewhere.} \end{cases}$$

Show that the random variables X_1 and X_2 are stochastically dependent.

Solution The marginal pdfs are given by

$$f(x_1) = \int_{-\infty}^{\infty} dx_2 f(x_1, x_2) = \int_0^1 dx_2\, (x_1 + x_2) = x_1 + \frac{1}{2}$$
$$f(x_2) = \int_{-\infty}^{\infty} dx_1 f(x_1, x_2) = \int_0^1 dx_1\, (x_1 + x_2) = x_2 + \frac{1}{2}.$$

Obviously, $f(x_1, x_2) \neq f(x_1)f(x_2)$, thus the two random variables X_1 and X_2 are stochastically dependent.

Problem 1.54 Show that the random variables X and Y with joint pdf

$$f(x, y) = \begin{cases} 2e^{-x-y} & \text{for } 0 < x < y, 0 < y < \infty \\ 0 & \text{elsewhere} \end{cases}$$

are stochastically dependent.

Solution The marginal pdfs are given by

$$f(x) = \int_x^{\infty} dy\, 2e^{-x-y} = 2e^{-2x} \qquad 0 < x < \infty$$

$$f(y) = \int_0^{y} dx\, 2e^{-x-y} = 2e^{-y}\left[1 - e^{-y}\right] \qquad 0 < y < \infty$$

Obviously, $f(x, y) \neq f(x)f(y)$, which means that the random variables X and Y are stochastically dependent.

Problem 1.55 Consider the joint pdf of two random variables X and Y,

$$f(x, y) \begin{cases} \frac{x(1+3y^2)}{4} & \text{for } 0 < x < 2, 0 < y < 1 \\ 0 & \text{elsewhere.} \end{cases}$$

Are the random variables X and Y stochastically independent? Compute $f(x|y)$ and hence $P(1/4 < X < 1/2 | Y = 3)$.

Solution The marginal pdfs are given by

$$f(x) = \int_0^1 dy \frac{x(1 + 3y^2)}{4} = \frac{x}{4} \int_0^1 dy\,(1 + 3y^2) = \frac{x}{4}\left[y + y^3\right]_0^1 = \frac{x}{2}$$

for $0 < x < 2$, and

$$f(y) = \int_0^2 dx \frac{x(1 + 3y^2)}{4} = \frac{(1 + 3y^2)}{4} \int_0^2 dx\, x = \frac{(1 + 3y^2)}{2}$$

for $0 < y < 1$. Obviously, $f(x, y) = f(x)f(y)$, thus the two random variables are stochastically independent. The conditional pdf for the random variable X given $Y = y$ is

$$f(x|y) = \frac{f(x, y)}{f(y)} = \frac{x(1 + 3y^2)}{4} \frac{2}{(1 + 3y^2)} = \frac{x}{2}$$

for $0 < x < 1$. Obviously, the conditional pdf does not depend on the variable y, thus

$$P(1/4 < X < 1/2|Y = 3) = \int_{1/4}^{1/2} dx f(x|y = 3) = \left[\frac{x^2}{4}\right]_{1/4}^{1/2} = \frac{3}{64}.$$

Problem 1.56 The random variables X and Y have joint pdf

$$f(x, y) = \begin{cases} 4x(1 - y) & \text{for } 0 < x < 1, 0 < y < 1 \\ 0 & \text{elsewhere.} \end{cases}$$

Find $P(0 < X < 1/3, 0 < Y < 1/3)$.

Solution The probability is easily calculated by

$$P(0 < X < 1/3, 0 < Y < 1/3) = \int_0^{1/3} dx \int_0^{1/3} dy\, 4x(1 - y) = \frac{5}{81}.$$

Problem 1.57 Let X_1, X_2, and X_3 be three stochastically independent random variables, each with pdf

$$f(x) = \begin{cases} e^{-x} & \text{for } 0 < x \\ 0 & \text{elsewhere.} \end{cases}$$

Find $P(X_1 < 2, 1 < X_2 < 3, X_3 > 2)$.

Solution Since the pdfs of each variable are stochastically independent, we find easily the joint pdf as

$$f(x_1, x_2, x_3) = f(x_1)f(x_2)f(x_3) = e^{-x_1-x_2-x_3}.$$

The probability is then given by

$$\begin{aligned} &P(X_1 < 2, 1 < X_2 < 3, X_3 > 2) \\ &= \int_0^2 dx_1 \int_1^3 dx_2 \int_2^\infty dx_3\, e^{-x_1-x_2-x_3} \\ &= \int_0^2 dx_1 e^{-x_1} \int_1^3 dx_2 e^{-x_2} \int_2^\infty dx_3\, e^{-x_3} = \left[1 - \frac{1}{e^2}\right]\left[\frac{1}{e} - \frac{1}{e^3}\right]\frac{1}{e^2} \\ &\approx 0.037. \end{aligned}$$

Problem 1.58 Show that the random variables X and Y with joint pdf

$$f(x,y) \begin{cases} e^{-x-y} & \text{for } 0 < x < \infty, 0 < y < \infty \\ 0 & \text{elsewhere} \end{cases}$$

are stochastically independent and that

$$E\left(e^{t(X+Y)}\right) = (1-t)^{-2}, \qquad t < 1.$$

Solution The marginal pdfs are

$$f(x) = \int_0^\infty dy\, e^{-x-y} = e^{-x} \qquad 0 < x < \infty$$

$$f(y) = \int_0^\infty dx\, e^{-x-y} = e^{-y} \qquad 0 < y < \infty.$$

Since $f(x,y) = f(x)f(y)$ both variables are stochastically independent. We further find

$$E\left(e^{t(X+Y)}\right) = \int_{-\infty}^\infty dx \int_{-\infty}^\infty dy\, e^{t(x+y)} f(x,y) = \int_0^\infty dx \int_0^\infty dy\, e^{t(x+y)} e^{-x-y}$$

$$= \int_0^\infty dx\, e^{-x(1-t)} \int_0^\infty dy\, e^{-y(1-t)} = \frac{1}{(1-t)^2} \qquad \text{for } t < 1.$$

Note that the integration converges only for $t < 1$.

Problem 1.59 Show that the random variables X and Y with joint pdf

$$f(x,y) = \begin{cases} 12xy(1-y) & \text{for } 0 < x < 1, 0 < y < 1 \\ 0 & \text{elsewhere} \end{cases}$$

are stochastically independent.

Solution The marginal pdfs are

$$f(x) = \int_0^1 dy\, 12xy(1-y) = 2x \qquad 0 < x < 1$$

$$f(y) = \int_0^1 dx\, 12xy(1-y) = 6y(1-y) \qquad 0 < y < 1.$$

Obviously, $f(x,y) = f(x)f(y)$, which means that the random variables X and Y are stochastically independent.

1.8 Particular Distributions

1.8.1 The Binomial Distribution

The pdf of the binomial distribution is given by

$$f(x) = \binom{n}{x} p^x (1-p)^{n-x}, \tag{1.28}$$

where the binomial coefficient describes the number of combinations (see also Problem 1.36) and is given by

$$C_x^n = \binom{n}{x} = \frac{n!}{x!(n-x)!}. \tag{1.29}$$

The mean and variance are given by $\mu = np$ and $\sigma^2 = np(1-p)$. The binomial distribution is commonly denoted by $B(n, p)$. The moment-generating function of the binomial distribution is defined by

$$M(t) = \left[(1-p) + pe^t\right]^n. \tag{1.30}$$

The *binomial theorem* can be written as

$$(x+y)^n = \sum_{k=0}^{n} \binom{n}{k} x^{n-k} y^k. \tag{1.31}$$

Problem 1.60 If the moment-generating function of a random variable X is

$$M(t) = \left(\frac{1}{3} + \frac{2}{3}e^t\right)^5,$$

find $P(X = 2 \text{ or } 3)$.

Solution By comparing the moment-generating function with the definition (1.30) above we find $n = 5$ and $p = 2/3$. The binomial pdf is then given by

$$f(x) = \binom{5}{x} \left(\frac{2}{3}\right)^x \left(\frac{1}{3}\right)^{5-x}.$$

The first factor is the binomial coefficient (1.29). The probability of $P(X = 2 \text{ or } 3) = P(2) + P(3)$ is given by

$$P(2) + P(3) = \sum_{x=2,3} \binom{5}{x} \left(\frac{2}{3}\right)^x \left(\frac{1}{3}\right)^{5-x} = \frac{40}{81}.$$

Problem 1.61 The moment-generating function of a random variable X is

$$M(t) = \left(\frac{2}{3} + \frac{1}{3}e^t\right)^9.$$

Show that

$$P(\mu - 2\sigma < X < \mu + 2\sigma) = \sum_{x=1}^{5} \binom{9}{x} \left(\frac{1}{3}\right)^x \left(\frac{2}{3}\right)^{9-x}.$$

Solution If we compare the moment-generating function above with the definition given by Eq. (1.30) we find $p = 1/3$ and $n = 9$. With $\mu = np$ and $\sigma^2 = np(1-p)$ we find $\mu = 3$ and $\sigma^2 = 2$, so that $\mu + 2\sigma \approx 5.8$ and $\mu - 2\sigma \approx 0.2$. Since the binomial distribution is discrete we find

$$P(\mu - 2\sigma < X < \mu + 2\sigma) = \sum_{x=1}^{5} \binom{9}{x} \left(\frac{1}{3}\right)^x \left(\frac{2}{3}\right)^{9-x}.$$

Problem 1.62 The probability that a patient recovers from heart surgery is 0.4. If 15 people have had surgery, what is the probability that (A) at least 10 survive (B) from 3 to 8 survive (C) exactly 5 survive? (D) Using Chebyshev's inequality, find and interpret the interval $\mu \pm 2\sigma$.

Solution The general form of the binomial distribution is given by Eq. (1.28), where the binomial coefficient is defined by Eq. (1.29). We use the following notation: The probability of surviving is $p = 0.4$. Consequently, the probability of dying is $1 - p = 0.6$. The experiment comprises $n = 15$ people, so that the pdf can be written as

$$f(x) = \binom{15}{x} (0.4)^x (0.6)^{15-x}. \tag{1.32}$$

Figure 1.15 shows this particular distribution. We further find:

A. The probability that at least 10 people survive is given by

$$P(X \geq 10) = \sum_{x=10}^{15} f(x) = 0.0338,$$

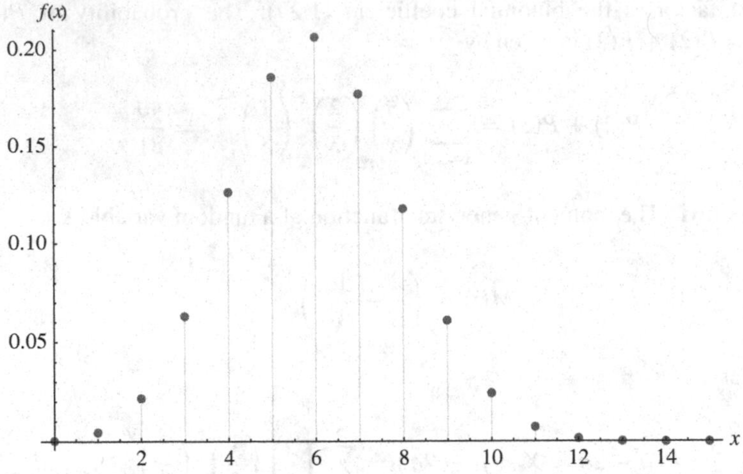

Fig. 1.15 Shown is the binomial distribution as given in Eq. (1.32)

which is the sum of the probabilities that exactly 10, 11, 12, 13, 14, and 15 people survive.

B. The probability that 3–8 people survive is given by

$$P(3 \leq X \leq 8) = \sum_{3}^{8} f(x) = 0.8778,$$

the sum of the probabilities that exactly 3, 4, 5, 6, 7, and 8 people survive.

C. The probability that exactly 5 people survive is given by

$$P(X = 5) = f(5) = 0.1859.$$

D. Next, we find $\mu = np = 15 \cdot 0.4 = 6$ and $\sigma^2 = np(1-p) = 36/10$, so that $\mu - 2\sigma = 6 - 12/\sqrt{10} = 2.2$ and $\mu + 2\sigma = 6 + 12/\sqrt{10} = 9.8$. With Chebyshev's inequality, see Eq. (1.20) from Problem 1.35, we find with $n = 2$

$$P(\mu - n\sigma < X < \mu + n\sigma) \geq 1 - \frac{1}{n^2}$$

$$P\,(3 \leq X \leq 9) \geq \frac{3}{4}.$$

The probability that between 3 and 9 people survive is *at least* 3/4. In fact the exact probability is

$$P\,(3 \leq X \leq 9) = \sum_{3}^{9} f(x) = 0.9391.$$

Problem 1.63 If the random variable X has a binomial distribution, $B(n, p)$, with parameters n and p, show that

$$E\left(\frac{X}{n}\right) = p \quad \text{and} \quad E\left[\left(\frac{X}{n} - p\right)^2\right] = \frac{p(1-p)}{n}.$$

Solution We remember that

$$E(X) = \mu = np \qquad E(X^2) = np + n(n-1)p^2.$$

Therefore, we find

$$E\left(\frac{X}{n}\right) = \frac{1}{n}E(X) = \frac{np}{n} = p$$

and

$$\begin{aligned} E\left[\left(\frac{X}{n} - p\right)^2\right] &= E\left(\frac{X^2}{n^2} - \frac{2p}{n}X + p^2\right) = \frac{1}{n^2}E(X^2) - \frac{2p}{n}E(X) + p^2 \\ &= \frac{np + n(n-1)p^2}{n^2} - \frac{2p^2 n}{n} + p^2 \\ &= \frac{np + n^2p^2 - np^2}{n^2} - p^2 = \frac{p(1-p)}{n}. \end{aligned}$$

1.8.2 The Poisson Distribution

The Poisson distribution is given by

$$f(x) = \frac{p^x e^{-p}}{x!}. \tag{1.33}$$

The mean and variance are given by $p = \mu = \sigma^2 > 0$. The general form of the moment-generating function for a Poisson distribution is

$$M(t) = \sum_x e^{tx} f(x) = \exp\left[p(e^t - 1)\right]. \tag{1.34}$$

Problem 1.64 If the random variable X has a Poisson distribution such that $P(X = 1) = P(X = 2)$, find $P(X = 4)$.

Solution The probability of $P(X = 1)$ is given by $f(1)$ and the probability of $P(X = 2)$ is given by $f(2)$. Since $P(X = 1) = P(X = 2)$, we calculate

$$f(1) = f(2) \qquad \Longrightarrow \qquad \frac{p^1 e^{-p}}{1!} = \frac{p^2 e^{-p}}{2!},$$

which leads to $p = 2$. The Poisson distribution for the random variable X is then given by

$$f(x) = \frac{2^x e^{-2}}{x!}.$$

The probability of $P(X = 4)$ is then

$$P(X = 4) = f(4) = \frac{2^4 e^{-2}}{4!} = \frac{2}{3}e^{-2} \approx 0.09.$$

Problem 1.65 Given that $M(t) = \exp[4(e^t - 1)]$ is the moment-generating function of a random variable X, show that $P(\mu - 2\sigma < X < \mu + \sigma) = 0.931$.

Solution By comparing $M(t) = \exp[4(e^t - 1)]$ with the general form (1.34) we find $p = 4$. Since the mean and the variance for a Poisson distribution are given by $\mu = \sigma^2 = p$, we find immediately

$$P(\mu - 2\sigma < X < \mu + 2\sigma) = P(0 < X < 8) = e^{-4} \sum_{x=1}^{7} \frac{4^x}{x!} \approx 0.931.$$

Note that the random variable X is larger than 0 but smaller than 8, i.e., the sum extends from 1 to 7.

Problem 1.66 Suppose that during a given rush hour Wednesday, the number of accidents on a certain stretch of highway has a Poisson distribution with mean 0.7. What is the probability that there will be at least three accidents on that stretch of highway at rush hour on Wednesday?

Solution The Poisson distribution is given by Eq. (1.33), where the mean and the variance corresponds to p, i.e., $p = \mu = \sigma^2$. The Poisson distribution with mean $\mu = 0.7$ can, therefore, be written as

$$f(x) = \frac{0.7^x e^{-0.7}}{x!},$$

where x denotes the number of accidents. Since $P(X \geq 3) = 1 - P(X < 3)$, we find

$$P(X \geq 3) = 1 - \sum_{0}^{2} f(x) = 1 - \left[\frac{0.7^0 e^{-0.7}}{0!} + \frac{0.7^1 e^{-0.7}}{1!} + \frac{0.7^2 e^{-0.7}}{2!}\right]$$
$$= 1 - [0.4966 + 0.3476 + 0.1216] = 1 - 0.9656$$
$$= 0.0341.$$

There is a 3.4 % probability that at least three accidents occur on that stretch of highway.

Problem 1.67 Compute the measures of skewness and kurtosis of the Poisson distribution with mean μ.

Solution The skewness and kurtosis are given by

$$S = \frac{E[(X-\mu)^3]}{\sigma^3} \qquad K = \frac{E[(X-\mu)^4]}{\sigma^4} \tag{1.35}$$

where the third moment is given by Eq. (1.21) and the fourth moment by Eq. (1.22). We know that the mean and the variance for the Poisson distribution is given by $p = \mu = \sigma^2$, so that the third and fourth moment can also be written as

$$E[(X-p)^3] = E[X^3] - 3pE[X^2] + 2p^3$$
$$E[(X-p)^4] = E[X^4] - 4pE[X^3] + 6p^2 E[X^2] - 3p^4.$$

The moment-generating function for the Poisson distribution is given by Eq. (1.34). The first four derivatives are

$$M'(t) = pe^t M(t)$$
$$M''(t) = pe^t M(t) + p^2 e^{2t} M(t)$$
$$M'''(t) = pe^t M(t) + 3p^2 e^{2t} M(t) + p^3 e^{3t} M(t)$$
$$M^4(t) = pe^t M(t) + 7p^2 e^{2t} M(t) + 6p^3 e^{3t} M(t) + p^4 e^{4t} M(t).$$

With $M(t=0) = 1$ the expectations are then given by

$$E[X] = M'(t=0) = p$$
$$E[X^2] = M''(t=0) = p + p^2$$
$$E[X^3] = M'''(t=0) = p + 3p^2 + p^3$$
$$E[X^4] = M^4(t=0) = p + 7p^2 + 6p^3 + p^4.$$

The third and fourth moment is then

$$E[(X-p)^3] = p$$
$$E[(X-p)^4] = p + 3p^2.$$

The skewness and kurtosis are then given by

$$S = \frac{p}{p^{3/2}} = \frac{1}{\sqrt{p}} \qquad K = \frac{p+3p^2}{p^2} = \frac{1}{p} + 3.$$

Problem 1.68 Suppose the random variables X and Y have the joint pdf

$$f(x,y) = \begin{cases} \frac{e^{-2}}{x!(y-x)!} & \text{for } y = 0,1,2,3,\ldots; \text{and } x = 0,1,2,\ldots,y \\ 0 & \text{elsewhere.} \end{cases}$$

Find the moment-generating function $M(t_1, t_2)$ of the joint pdf. Compute the means, variances, and correlation coefficient of X and Y. Determine the conditional mean $E(X|y)$.

Solution The moment-generating function is given by

$$M(t_1,t_2) = \sum_{y=0}^{\infty}\sum_{x=0}^{\infty} e^{xt_1}e^{yt_2}f(x,y) = \sum_{y=0}^{\infty}\sum_{x=0}^{y} e^{xt_1}e^{yt_2}\frac{e^{-2}}{x!(y-x)!}.$$

Note that the sum over the variable x extends from 0 to y and not to infinity (compare with the pdf)! Multiplying this expression by $y!/y!$ we find

$$M(t_1,t_2) = e^{-2}\sum_{y=0}^{\infty}\frac{e^{yt_2}}{y!}\sum_{x=0}^{y}\frac{y!}{x!(y-x)!}e^{xt_1} = e^{-2}\sum_{y=0}^{\infty}\frac{e^{yt_2}}{y!}\sum_{x=0}^{y}\binom{y}{x}e^{xt_1},$$

where we used the definition of the binomial coefficient (1.29). According to the binomial theorem (1.31) the summation over x yields $(1+e^{t_1})^y$ and we obtain

$$M(t_1,t_2) = e^{-2}\sum_{y=0}^{\infty}\frac{e^{yt_2}}{y!}\left(1+e^{t_1}\right)^y = e^{-2}\sum_{y=0}^{\infty}\frac{\left[e^{t_2}+e^{t_1+t_2}\right]^y}{y!}.$$

With the series for the exponential function we obtain the moment-generating function

$$M(t_1,t_2) = e^{\left(e^{t_2}+e^{t_1+t_2}-2\right)}.$$

The means of the random variables X and Y are given by the first derivation with respect to t_1 and t_2 respectively. For the derivations we find therefore

$$\frac{\partial M(t_1,t_2)}{\partial t_1} = e^{(t_1+t_2)} e^{(e^{t_2}+e^{t_1+t_2}-2)} \quad \Rightarrow \quad \mu_x = \frac{\partial M(0,0)}{\partial t_1} = 1$$

and

$$\frac{\partial M(t_1,t_2)}{\partial t_2} = \left[e^{t_2} + e^{(t_1+t_2)}\right] e^{(e^{t_2}+e^{t_1+t_2}-2)} \quad \Rightarrow \quad \mu_y = \frac{\partial M(0,0)}{\partial t_2} = 2.$$

For the variance we need to calculate the second derivation. With $\mu_x = 1$ we obtain

$$\frac{\partial^2 M(t_1,t_2)}{\partial t_1^2} = \left[1 + e^{(t_1+t_2)}\right] e^{(t_1+t_2)} e^{(e^{t_2}+e^{t_1+t_2}-2)}$$

$$\Rightarrow \quad \sigma_x^2 = \frac{\partial^2 M(0,0)}{\partial t_1^2} - \mu_x^2 = 2 - 1 = 1$$

and with $\mu_y = 2$

$$\frac{\partial^2 M(t_1,t_2)}{\partial t_2^2} = \left[1 + \left(e^{t_2} + e^{(t_1+t_2)}\right)\right]\left(e^{t_2} + e^{(t_1+t_2)}\right) e^{(e^{t_2}+e^{t_1+t_2}-2)}$$

$$\Rightarrow \quad \sigma_y^2 = \frac{\partial^2 M(0,0)}{\partial t_2^2} - \mu_y^2 = 6 - 4 = 2.$$

For the correlation coefficient we calculate first

$$\frac{\partial^2 M(t_1,t_2)}{\partial t_1 \partial t_2} = \left[1 + \left(e^{t_2} + e^{(t_1+t_2)}\right)\right] e^{(t_1+t_2)} e^{(e^{t_2}+e^{t_1+t_2}-2)}$$

$$\Rightarrow \quad \frac{\partial^2 M(0,0)}{\partial t_1 \partial t_2} = E(XY) = 3.$$

The correlation coefficient is then given by (compare with Eq. (1.26) in Problem 1.51)

$$\rho = \frac{E(XY) - \mu_x \mu_y}{\sigma_x \sigma_y} = \frac{3 - 1 \cdot 2}{\sqrt{2}} = \frac{1}{\sqrt{2}}.$$

For the conditional mean $E(X|y)$ we need to calculate first the conditional pdf $f(x|y)$, which is given by $f(x|y) = f(x,y)/f(y)$, where the marginal pdf of Y is given by $f(y) = \sum_{x=0}^{y} f(x,y)$

$$f(y) = \sum_{x=0}^{y} \frac{e^{-2}}{x!(y-x)!} = \frac{e^{-2}}{y!} \sum_{x=0}^{y} \frac{y!}{x!(y-x)!} = \frac{e^{-2}}{y!} \sum_{x=0}^{y} \binom{y}{x} = \frac{e^{-2}2^y}{y!},$$

where we again multiplied by $y!/y!$ and where we also used the fact that

$$\sum_{x=0}^{y} \binom{y}{x} = 2^y.$$

This can easily be verified by using the binomial theorem (1.31). The conditional pdf is then given by

$$f(x|y) = \frac{f(x,y)}{f(y)} = \frac{e^{-2}}{x!(y-x)!} \frac{y!}{e^{-2}2^y} = 2^{-y} \frac{y!}{x!(y-x)!}.$$

The conditional mean is then calculated by

$$E(X|y) = \sum_{x=0}^{y} x f(x|y) = \sum_{x=0}^{y} x 2^{-y} \frac{y!}{x!(y-x)!}.$$

With the binomial theorem (1.31) we find

$$E(X|y) = 2^{-y} \sum_{x=0}^{y} x \binom{y}{x} = 2^{-y} y 2^{y-1} = \frac{y}{2},$$

where we used $\sum_{k=0}^{n} k \binom{n}{k} = n2^{n-1}$.

1.8.3 The Normal or Gaussian Distribution

The Gaussian distribution is given by

$$f(x) = \frac{1}{\sigma\sqrt{2\pi}} e^{-\frac{(x-\mu)^2}{2\sigma^2}}. \tag{1.36}$$

(continued)

and commonly denoted by $n(\mu, \sigma^2)$. The moment-generating function is given by

$$M(t) = \exp\left(\mu t + \frac{\sigma^2 t^2}{2}\right). \tag{1.37}$$

As a special case, consider the probability

$$P(a < X < b) = N\left(\frac{b-\mu}{\sigma}\right) - N\left(\frac{a-\mu}{\sigma}\right), \tag{1.38}$$

where $N(x)$ is given by the integral

$$N(x) = \frac{1}{\sqrt{2\pi}} \int_{-\infty}^{x} e^{-y^2/2} dy, \tag{1.39}$$

which is based on the distribution $n(0, 1)$, i.e., a normal distribution with zero mean and a variance of 1.

Problem 1.69 If Eq. (1.39) holds, show that $N(-x) = 1 - N(x)$.

Solution From the definition of the probability distribution function we know that

$$N(-x) = \frac{1}{\sqrt{2\pi}} \int_{-\infty}^{-x} e^{-y^2/2} dy. \tag{1.40}$$

We consider now $N(x)$ with the substitution $z = -y$ with $dz = -dy$ and obtain

$$N(x) = -\frac{1}{\sqrt{2\pi}} \int_{\infty}^{-x} e^{-z^2/2} dz = \frac{1}{\sqrt{2\pi}} \int_{-x}^{\infty} e^{-z^2/2} dz. \tag{1.41}$$

From the definition of the pdf we know that

$$1 = \frac{1}{\sqrt{2\pi}} \int_{-\infty}^{\infty} e^{-y^2/2} dy = \frac{1}{\sqrt{2\pi}} \int_{-\infty}^{-x} e^{-y^2/2} dy + \frac{1}{\sqrt{2\pi}} \int_{-x}^{\infty} e^{-y^2/2} dy$$
$$= N(-x) + N(x), \tag{1.42}$$

where we substituted the results from Eqs. (1.40) and (1.41). It follows that $N(-x) = 1 - N(x)$.

Problem 1.70 If X is $n(75, 100)$, find $P(X < 60)$ and $P(70 < X < 100)$.

Solution We can immediately deduce that $\mu = 75$ and $\sigma^2 = 100$. The distribution function is then given by

$$f(x) = \frac{1}{10\sqrt{2\pi}} e^{-\frac{(x-75)^2}{200}}.$$

For the probabilities we obtain

$$P(X < 60) = \int_{-\infty}^{60} f(x)\,dx = 0.067$$
$$P(70 < X < 100) = \int_{70}^{100} f(x)\,dx = 0.69.$$

Problem 1.71 If X is $n(\mu, \sigma^2)$, find a so that $P(-a < (X - \mu)/\sigma < a) = 0.9$.

Solution From definition (1.38) we know that

$$\begin{aligned} P\left(-a < \frac{X-\mu}{\sigma} < a\right) &= P(\mu - a\sigma < X < \mu + a\sigma) \\ &= N(a) - N(-a) \\ &= 2N(a) - 1, \end{aligned}$$

where we used $N(-x) = 1 - N(x)$. Since the probability is $P(-a < (X - \mu)/\sigma < a) = 0.9$ we find $N(a) = 0.95$, so that according to Eq. (1.39),

$$N(a) = \frac{1}{\sqrt{2\pi}} \int_{-\infty}^{a} e^{-y^2/2} dy = \frac{1}{2} + \frac{1}{2} \operatorname{erf}\left(\frac{a}{\sqrt{2}}\right) \stackrel{!}{=} 0.95.$$

The error function is solved numerically and we find $a = 1.645$.

Problem 1.72 If X is $n(\mu, \sigma^2)$, show that $E(|X - \mu|) = \sigma\sqrt{2/\pi}$.

Solution The expectation of $E(|X - \mu|)$ is given by

$$E(|X-\mu|) = \frac{1}{\sigma\sqrt{2\pi}} \int_{-\infty}^{\infty} |x-\mu| e^{-\frac{(x-\mu)^2}{2\sigma^2}}\,dx.$$

The integral has to be split up, where $x < \mu$ in the first and $x > \mu$ in the second integral, so that

$$E(|X-\mu|) = \frac{1}{\sigma\sqrt{2\pi}} \left[\int_{-\infty}^{\mu} (\mu - x) e^{-\frac{(x-\mu)^2}{2\sigma^2}}\,dx + \int_{\mu}^{\infty} (x-\mu) e^{-\frac{(x-\mu)^2}{2\sigma^2}}\,dx\right].$$

In the first integral we substitute $z = (\mu - x)/\sigma$ and in the second integral we substitute $z = (x - \mu)/\sigma$. We obtain

$$E(|X-\mu|) = \frac{\sigma}{\sqrt{2\pi}}\left[-\int_{\infty}^{0} z e^{-\frac{z^2}{2}}\,dz + \int_{0}^{\infty} z e^{-\frac{z^2}{2}}\,dz\right].$$

By swapping the limits of the first integral it follows with $\int_0^\infty z\exp\left(-z^2/2\right)\,dz = 1$ that

$$E(|X-\mu|) = \frac{2\sigma}{\sqrt{2\pi}}\int_0^\infty z e^{-\frac{z^2}{2}}\,dz = \sqrt{\frac{2}{\pi}}\sigma.$$

Problem 1.73 Show that the pdf $n(\mu, \sigma^2)$ has points of inflection at $x_i = \mu \pm \sigma$.

Solution For inflection points we have to show as a necessary condition that $d^2f(x)/dx^2 = 0$. Thus,

$$\begin{aligned}
f(x) &= \frac{1}{\sigma\sqrt{2\pi}} e^{-\frac{(x-\mu)^2}{2\sigma^2}} \\
\frac{df(x)}{dx} &= -\frac{1}{\sigma^3\sqrt{2\pi}}(x-\mu)e^{-\frac{(x-\mu)^2}{2\sigma^2}} \\
\frac{d^2f(x)}{dx^2} &= \frac{1}{\sigma^3\sqrt{2\pi}}\left[-1 + \frac{(x-\mu)^2}{\sigma^2}\right] e^{-\frac{(x-\mu)^2}{2\sigma^2}} \stackrel{!}{=} 0.
\end{aligned}$$

We consider only the term in square brackets and obtain

$$(x-\mu)^2 = \sigma^2 \quad \Longrightarrow \quad x^2 - 2\mu x + \mu^2 - \sigma^2 = 0.$$

This quadratic equation can easily be solved and we obtain two solutions $x_i = \mu \pm \sigma$. The sufficient condition requires $d^3f(x)/dx^3 \neq 0$ at the points of inflection, so that

$$\frac{d^3f(x)}{dx^3} = -\frac{1}{\sigma^5\sqrt{2\pi}}(x-\mu)\left[-3 + \frac{(x-\mu)^2}{\sigma^2}\right] e^{-\frac{(x-\mu)^2}{2\sigma^2}}.$$

At the point of inflection we have

$$\left.\frac{d^3f(x)}{dx^3}\right|_{x_i} = \pm\frac{2}{\sigma^4\sqrt{2\pi}} e^{-\frac{1}{2}} \neq 0,$$

since $\sigma \neq 0$ per definition. If $\sigma = 0$ the distribution function $f(x)$ is a delta distribution (see Problem 1.77) and the consideration here is obsolete.

Problem 1.74 Suppose a random variable X has the pdf

$$f(x) = \begin{cases} \frac{2}{\sqrt{2\pi}} e^{-x^2/2} & \text{for } 0 < x < \infty \\ 0 & \text{elsewhere.} \end{cases}$$

Find the mean and the variance of X.

Solution The mean is given by

$$\mu = E(X) = \frac{2}{\sqrt{2\pi}} \int_0^\infty dx\, x e^{-x^2/2} = \sqrt{\frac{2}{\pi}},$$

since $\int_0^\infty dx\, x \exp\left(-x^2/2\right) = 1$. The variance is given by

$$\sigma^2 = E(X^2) - \mu^2 = \frac{2}{\sqrt{2\pi}} \int_0^\infty dx\, x^2 e^{-x^2/2} - \frac{2}{\pi}. \tag{1.43}$$

Consider an arbitrary random variable with the Gaussian pdf $n(0, 1)$, from which we know that the mean is zero and the variance is one. The pdf of that random variable can be found as

$$f(x) = \frac{1}{\sqrt{2\pi}} e^{-\frac{x^2}{2}}.$$

Since the mean is zero ($\mu = 0$), the expectation for X^2 is equal to the variance $\sigma = 1$ and therefore

$$1 \stackrel{!}{=} \sigma^2 = E(X^2) = \frac{1}{\sqrt{2\pi}} \int_{-\infty}^\infty dx\, x^2 e^{-\frac{x^2}{2}} = \frac{2}{\sqrt{2\pi}} \int_0^\infty dx\, x^2 e^{-\frac{x^2}{2}},$$

since the integrand is an even function. If we compare this result with Eq. (1.43) we find immediately

$$\sigma^2 = E(X^2) - \mu^2 = 1 - \frac{2}{\pi}.$$

Problem 1.75 Let X_1 and X_2 be two stochastically independent normally distributed random variables with means μ_1 and μ_2 and variances σ_1 and σ_2. Show that $X_1 + X_2$ is normally distributed with mean $(\mu_1 + \mu_2)$ and variance $\sigma_1^2 + \sigma_2^2$. (Hint: use the uniqueness of the moment-generating function.)

Solution The moment-generating functions for the normally distributed random variables X_1 and X_2 can be written as

$$M_{X_1}(t) = E\left[e^{x_1 t}\right] = \exp\left(\mu_{x_1} t + \frac{\sigma_{x_1}^2 t^2}{2}\right)$$

$$M_{X_2}(t) = E\left[e^{x_2 t}\right] = \exp\left(\mu_{x_2} t + \frac{\sigma_{x_2}^2 t^2}{2}\right).$$

Let's introduce a third moment-generating function for the normally distributed random variable $X_1 + X_2$, with mean $\mu_{x_1+x_2}$ and variance $\sigma_{x_1+x_2}$,

$$M_{X_1+X_2}(t) = E\left[e^{(x_1+x_2)t}\right] = E\left[e^{x_1 t} e^{x_2 t}\right] = \exp\left(\mu_{x_1+x_2} t + \frac{\sigma_{x_1+x_2}^2 t^2}{2}\right).$$

Since the random variables X_1 and X_2 are stochastically independent, we find for the expectation

$$E\left[e^{x_1 t} e^{x_2 t}\right] = E\left[e^{x_1 t}\right] \cdot E\left[e^{x_2 t}\right].$$

This means, that the moment-generating function for the random variable $X_1 + X_2$ can also be written as

$$\begin{aligned} M_{X_1+X_2}(t) &= M_{X_1}(t) \cdot M_{X_2}(t) \\ &= \exp\left(\mu_{x_1} t + \frac{\sigma_{x_1}^2 t^2}{2}\right) \cdot \exp\left(\mu_{x_2} t + \frac{\sigma_{x_2}^2 t^2}{2}\right) \\ &= \exp\left((\mu_{x_1} + \mu_{x_2}) t + (\sigma_{x_1}^2 + \sigma_{x_2}^2) \frac{t^2}{2}\right). \end{aligned}$$

Obviously, $M_{X_1+X_2}(t)$ is the moment-generating function of a normal distribution. Thus, the new random variable $X_1 + X_2$ is also normally distributed with mean and variance given by

$$\begin{aligned} \mu_{x_1+x_2} &= \mu_{x_1} + \mu_{x_2} \\ \sigma_{x_1+x_2} &= \sigma_{x_1}^2 + \sigma_{x_2}^2. \end{aligned}$$

Problem 1.76 Compute $P(1 < X^2 < 9)$ if X is $n(1, 4)$.

Solution From $n(1, 4)$ we deduce that $\mu = 1$ and $\sigma = 2$. It is

$$\begin{aligned} &P(1 < X^2 < 9) \\ &= P(1 < X < 3) + P(-3 < X < -1) \\ &= P(X < 3) - P(X < 1) + P(X < -1) - P(X < -3) \end{aligned}$$

$$
\begin{aligned}
&= P\left(\frac{X-\mu}{\sigma} < \frac{3-\mu}{\sigma}\right) - P\left(\frac{X-\mu}{\sigma} < \frac{1-\mu}{\sigma}\right) \\
&+P\left(\frac{X-\mu}{\sigma} < \frac{-1-\mu}{\sigma}\right) - P\left(\frac{X-\mu}{\sigma} < \frac{-3-\mu}{\sigma}\right) \\
&= N\left(\frac{3-\mu}{\sigma}\right) - N\left(\frac{1-\mu}{\sigma}\right) + N\left(\frac{-1-\mu}{\sigma}\right) - N\left(\frac{-3-\mu}{\sigma}\right).
\end{aligned}
$$

By substituting $\mu = 1$ and $\sigma = 2$ we obtain

$$
\begin{aligned}
P(1 < X^2 < 9) &= N\left(\frac{3-1}{2}\right) - N\left(\frac{1-1}{2}\right) + N\left(\frac{-1-1}{2}\right) - N\left(\frac{-3-1}{2}\right) \\
&= N(1) - N(0) + N(-1) - N(-2) \\
&= 0.8413 - 0.5 + 0.1587 - 0.0228 \\
&= 0.4772,
\end{aligned}
$$

where the values for the probability distribution were taken from tables.

Problem 1.77 Suppose the random variable X is normally distributed with $n(\mu, \sigma^2)$. What will the distribution be if $\sigma^2 = 0$?

Solution The Gaussian pdf is given by

$$
f(x) = \frac{1}{\sigma\sqrt{2\pi}} e^{-\frac{(x-\mu)^2}{2\sigma^2}}.
$$

The limit for $\sigma \to 0$ yields the delta distribution,

$$
\lim_{\sigma\to 0} f(x) = \lim_{\sigma\to 0} \frac{1}{\sigma\sqrt{2\pi}} e^{-\frac{(x-\mu)^2}{2\sigma^2}} = \delta(x-\mu),
$$

with $F(X) = 1$ for $x \geq \mu$ and $F(X) = 0$ for $x < \mu$.

1.9 The Central Limit Theorem

Suppose that X_i, with $i = 1, 2, 3, \ldots, n$ is a random sample from a distribution that has mean μ and variance σ^2. Then the random variable

(continued)

$$Y_n = \frac{\sum_{i=1}^{n} X_i - n\mu}{\sqrt{n}\sigma} = \sqrt{n}\frac{\bar{X}_n - \mu}{\sigma} \tag{1.44}$$

has a limiting distribution that is normal with mean 0 and variance 1.

Problem 1.78 Compute an approximate probability that the mean $\bar{X}_n$ of a random sample of size 15 from a distribution having pdf

$$f(x) = \begin{cases} 3x^2 & \text{for } 0 < x < 1 \\ 0 & \text{elsewhere} \end{cases}$$

is between $3/5$ and $4/5$.

Solution First, we calculate the mean and variance of the pdf,

$$\mu = E(X) = \int_0^1 dx\, 3x^3 = \frac{3}{4}$$

$$\sigma^2 = E(X^2) - \mu^2 = \int_0^1 dx\, 3x^4 - \frac{9}{16} = \frac{3}{5} - \frac{9}{16} = \frac{3}{80}.$$

Now we search for an approximate probability that the mean $\bar{X}_n$ of a random sample of size $n = 15$ is between $3/5$ and $4/5$. In other words, we search for the probability $P(3/5 < \bar{X}_n < 4/5)$. In order to do so we introduce the new variable Y_n, given by Eq. (1.44), so that the expression for our probability transforms as

$$P\left(\frac{3}{5} < \bar{X} < \frac{4}{5}\right) = P\left[\frac{\sqrt{n}(3/5 - \mu)}{\sigma} < \frac{\sqrt{n}(\bar{X} - \mu)}{\sigma} < \frac{\sqrt{n}(4/5 - \mu)}{\sigma}\right].$$

Basically, we subtracted the mean μ from each parameter in the expression and multiplied the results by $\sqrt{n}/\sigma$. Obviously, the parameter in the middle corresponds to the new variable Y_n. Substituting now the mean and variance we obtain

$$P\left(\frac{3}{5} < \bar{X} < \frac{4}{5}\right) = P\left[\frac{\sqrt{15}(3/5 - 3/4)}{\sqrt{3/80}} < Y_n < \frac{\sqrt{15}(4/5 - 3/4)}{\sqrt{3/80}}\right].$$

Since the new random variable Y_n is normally distributed with $\mu = 0$ and $\sigma^2 = 1$ we can use the Eq. (1.38) and obtain

$$P\left(\frac{3}{5} < \bar{X} < \frac{4}{5}\right) = P(-3 < Y_n < 1) = N(1) - N(-3) = 0.84,$$

where the values for $N(1)$ and $N(-3)$ were taken from tables.

Problem 1.79 Let Y be a binomial distribution $B(72, 1/3)$. Approximate $P(22 \leq Y \leq 28)$.

Solution It follows immediately that $n = 72$ and $p = 1/3$. Therefore, the random variable Y has a binomial distribution with mean $\mu = np = 24$ and variance $\sigma^2 = np(1-p) = 16$. For large sample sizes (besides other restrictions) the binomial distribution is approximated very well by a normal distribution with the same mean and variance.

Let X be the random variable of a normal distribution with $n(\mu, \sigma^2)$, then

$$P(22 \leq Y \leq 28) \approx P(21.5 < X < 28.5),$$

where we used the convention of taking 0.5 above and below the limiting discrete value. On using Eq. (1.38) we find with $\mu = 24$ and $\sigma = 4$

$$\begin{aligned} P(22 \leq Y \leq 28) &\approx N\left(\frac{28.5 - \mu}{\sigma}\right) - N\left(\frac{21.5 - \mu}{\sigma}\right) \\ &\approx N(1.125) - N(-0.625) = N(1.125) - 1 + N(0.625) \\ &= 0.87 - 1 + 0.734 = 0.604. \end{aligned}$$

1.10 The Language of Fluid Turbulence

Problem 1.80 Show that the joint covariance is not symmetric in the time lag τ, i.e., that $R_{uv}(\tau) = R_{vu}(-\tau)$.

Solution The joint covariance is given by

$$R_{uv}(\tau) = \langle u(t)v(t+\tau) \rangle = \langle u(t' - \tau)v(t') \rangle = \langle v(t')u(t' - \tau) \rangle = R_{vu}(-\tau),$$

where we used the new variable $t' = t + \tau$.

Problem 1.81 Show that the joint covariance for u and a time derivative of v satisfies $R_{u\dot{v}}(\tau) = \partial R_{vu}(-\tau)/\partial \tau$.

Solution The joint covariance is given by

$$\begin{aligned} R_{u\dot{v}}(\tau) = \left\langle u(t)\frac{\partial}{\partial \tau}v(t+\tau) \right\rangle &= \frac{\partial}{\partial \tau}\langle u(t' - \tau)v(t') \rangle \\ &= \frac{\partial}{\partial \tau}\langle v(t')u(t' - \tau) \rangle = \frac{\partial}{\partial \tau}R_{vu}(-\tau), \end{aligned}$$

where we used the new variable $t' = t + \tau$.

Problem 1.82 Show that the co-spectrum and quadrature spectrum may be expressed as integrals

$$Co_{uv}(\omega) = \frac{1}{2\pi}\int_0^\infty [R_{uv}(\tau) + R_{uv}(-\tau)]\cos(\omega\tau)\,d\tau,$$

$$Qu_{uv}(\omega) = \frac{1}{2\pi}\int_0^\infty [R_{uv}(\tau) - R_{uv}(-\tau)]\sin(\omega\tau)\,d\tau.$$

Solution The joint or cross-spectral density of the joint pair of random functions u and v is given by

$$S_{uv}(\omega) = \frac{1}{2\pi}\int_{-\infty}^{\infty} d\tau\, e^{i\omega\tau} R_{uv}(\tau) = Co_{uv}(\omega) + iQu_{uv}(\omega),$$

where the co-spectrum is the real part of the cross-spectral density, $Co_{uv}(\omega) = \mathfrak{Re}\,(S_{uv}(\omega))$, and the quadrature spectrum is the imaginary part, $Qu_{uv}(\omega) = \mathfrak{Im}\,(S_{uv}(\omega))$. The integral can also be written in the form

$$\begin{aligned}
S_{uv}(\omega) &= \frac{1}{2\pi}\int_{-\infty}^{\infty} d\tau\, e^{i\omega\tau} R_{uv}(\tau) \\
&= \frac{1}{2\pi}\int_0^{\infty} d\tau\, e^{i\omega\tau} R_{uv}(\tau) + \frac{1}{2\pi}\int_{-\infty}^{0} d\tau\, e^{i\omega\tau} R_{uv}(\tau) \\
&= \frac{1}{2\pi}\int_0^{\infty} d\tau\, e^{i\omega\tau} R_{uv}(\tau) + \frac{1}{2\pi}\int_0^{\infty} d\tau\, e^{-i\omega\tau} R_{uv}(-\tau) \\
&= \frac{1}{2\pi}\int_0^{\infty} d\tau\, \left[e^{i\omega\tau} R_{uv}(\tau) + e^{-i\omega\tau} R_{uv}(-\tau)\right],
\end{aligned}$$

where we used the substitution $\tau = -\tau$ in the second step. Bear in mind that $e^{\pm i\omega\tau} = \cos(\omega\tau) \pm i\sin(\omega\tau)$, and therefore $\mathfrak{Re}\left(e^{\pm i\omega\tau}\right) = \cos(\omega\tau)$ and $\mathfrak{Im}\left(e^{\pm i\omega\tau}\right) = \pm\sin(\omega\tau)$. Since $R_{uv}(\pm\tau)$ is a real function it follows that

$$\begin{aligned}
Co_{uv}(\omega) &= \mathfrak{Re}\,[S_{uv}(\omega)] \\
&= \mathfrak{Re}\left[\frac{1}{2\pi}\int_0^{\infty} d\tau\, \left[e^{i\omega\tau} R_{uv}(\tau) + e^{-i\omega\tau} R_{uv}(-\tau)\right]\right] \\
&= \frac{1}{2\pi}\int_0^{\infty} d\tau\, \left[\mathfrak{Re}\left[e^{i\omega\tau}\right] R_{uv}(\tau) + \mathfrak{Re}\left[e^{-i\omega\tau}\right] R_{uv}(-\tau)\right] \\
&= \frac{1}{2\pi}\int_0^{\infty} d\tau\, [\cos(\omega\tau) R_{uv}(\tau) + \cos(\omega\tau) R_{uv}(-\tau)] \\
&= \frac{1}{2\pi}\int_0^{\infty} d\tau\, [R_{uv}(\tau) + R_{uv}(-\tau)]\cos(\omega\tau)
\end{aligned}$$

and similarly

$$
\begin{aligned}
Qu_{uv}(\omega) &= \Im\mathrm{m}\,[S_{uv}(\omega)] \\
&= \Im\mathrm{m}\left[\frac{1}{2\pi}\int_0^\infty d\tau\,\left[e^{i\omega\tau}R_{uv}(\tau) + e^{-i\omega\tau}R_{uv}(-\tau)\right]\right] \\
&= \frac{1}{2\pi}\int_0^\infty d\tau\,\left[\Im\mathrm{m}\left[e^{i\omega\tau}\right]R_{uv}(\tau) + \Im\mathrm{m}\left[e^{-i\omega\tau}\right]R_{uv}(-\tau)\right] \\
&= \frac{1}{2\pi}\int_0^\infty d\tau\,[\sin(\omega\tau)R_{uv}(\tau) - \sin(\omega\tau)R_{uv}(-\tau)] \\
&= \frac{1}{2\pi}\int_0^\infty d\tau\,[R_{uv}(\tau) - R_{uv}(-\tau)]\sin(\omega\tau).
\end{aligned}
$$

Problem 1.83 By introducing the *coherence*

$$
Coh_{uv}(\omega) = \sqrt{\frac{Co_{uv}^2 + Qu_{uv}^2}{|S_{vv}||S_{uu}|}}
$$

and the phase $\theta_{uv}(\omega) = \arg(S_{uv})$ (the argument of S_{uv}), show that the joint or cross spectral density can be expressed in terms of its magnitude and argument,

$$
S_{uv}(\omega) = \sqrt{|S_{uu}(\omega)||S_{vv}(\omega)|}\;Coh_{uv}(\omega)\;e^{i\theta_{uv}(\omega)}.
$$

Solution Any complex number z can be rewritten in a polar form and, by using Euler's formula, expressed by

$$
z = a + ib = |z|\,[\cos\phi + i\sin\phi] = |z|e^{i\phi},
$$

where the argument $\phi = \arg z = \arctan(b/a)$. The complex cross spectral density S_{uv} can then be written as

$$
S_{uv} = |S_{uv}|e^{i\arg S_{uv}}.
$$

The argument of S_{uv} is given by $\theta_{uv}(\omega) = \arg(S_{uv})$. The absolute value is given by

$$
|S_{uv}| = |Co_{uv}(\omega) + iQu_{uv}(\omega)| = \sqrt{Co_{uv}^2(\omega) + Qu_{uv}^2(\omega)},
$$

so that

$$
S_{uv}(\omega) = \sqrt{|S_{uu}(\omega)||S_{vv}(\omega)|}\;Coh_{uv}(\omega)\;e^{i\theta_{uv}(\omega)}
$$

where we used the definition of the coherence given above.

Problem 1.84 Show that if $u' = \partial u/\partial x$, then $\phi_{u'u'} = k^2\phi_{uu}$.

Solution We have

$$R_{u'u'} = \left\langle \frac{\partial u(x)}{\partial x}\frac{\partial u(x')}{\partial x'} \right\rangle = \frac{\partial^2}{\partial x \partial x'}\langle u(x)u(x')\rangle = \frac{\partial^2}{\partial x \partial x'}R_{uu}.$$

Introducing the new variable $\xi = x' - x$, we find

$$\frac{\partial}{\partial x'} = \frac{\partial \xi}{\partial x'}\frac{\partial}{\partial \xi} = \frac{\partial}{\partial \xi} \quad \text{and} \quad \frac{\partial}{\partial x} = \frac{\partial \xi}{\partial x}\frac{\partial}{\partial \xi} = -\frac{\partial}{\partial \xi},$$

so that

$$R_{u'u'}(\xi) = -\frac{\partial^2}{\partial \xi^2}R_{uu}(\xi).$$

The (1D) wavenumber spectrum is given by

$$\phi_{u'u'}(k) = \frac{1}{2\pi}\int_{-\infty}^{\infty} e^{-ik\xi}R_{u'u'}(\xi)d\xi = -\frac{1}{2\pi}\int_{-\infty}^{\infty} e^{-ik\xi}\frac{\partial^2}{\partial \xi^2}R_{uu}(\xi)d\xi.$$

Using integration by parts twice we find

$$\phi_{u'u'}(k) = k^2\frac{1}{2\pi}\int_{-\infty}^{\infty} e^{-ik\xi}R_{uu}(\xi)d\xi = k^2\phi_{uu}(k),$$

using that $R_{uu}(\xi) = 0$ and $\partial R_{uu}(\xi)/\partial \xi = 0$ for $\xi = \pm\infty$.

Problem 1.85 Show that an exponentially decaying covariance

$$R_{uu}(\tau) = \langle u(0)u(\tau)\rangle = Ce^{-\gamma|\tau|} \tag{1.45}$$

yields a Lorentz distribution for the power spectral density,

$$S_{uu}(\omega) = \frac{1}{\pi}\frac{C\gamma}{\omega^2 + \gamma^2}. \tag{1.46}$$

Solution The joint or cross-spectral density of the joint pair of random functions is given by

$$S_{uu}(\omega) = \frac{1}{2\pi}\int_{-\infty}^{\infty} d\tau\, e^{i\omega\tau}R_{uu}(\tau) = \frac{C}{2\pi}\int_{-\infty}^{\infty} d\tau\, e^{i\omega\tau}e^{-\gamma|\tau|}.$$

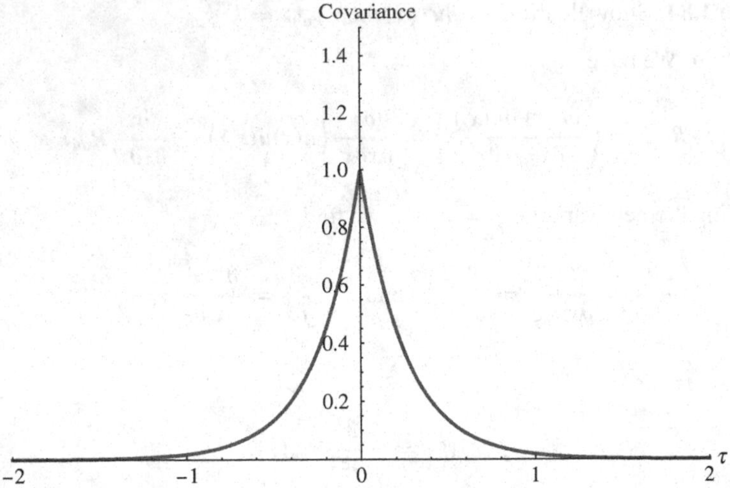

Fig. 1.16 Shown is the covariance $\langle u(\tau)u(0)\rangle$ calculated from Eq. (1.45) for $\gamma = 4$ and $C = 1$

As an example, the covariance is plotted in Fig. 1.16 for $\gamma = 4$ and $C = 1$. With $|\tau| = \tau$ for $\tau > 0$ and $|\tau| = -\tau$ for $\tau < 0$ we split the integral into

$$S_{uu}(\omega) = \frac{C}{2\pi}\int_0^\infty d\tau\, e^{(i\omega-\gamma)\tau} + \frac{C}{2\pi}\int_{-\infty}^0 d\tau\, e^{(i\omega\tau+\gamma)\tau}.$$

By substituting $\tau = -\tau$ in the second integral we obtain

$$\begin{aligned} S_{uu}(\omega) &= \frac{C}{2\pi}\int_0^\infty d\tau\, e^{-(-i\omega+\gamma)\tau} + \frac{C}{2\pi}\int_0^\infty d\tau\, e^{-(i\omega\tau+\gamma)\tau} \\ &= \frac{C}{2\pi}\left[\frac{1}{(-i\omega+\gamma)} + \frac{1}{(i\omega+\gamma)}\right] = \frac{1}{\pi}\frac{C\gamma}{\omega^2+\gamma^2}. \end{aligned}$$

As an example, the spectral density is plotted in Fig. 1.17 for $\gamma = 4$ and $C = 1$.

Problem 1.86 Consider now an additional periodic component to $u(t)$, say $u(t) = v(t) + Ae^{-i\omega_0 t}$, with $\langle v(t)\rangle = 0$ and $\langle v(0)v(\tau)\rangle = Ce^{-\gamma|\tau|}$. Show that

$$\langle u(\tau)u(0)\rangle = Ce^{-\gamma|\tau|} + A^2 e^{-i\omega_0\tau} \tag{1.47}$$

and

$$S_{uu}(\omega) = \frac{1}{\pi}\frac{C\gamma}{\omega^2+\gamma^2} + A^2\delta(\omega-\omega_0). \tag{1.48}$$

Sketch the covariance and the power spectral density.

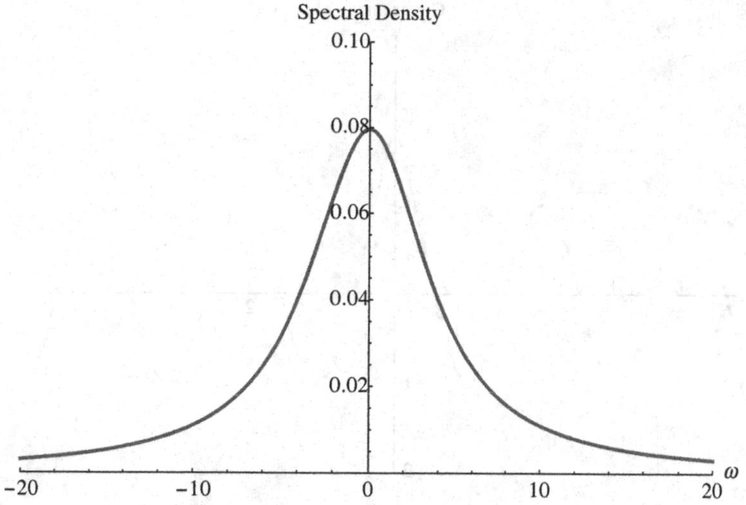

Fig. 1.17 Shown is the spectral density $S_{uu}(\omega)$ calculated from Eq. (1.46) for $\gamma = 4$ and $C = 1$

Solution For calculating the autocovariance we consider first

$$u(t)u(t') = v(t)v(t') + Av(t)e^{-i\omega_0 t'} + Av(t')e^{-i\omega_0 t} + A^2 e^{-i\omega_0 (t+t')}.$$

By averaging over $u(t)u(t')$ we obtain

$$\langle u(t)u(t') \rangle = \langle v(t)v(t') \rangle + A^2 e^{-i\omega_0 (t+t')},$$

since $< Av(t)e^{-i\omega_0 t'} >= A \langle v(t) \rangle e^{-i\omega_0 t'} = 0$. By substituting $\tau = t' - t$ and by setting $t = 0$ (because the homogeneous autocovariance depends only on the time difference τ), we find

$$\langle u(0)u(\tau) \rangle = Ce^{-\gamma|\tau|} + A^2 e^{-i\omega_0 \tau},$$

where we used $\langle v(0)v(\tau) \rangle = Ce^{-\gamma|\tau|}$. Shown in Fig. 1.18 is the real part of Eq. (1.47) for $\gamma = 4, C = 1, A = 1$, and $\omega_0 = 3$. The Fourier transform of the autocovariance is then given by

$$\begin{aligned} S_{uu}(\omega) &= \frac{1}{2\pi} \int_{-\infty}^{\infty} d\tau \, e^{i\omega\tau} \langle u(0)u(\tau) \rangle \\ &= \frac{C}{2\pi} \int_{-\infty}^{\infty} d\tau \, e^{i\omega\tau} e^{-\gamma|\tau|} + \frac{A^2}{2\pi} \int_{-\infty}^{\infty} d\tau \, e^{i(\omega-\omega_0)\tau}. \end{aligned}$$

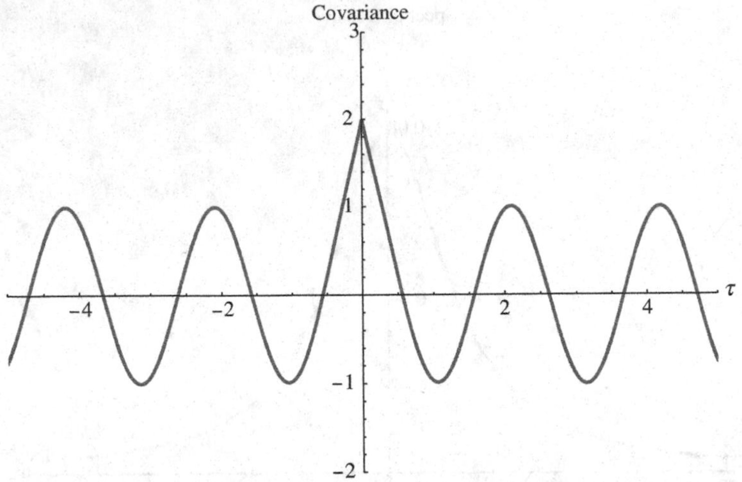

Fig. 1.18 Shown is the covariance $\langle u(\tau)u(0) \rangle$ calculated from Eq. (1.47) for $\gamma = 4, C = 1, A = 1$, and $\omega_0 = 3$

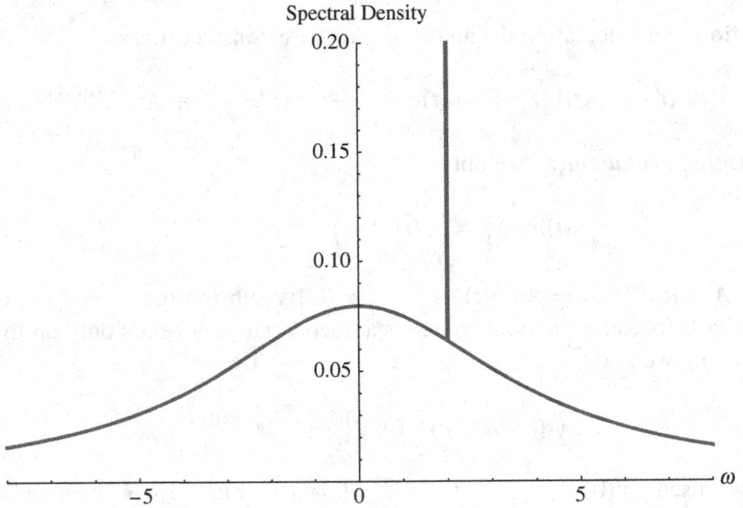

Fig. 1.19 Shown is the spectral density $S_{uu}(\omega)$ calculated from Eq. (1.48) for $\gamma = 4, C = 1, A = 1$, and $\omega_0 = 3$

The first integral has been evaluated in the previous problem. The second integral yields the delta function with $2\pi\delta(x) = \int_{-\infty}^{\infty} \exp(ixt)dt$, so that

$$S_{uu}(\omega) = \frac{1}{\pi}\frac{C\gamma}{\omega^2 + \gamma^2} + A^2\delta(\omega - \omega_0).$$

Figure 1.19 shows the spectral density for $\gamma = 4, C = 1, A = 1$, and $\omega_0 = 3$.

Chapter 2
The Boltzmann Transport Equation

2.1 Derivation of the Boltzmann Transport Equation

The non-relativistic Boltzmann equation is given by

$$\frac{\partial f}{\partial t} + \boldsymbol{v} \cdot \nabla_r f + \frac{\boldsymbol{F}}{m} \cdot \nabla_v f = \left(\frac{\delta f}{\delta t}\right)_{coll}, \tag{2.1}$$

where $\boldsymbol{r}$, $\boldsymbol{v}$, and t are *independent* variables.

Problem 2.1 Show that the Boltzmann equation (2.1) is invariant with respect to Galilean transformations.

Solution For simplicity let us consider a Cartesian inertial system K with (orthogonal) axes x, y, z and an inertial system K' with axes x', y', z' that is moving with speed w in positive x-direction with respect to K. The Galilean transformation is then given by

$$t' = t \qquad \boldsymbol{r}'(t) = \boldsymbol{r} - wt\boldsymbol{e}_x \qquad \boldsymbol{v}' = \boldsymbol{v} - w\boldsymbol{e}_x \qquad \boldsymbol{a}' = \boldsymbol{a},$$

where $\boldsymbol{e}_x$ is the unit vector in x-direction. In particular, for each coordinate we find

$$\begin{aligned} t' &= t \qquad & x'(t) &= x - wt \qquad & v'_x &= v_x - w \\ & & y' &= y & v'_y &= v_y \\ & & z' &= z & v'_z &= v_z. \end{aligned}$$

A. Dosch, G.P. Zank, *Transport Processes in Space Physics and Astrophysics*, Lecture Notes in Physics 918, DOI 10.1007/978-3-319-24880-6_2

Note that $\boldsymbol{r}$, $\boldsymbol{v}$, and t are independent variables, while the new coordinate $\boldsymbol{r}'$ depends on time t. More precisely, x' depends on time t.

Consider now the change in variables, i.e., $f(\boldsymbol{r}, \boldsymbol{v}, t) \to f(\boldsymbol{r}'(t), \boldsymbol{v}', t)$. Thus, we have the following transformations:

A. Since the new spatial coordinate depends on time we have to rewrite the time derivative and obtain

$$\frac{\partial f}{\partial t} \Rightarrow \frac{\partial f}{\partial t} + \frac{\partial f}{\partial \boldsymbol{r}'} \cdot \frac{\partial \boldsymbol{r}'}{\partial t} = \frac{\partial f}{\partial t} - w\boldsymbol{e}_x \cdot \nabla_{r'} f,$$

where $\partial f / \partial \boldsymbol{r}' = \nabla_{r'}$ and $\partial \boldsymbol{r}' / \partial t = -w\boldsymbol{e}_x$.

B. For the gradient of f we find

$$\nabla_r f = \frac{\partial f}{\partial x} + \frac{\partial f}{\partial y} + \frac{\partial f}{\partial z} = \frac{\partial f}{\partial x'}\frac{\partial x'}{\partial x} + \frac{\partial f}{\partial y'}\frac{\partial y'}{\partial y} + \frac{\partial f}{\partial z'}\frac{\partial z'}{\partial z}$$
$$= \frac{\partial f}{\partial x'} + \frac{\partial f}{\partial y'} + \frac{\partial f}{\partial z'} = \nabla_{r'} f.$$

Therefore, $\nabla_r = \nabla_{r'}$.

C. Similarly we find

$$\nabla_v f = \frac{\partial f}{\partial v_x} + \frac{\partial f}{\partial v_y} + \frac{\partial f}{\partial v_z} = \frac{\partial f}{\partial v_x'}\frac{\partial v_x'}{\partial v_x} + \frac{\partial f}{\partial v_y'}\frac{\partial v_y'}{\partial v_y} + \frac{\partial f}{\partial v_z'}\frac{\partial v_z'}{\partial v_z}$$
$$= \frac{\partial f}{\partial v_x'} + \frac{\partial f}{\partial v_y'} + \frac{\partial f}{\partial v_z'} = \nabla_{v'} f.$$

It follows that $\nabla_v = \nabla_{v'}$. Moreover, since $\boldsymbol{a}' = \boldsymbol{a}$ we have $\boldsymbol{F}' = \boldsymbol{F}$.

Substituting the above results, the Boltzmann equation reads

$$-w\boldsymbol{e}_x \nabla_{r'} f + \frac{\partial f}{\partial t} + \left[\boldsymbol{v}' + w\boldsymbol{e}_x\right] \cdot \nabla_{r'} f + \frac{\boldsymbol{F}'}{m} \cdot \nabla_{v'} f = \left(\frac{\delta f}{\delta t}\right)_{coll}'$$
$$\Rightarrow \quad \frac{\partial f}{\partial t} + \boldsymbol{v}' \cdot \nabla_{r'} f + \frac{\boldsymbol{F}'}{m} \cdot \nabla_{v'} f = \left(\frac{\delta f}{\delta t}\right)_{coll}'.$$

Note that $\boldsymbol{v}$ was replaced by $\boldsymbol{v}' + w\boldsymbol{e}_x$. Therefore, the Boltzmann equation is invariant under Galilean transformation, i.e., it retains its form.

Problem 2.2 Show that the Boltzmann equation (2.1) transforms into the mixed phase space coordinate form

$$\frac{\partial f}{\partial t} + (u_i + c_i)\frac{\partial f}{\partial r_i} - \left(\frac{\partial u_i}{\partial t} + (u_j + c_j)\frac{\partial u_i}{\partial r_j} - \frac{F_i}{m}\right)\frac{\partial f}{\partial c_i} = \left(\frac{\delta f}{\delta t}\right)_{coll}, \tag{2.2}$$

where we used Einstein's summation convention, i.e., we sum over double indices.

Solution We rewrite the Boltzmann equation as

$$\frac{\partial f}{\partial t} + v_i \frac{\partial f}{\partial r_i} + \frac{F_i}{m} \frac{\partial f}{\partial v_i} = \left(\frac{\delta f}{\delta t} \right)_{coll}.$$

The velocity $\boldsymbol{v}$ of a particle can also be described with respect to the bulk flow velocity $\boldsymbol{u}(\boldsymbol{r}, t)$, through $\boldsymbol{v} = \boldsymbol{c}(\boldsymbol{r}, t) + \boldsymbol{u}(\boldsymbol{r}, t)$, where $\langle \boldsymbol{v} \rangle = \boldsymbol{u}(\boldsymbol{r}, t)$ and $\langle \boldsymbol{c}(\boldsymbol{r}, t) \rangle = 0$. Here, $\boldsymbol{c}$ is the random velocity and sometimes also called the peculiar or thermal velocity. The random velocity can then be written as $\boldsymbol{c}(\boldsymbol{r}, t) = \boldsymbol{v} - \boldsymbol{u}(\boldsymbol{r}, t)$, where each component can be described by

$$c_i(\boldsymbol{r}, t) = v_i - u_i(\boldsymbol{r}, t). \tag{2.3}$$

Note that each component c_i depends on all spatial coordinates and time.

Sometimes it might be more convenient to express the phase space distribution f in terms of $\boldsymbol{r}, \boldsymbol{c}$, and t instead of $\boldsymbol{r}, \boldsymbol{v}$, and t. This means, that one has to introduce a *mixed* phase space, where the configuration space coordinate is inertial, but the velocity-space coordinate system is an accelerated system because it is tied to the instantaneous local bulk velocity. (Taken from [1].)

In this case one has to transfer the Boltzmann equation into the new *mixed* phase space coordinate system. The pdf $f(\boldsymbol{r}, \boldsymbol{v}, t)$ transforms then into $f(\boldsymbol{r}, \boldsymbol{c}, t)$. By replacing the independent variable $\boldsymbol{v}$ with the random velocity $\boldsymbol{c}$, one has to take into account that the new variable $\boldsymbol{c}(\boldsymbol{r}, t)$ depends on $\boldsymbol{r}$ and t. Similar to the preceding problem we have to transform the derivatives accordingly, which is done in the following.

- The time derivative transforms into

$$\frac{\partial f}{\partial t} \Longrightarrow \frac{\partial f}{\partial t} + \frac{\partial f}{\partial c_i} \frac{\partial c_i}{\partial t}. \tag{2.4}$$

The time derivative of the new variable $c_i(\boldsymbol{r}, t)$ can be written as (because it depends on $\boldsymbol{u}$, which depends on time t)

$$\frac{\partial c_i}{\partial t} = \frac{\partial c_i}{\partial u_i} \frac{\partial u_i}{\partial t} = -\frac{\partial u_i}{\partial t},$$

where we used $\partial c_i / \partial u_i = -1$. In this case Eq. (2.4) becomes

$$\frac{\partial f}{\partial t} \Longrightarrow \frac{\partial f}{\partial t} - \frac{\partial u_i}{\partial t} \frac{\partial f}{\partial c_i}. \tag{2.5}$$

- The derivative with respect to the spatial coordinates can be written as

$$v_i \frac{\partial f}{\partial r_i} \Longrightarrow v_i \frac{\partial f}{\partial r_i} + v_i \frac{\partial f}{\partial c_j} \frac{\partial c_j}{\partial r_i}. \tag{2.6}$$

Note that we use a different index for $\boldsymbol{c}$, since each component c_j depends on all spatial coordinates r_i. By using Eq. (2.3) we find

$$\frac{\partial c_j}{\partial r_i} = \frac{\partial c_j}{\partial u_j}\frac{\partial u_j}{\partial r_i} = -\frac{\partial u_j}{\partial r_i},$$

where we again used $\partial c_j/\partial u_j = -1$. In this case we find for relation (2.6)

$$v_i\frac{\partial f}{\partial r_i} \Longrightarrow v_i\frac{\partial f}{\partial r_i} - v_i\frac{\partial f}{\partial c_j}\frac{\partial u_j}{\partial r_i} = v_i\frac{\partial f}{\partial r_i} - v_j\frac{\partial f}{\partial c_i}\frac{\partial u_i}{\partial r_j}. \tag{2.7}$$

Note that we swapped the indices i and j for the velocity v and the random velocity c in the last term on the right side. We do that, because we want to summarize the result in terms of $\partial f/\partial c_i$, and not $\partial f/\partial c_j$. According to Einstein's summation convention, swapping the indices has no influence on the summation. Substituting $v_{i,j}$ by Eq. (2.3) we find

$$v_i\frac{\partial f}{\partial r_i} \Rightarrow (u_i + c_i)\frac{\partial f}{\partial r_i} - (u_j + c_j)\frac{\partial u_i}{\partial r_j}\frac{\partial f}{\partial c_i}.$$

- Finally, we transform

$$\frac{\partial f}{\partial v_i} \Longrightarrow \frac{\partial f}{\partial c_i}\frac{\partial c_i}{\partial v_i} = \frac{\partial f}{\partial c_i},$$

  where we used $\partial c_i/\partial v_i = 1$ according to Eq. (2.3).

By summarizing all transformations we obtain Eq. (2.2).

Problem 2.3 Find the general solution to the Boltzmann equation (2.1) in the absence of collisions, i.e., $(\delta f/\delta t)_{coll} = 0$. Derive the general solution for the case that the force $\boldsymbol{F} = 0$.

Solution We consider first the force-free case with $\boldsymbol{F} = 0$ and then the case where $F = const. \neq 0$.

- For simplicity we use the one-dimensional Boltzmann equation

$$\frac{\partial f}{\partial t} + v_x\frac{\partial f}{\partial x} = 0$$

  and deduce the three-dimensional case from that result. Note that the function f depends on the independent variables x and t, i.e., $f = f(x,t)$. We solve this partial differential equation by using the method of characteristics. Therefore, we parameterize x and t through the parameter s, so that $f = f(t(s), x(s))$.

The *system of characteristics* (system of ordinary differential equations, ODE) and the initial conditions (for $s = 0$) are given by

$$\frac{dt(s)}{ds} = 1 \qquad t(0) = 0 \tag{2.8a}$$

$$\frac{dx(s)}{ds} = v_x \qquad x(0) = x_0 \tag{2.8b}$$

$$\frac{df(s)}{ds} = 0 \qquad f(0) = f(x_0). \tag{2.8c}$$

The solutions to the differential equations (2.8a)–(2.8c) are given by

$$\begin{array}{lll} t(s) = s + c_1 & c_1 = 0 & \quad t(s) = s \\ x(s) = v_x s + c_2 & c_2 = x_0 \quad \Rightarrow & \quad x(s) = v_x s + x_0 \\ f(s) = c_3 & c_3 = f(x_0) & \quad f(s) = f(x_0), \end{array}$$

and are referred to as the characteristic curve or simply the *characteristic*. In particular, we find immediately

$$s = t$$

$$x(t) = v_x t + x_0 \qquad x_0 = x - v_x t.$$

The characteristics are therefore curves which go through the point x_0 at time $t = 0$ into the direction of v_x. The solution of the Boltzmann equation with initial condition $f(x, t = 0) = f(x_0)$ can, therefore, be written as

$$f(x, t) = f(x_0) = G(x - v_x t),$$

which means that the solution is constant along the characteristic. One can interpret this solution in the way, that the initial profile $f(x_0)$ is transported with velocity v_x without changing the form of that profile. Going back to the three-dimensional case the solution is simply given by

$$f(\boldsymbol{x}, t) = f(\boldsymbol{x}_0) = G(\boldsymbol{x} - \boldsymbol{v}t).$$

- For the (1D) collisionless Boltzmann equation with $F = const. \neq 0$ we have

$$\frac{\partial f}{\partial t} + v_x \frac{\partial f}{\partial x} + a_x \frac{\partial f}{\partial v_x} = 0.$$

Here, the function f depends on x, v_x and t, i.e., $f = f(x, v_x, t)$. The system of characteristics with the initial conditions is given by

$$\frac{dt(s)}{ds} = 1 \qquad t(0) = 0 \tag{2.9a}$$

$$\frac{dx(s)}{ds} = v_x \qquad x(0) = x_0 \tag{2.9b}$$

$$\frac{dv_x(s)}{ds} = a_x \qquad v_x(0) = v_{x0} \tag{2.9c}$$

$$\frac{df(s)}{ds} = 0 \qquad f(0) = f(x_0, v_{x0}), \tag{2.9d}$$

where a_x is the acceleration in x direction. We start by solving the characteristic equation (2.9c)

$$\frac{dv_x(s)}{ds} = a_x \quad \Longrightarrow \quad v_x(s) = a_x s + c_1 \quad \Longrightarrow \quad v_x(s) = a_x s + v_{x0},$$

where we used the initial condition $v_x(s = 0) = v_{x0}$. The second differential equation (2.9b) can then be written as

$$\frac{dx(s)}{ds} = v_x = a_x s + v_{x0},$$

where we substituted the solution of the first differential equation. The solution to this differential equation is given by

$$x(s) = \frac{a_x}{2}s^2 + v_{x0}s + c_2 \quad \Longrightarrow \quad x(s) = \frac{a_x}{2}s^2 + v_{x0}s + x_0,$$

where we used the above initial condition. Obviously, we also find

$$t(s) = s + c_3 \quad \Longrightarrow \quad t(s) = s.$$

Therefore, the characteristics are

$$x(t) = \frac{a_x}{2}t^2 + v_{x0}t + x_0 \quad \Longrightarrow \quad x_0 = x - \frac{a_x}{2}t^2 - v_{x0}t$$

$$v_x(t) = a_x t + v_{x0} \quad \Longrightarrow \quad v_{x0} = v_x - a_x t. \tag{2.10}$$

The function f is constant along the characteristic described by Eq. (2.10),

$$f(x, v_x, t) = f(x_0, v_{x0}) = G\left[\left(x - \frac{a_x}{2}t^2 - v_{x0}t\right), (v_x - a_x t)\right].$$

We can easily test, that the function G solves the Boltzmann equation. We use

$$f(x, v_x, t) = G(\Phi, \Theta),$$

where

$$\Phi = x - \frac{a_x}{2}t^2 - v_{x0}t, \qquad \Theta = v_x - a_x t.$$

We find for the derivatives (for simplicity we abbreviate derivatives as $\partial G/\partial t = G_t$ and $\partial G/\partial \Phi = G_\Phi$):

$$\frac{\partial f}{\partial t} = G_\Phi \Phi_t + G_\Theta \Theta_t = G_\Phi \left(-a_x t - v_{x0}\right) + G_\Theta \left(-a_x\right)$$
$$v_x \frac{\partial f}{\partial x} = v_x G_\Phi \Phi_x = v_x G_\Phi$$
$$a_x \frac{\partial f}{\partial v_x} = a_x G_\Theta \Theta_{v_x} = a_x G_\Theta.$$

Adding all results up we find (since $-a_x t - v_{x0} + v_x = 0$, see Eq. (2.10))

$$\frac{\partial f}{\partial t} + v_x \frac{\partial f}{\partial x} + a_x \frac{\partial f}{\partial v_x} = G_\Phi \left(-a_x t - v_{x0} + v_x\right) + G_\Theta \left(-a_x + a_x\right) = 0.$$

2.2 The Boltzmann Collision Operator

Problem 2.4 Show that for $a \neq b$,

$$\delta\left((x-a)(x-b)\right) = \frac{1}{|a-b|}\left[\delta(x-a) + \delta(x-b)\right].$$

Solution For an arbitrary function $g(x)$ with roots[1] x_i and $g'(x_i) \neq 0$ the delta function is given by

$$\delta\left(g(x)\right) = \sum_i \frac{\delta(x - x_i)}{|\, g'(x_i)\,|}.$$

[1]A *root* or *zero*, x_i, of a function $g(x)$ is defined such that $g(x_i) = 0$, i.e., the function vanishes at x_i.

For the above function $g(x) = (x-a)(x-b) = x^2 - ax - bx + ab$ the roots are $x_1 = a$ and $x_2 = b$ and the derivative is given by $g'(x) = 2x - a - b$. We find

$$\delta\left((x-a)(x-b)\right) = \frac{1}{|2a-a-b|}\delta(x-a) + \frac{1}{|2b-a-b|}\delta(x-b)$$
$$= \frac{1}{|a-b|}\left[\delta(x-a) + \delta(x-b)\right].$$

Problem 2.5 Consider the relative motion of two particles P_1 and P_2 moving in each other's field of force with position vectors $\boldsymbol{r}_1$ and $\boldsymbol{r}_2$, and with masses m_1 and m_2, see Fig. 2.1. The particles are subject to the central forces $\boldsymbol{F}_1$ and $\boldsymbol{F}_2$, which are parallel to $\boldsymbol{r} = \boldsymbol{r}_1 - \boldsymbol{r}_2$ and depend, therefore, only on $r = |\boldsymbol{r}_1 - \boldsymbol{r}_2|$. Starting from the reduced mass equation of motion in polar coordinates,

$$M\left(\ddot{r} - r\dot{\theta}^2\right)\boldsymbol{e}_r + M\left(r\ddot{\theta} + 2\dot{r}\dot{\theta}\right)\boldsymbol{e}_\theta = -\frac{\partial V(r)}{\partial r}\boldsymbol{e}_r, \tag{2.11}$$

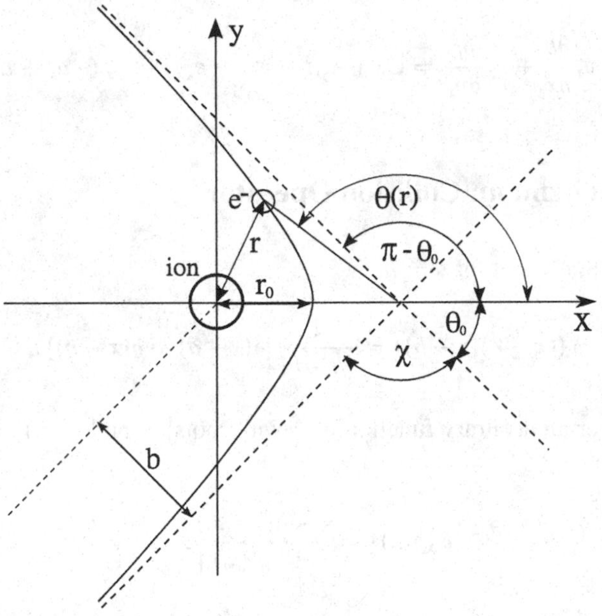

Fig. 2.1 Schematic of an electron (with charge eZ_1, where $Z_1 = -1$) scattering in the Coulomb field of an ion (with charge eZ_2, where $Z_2 > 0$). The trajectory of the electron is hyperbolic with eccentricity ϵ

complete the steps in the derivation of

$$\frac{d\theta}{dr} = \pm\frac{b}{r^2}\left[1 - \frac{b^2}{r^2} - \frac{2V(r)}{Mg^2}\right], \tag{2.12}$$

where θ is the angle of the incoming particle, b is the *impact parameter*, g is the constant relative speed, $M = m_1m_2/(m_1+m_2)$ is the relative (or reduced) mass, and $V(r)$ is the potential energy with $V(r=\infty) = 0$.

Solution We consider first the conservation laws for angular momentum and total energy. In the second part we derive the expression given by Eq. (2.12).

A. **Conservation of Angular Momentum.** Here we describe two alternatives. The first alternative is based on the *Lagrangian* and the conservation law for angular momentum is derived by considering the angular component (coordinate) of the Lagrangian. The second alternative is based on the *definition of the angular momentum*.

a. *Alternative 1*—Equation (2.11) describes the equations of motion in both directions, r and θ. To verify this equation we start with the Lagrangian, which is given by

$$\mathcal{L} = T - V,$$

where T is the kinetic energy and $V = V(r)$ is the potential energy. Note that in this particular case (polar coordinates) the kinetic energy $T = Mv^2/2$ consists of a radial and angular component, $T = T_r + T_\theta$, so that

$$\mathcal{L} = \frac{M}{2}v_r^2 + \frac{M}{2}v_\theta^2 - V(r) = \frac{M}{2}\dot{r}^2 + \frac{M}{2}r^2\dot{\theta}^2 - V(r),$$

where the radial and angular components of the velocity are given by $v_r = \dot{r}$ and $v_\theta = r\dot{\theta}$, with $v^2 = v_r^2 + v_\theta^2$. The equations of motion are then given by

$$\frac{d}{dt}\frac{\partial\mathcal{L}}{\partial\dot{q}_i} - \frac{\partial\mathcal{L}}{\partial q_i} = 0,$$

where the coordinate $q_i = r, \theta$. Let us consider first the radial component, where

$$\frac{d}{dt}\frac{\partial\mathcal{L}}{\partial\dot{r}} - \frac{\partial\mathcal{L}}{\partial r} = M\left(\ddot{r} - r\dot{\theta}^2\right) + \frac{\partial V(r)}{\partial r} = 0.$$

Note that this corresponds exactly to the radial component in Eq. (2.11). Similarly we obtain for the angular component

$$\frac{d}{dt}\frac{\partial\mathcal{L}}{\partial\dot{\theta}} - \frac{\partial\mathcal{L}}{\partial\theta} = M\frac{d}{dt}\left(r^2\dot{\theta}\right) = Mr\left(r\ddot{\theta} + 2\dot{r}\dot{\theta}\right) = 0, \tag{2.13}$$

since $\partial L/\partial\theta = 0$. For $r \neq 0$ we can divide by r. The remaining expression corresponds to the θ-component in Eq. (2.11). From Eq. (2.13) we can deduce that

$$r^2\dot{\theta} = const. = gb,$$

since its time derivative is zero. However, the latter relation can also be derived from the conservation law for the angular momentum L, which is described briefly in the following.

b. *Alternative 2*—The conservation law for angular momentum L is $dL/dt = 0$. With $\boldsymbol{v} = v_r\boldsymbol{e}_r + v_\theta\boldsymbol{e}_\theta$ and $\boldsymbol{r} = r\boldsymbol{e}_r$ we find

$$\boldsymbol{L} = \boldsymbol{r} \times \boldsymbol{p} = M\boldsymbol{r} \times \boldsymbol{v} = Mrv_r\boldsymbol{e}_r \times \boldsymbol{e}_r + Mrv_\theta\boldsymbol{e}_r \times \boldsymbol{e}_\theta$$
$$= Mr^2\dot{\theta}\boldsymbol{e}_r \times \boldsymbol{e}_\theta,$$

since $\boldsymbol{e}_r \times \boldsymbol{e}_r = 0$ and $v_\theta = r\dot{\theta}$. Since $\boldsymbol{e}_r \times \boldsymbol{e}_\theta$ is perpendicular to the plane of the particle motion, we simply write $L = mr^2\dot{\theta}$. Since the particle mass M is a constant we find

$$\frac{dL}{dt} = M\frac{d}{dt}\left(r^2\dot{\theta}\right) = 0 \quad \Rightarrow \quad r^2\dot{\theta} = const. = gb. \tag{2.14}$$

B. **Conservation of Total Energy.** The *conservation law for the total energy* can be derived by using the Hamiltonian,

$$\mathcal{H} = T + V.$$

With the above considerations we obtain

$$\mathcal{H} = \frac{M}{2}\dot{r}^2 + \frac{M}{2}r^2\dot{\theta}^2 + V(r). \tag{2.15}$$

The Hamiltonian $\mathcal{H}$ describes the total energy of the system. It can be shown that the total time derivative of the Hamiltonian equals the partial time derivative of $\mathcal{H}$,

$$\frac{d\mathcal{H}}{dt} = \frac{\partial\mathcal{H}}{\partial t}.$$

Further, if the Hamiltonian does not explicitly depend on time, the total time derivative vanishes and the total energy is conserved (constant). According to Eq. (2.15) the Hamiltonian is independent of time, i.e., $\partial\mathcal{H}/\partial t = 0$, and hence

$$\frac{d\mathcal{H}}{dt} = 0.$$

Therefore, the total energy is constant and given by

$$\frac{M}{2}\left(\dot{r}^2 + r^2\dot{\theta}^2\right) + V(r) = const. = \frac{M}{2}g^2. \tag{2.16}$$

We have now established that Eq. (2.11) describes the equations of motion for both the r and θ direction, and that the angular momentum is conserved. After deriving the conservation laws for the angular momentum and the total energy we proceed now with the derivation of Eq. (2.12).

Derivation of Eq. (2.12) We consider two particles P_1 and P_2. When particle P_1 approaches particle P_2 both the radial distance r and angle θ change with time t. Therefore, the time derivative of r is given by

$$\dot{r} = \frac{dr}{dt} = \frac{dr}{d\theta}\frac{d\theta}{dt} = \frac{dr}{d\theta}\dot{\theta},$$

where $\dot{\theta} = d\theta/dt$. Starting with Eq. (2.16) we substitute $\dot{r}$, multiply by 2, and divide by M and obtain

$$\left[\left(\frac{dr}{d\theta}\right)^2 \dot{\theta}^2 + r^2\dot{\theta}^2\right] = g^2 - \frac{2V(r)}{M},$$

where we moved the potential energy $V(r)$ to the right side. Substituting now $\dot{\theta}^2$ by Eq. (2.14) we obtain

$$\left(\frac{dr}{d\theta}\right)^2 \frac{g^2b^2}{r^4} = g^2 - \frac{2V(r)}{M} - \frac{g^2b^2}{r^2},$$

$$\left(\frac{dr}{d\theta}\right)^2 = \frac{r^4}{b^2}\left(1 - \frac{2V(r)}{Mg^2} - \frac{b^2}{r^2}\right).$$

By inverting and taking the square root, we find

$$\frac{d\theta}{dr} = \pm\frac{b}{r^2}\left[1 - \frac{2V(r)}{Mg^2} - \frac{b^2}{r^2}\right]^{-1/2}. \tag{2.17}$$

The negative root corresponds to an incoming particle since the radial coordinate decreases with time until reaching the point of closest approach. The positive root corresponds to an outgoing particle.

Problem 2.6 Consider the scattering of an electron with charge $q_1 = eZ_1$ (where $Z_1 = -1$) in the Coulomb field of an ion of charge $q_2 = eZ_2$ (where $Z_2 > 0$),

$$\boldsymbol{E} = \frac{eZ_2}{4\pi\epsilon_0}\frac{1}{r^2}\boldsymbol{e}_r,$$

where ϵ_0 is the permittivity of free space.

A. Show that

$$\frac{b^2}{rb_0} = 1 + \varepsilon \cos\theta, \tag{2.18}$$

where the eccentricity is given by

$$\epsilon \equiv \sqrt{1 + \frac{b^2}{b_0^2}}; \qquad b_0 = \frac{|Z_1 Z_2| e^2}{4\pi\epsilon_0 M g^2}. \tag{2.19}$$

B. Show that $\tan\theta_0 = b/b_0$ at the point of closest approach!
C. Show that the Coulomb or Rutherford scattering cross section is given by

$$\sigma = \frac{b_0^2}{4\sin^4\left(\frac{\chi}{2}\right)} = \left(\frac{Z_1 Z_2 e^2}{8\pi\epsilon_0 M g \sin^2\frac{\chi}{2}}\right)^2.$$

Solution

A. The electron experiences the (attractive) force

$$\boldsymbol{F}(r) = q_1 \boldsymbol{E}(r) = \frac{e^2 Z_1 Z_2}{4\pi\epsilon_0}\frac{1}{r^2}\boldsymbol{e}_r = -\frac{e^2|Z_1 Z_2|}{4\pi\epsilon_0}\frac{1}{r^2}\boldsymbol{e}_r \tag{2.20}$$

in the Coulomb field of the ion (see Fig. 2.1). Note that Z_1 and Z_2 have opposite signs (attractive force) and therefore the minus sign appears by taking the absolute value of Z_1 and Z_2. The potential energy at a particular distance r is then given by

$$V(r) = \int_r^\infty F(r')dr' = -\frac{e^2|Z_1 Z_2|}{4\pi\epsilon_0}\int_r^\infty \frac{1}{r'^2}dr' = -\frac{e^2|Z_1 Z_2|}{4\pi\epsilon_0}\frac{1}{r}, \tag{2.21}$$

where we used the fact that the potential energy for $r = \infty$ vanishes, i.e., $V(r = \infty) = 0$. Since the parameter b_0 is per definition a positive number, see Eq. (2.19), we find

$$\frac{V(r)}{Mg^2} = -\frac{b_0}{r}. \tag{2.22}$$

By substituting Eq. (2.22) into Eq. (2.17) from the previous problem we obtain

$$\frac{d\theta}{dr} = \pm\frac{b}{r^2}\left[1 + 2\frac{b_0}{r} - \frac{b^2}{r^2}\right]^{-1/2}. \tag{2.23}$$

Note that Eq. (2.17) was derived under a certain configuration in which the negative root denotes an incoming particle and the positive root denotes an outgoing particle. Now, according to Fig. 2.1 the coordinate system has been rotated so that the configuration is symmetric and each direction can refer to both an incoming or outgoing particle. We have to distinguish between the following two cases:

- a positive root refers to a particle moving into negative y direction
- a negative root refers to a particle moving into positive y direction.

This can easily be understood from Fig. 2.1. For a particle that moves in a negative y direction the angle $\theta(r)$ is always increasing (positive sign), while for a particle moving in a positive y direction the angle $\theta(r)$ is always decreasing (negative sign). The point of closest approach, r_0, is given by $d\theta/dr = 0$, which means that the square root of Eq. (2.23) has to be zero,

$$r_0^2 + 2b_0 r_0 - b^2 = 0. \tag{2.24}$$

The quadratic equation can easily be solved by

$$r_0 = -b_0 \pm \sqrt{b_0^2 + b^2}. \tag{2.25}$$

Since, per definition, the radial distance cannot be negative we have to choose the positive root, $r_0 = -b_0 + \sqrt{b_0^2 + b^2}$! By rearranging the root we can express the minimal distance in terms of the eccentricity (2.19),

$$r_0 = b_0(\epsilon - 1). \tag{2.26}$$

According to Fig. 2.1 we find that the angle under which the closest approach occurs is given by $\theta(r_0) = \pi$. (Note that the angle θ is taken anticlockwise from the positive x axis, the mathematical correct direction.) By considering a particle moving into negative y direction (starting from r_0) we can derive from Eq. (2.23) by choosing the positive root

$$\int_{\pi}^{\pi+\theta(r)} d\theta = \int_{r_0}^{r} \frac{b}{r'^2}\left[1 + 2\frac{b_0}{r'} - \frac{b^2}{r'^2}\right]^{-1/2} dr'.$$

With the substitution $u = b/r'$ we find (according to [2], Eq. (2.261))

$$\theta(r) = -\int_{b/r_0}^{b/r} du \frac{1}{\sqrt{1 + 2\frac{b_0}{b}u - u^2}} = -\arcsin\left(\frac{u - \frac{b_0}{b}}{\sqrt{1 + \frac{b_0^2}{b^2}}}\right)\Bigg|_{b/r_0}^{b/r}.$$

The denominator in the argument of the arcsin function can be expressed through the eccentricity,

$$\sqrt{1+\frac{b_0^2}{b^2}} = \frac{b_0}{b}\sqrt{1+\frac{b^2}{b_0^2}} = \frac{b_0}{b}\epsilon.$$

By using the identity $\arcsin y = \pi/2 - \arccos(y)$ we find

$$\theta(r) = -\arcsin\left(\frac{\frac{b}{b_0}u - 1}{\epsilon}\right)\Bigg|_{b/r_0}^{b/r} = \left[\arccos\left(\frac{\frac{b}{b_0}u - 1}{\epsilon}\right) - \frac{\pi}{2}\right]_{b/r_0}^{b/r}.$$

The angle θ is then given by

$$\theta(r) = \arccos\left(\frac{\frac{b^2}{b_0 r} - 1}{\epsilon}\right) - \arccos\left(\frac{\frac{b^2}{b_0 r_0} - 1}{\epsilon}\right).$$

Let us consider now the last term on the right side. More specifically, we want to calculate the argument of the arccos function. Therefore, we substitute r_0 (see Eq. 2.26) and we find

$$\frac{b^2}{b_0 r_0 \epsilon} - \frac{1}{\epsilon} = \frac{b^2}{b_0^2\epsilon(\epsilon-1)} - \frac{1}{\epsilon} = \frac{\epsilon^2-1}{\epsilon(\epsilon-1)} - \frac{1}{\epsilon} = \frac{\epsilon^2-1}{(\epsilon^2-\epsilon)} - \frac{(\epsilon-1)}{(\epsilon^2-\epsilon)} = 1,$$

where we used $b^2/b_0^2 = \epsilon^2 - 1$ in the second step, see also Eq. (2.19). Since $\arccos(1) = 0$ we find immediately

$$\theta(r) = \arccos\left[\frac{1}{\epsilon}\left(\frac{b^2}{rb_0} - 1\right)\right],$$

which leads to

$$\frac{b^2}{rb_0} = 1 + \epsilon\cos\theta. \tag{2.27}$$

B. From Fig. 2.1 we can deduce that for $r \to \infty$ the angle $\theta \to \pi \pm \theta_0$. Note that this is valid for both, a particle moving in positive *and* negative y-direction. Letting $r \to \infty$ in Eq. (2.27) we find that

$$0 = 1 + \epsilon\cos(\pi \pm \theta_0). \quad \Longrightarrow \quad \cos(\theta_0) = \frac{1}{\epsilon},$$

since $\cos(\pi \pm x) = -\cos(x)$. It is an easy matter to show then that

$$\tan(\theta_0) = \frac{\sqrt{1-\cos^2\theta_0}}{\cos\theta_0} = \sqrt{\epsilon^2 - 1} = \frac{b}{b_0}.$$

C. The cross section is defined by

$$\sigma = \frac{b}{\sin\chi}\left|\frac{db}{d\chi}\right|, \tag{2.28}$$

where χ is the scattering angle (i.e., the angle between the two asymptotes, which describe an incoming/outgoing particle). We know that the point of closest approach is defined by the angle θ_0 with (compare with Fig. 2.1)

$$\tan\theta_0 = \frac{b}{b_0} \quad \text{and} \quad \theta_0 = \frac{\pi}{2} - \frac{\chi}{2}.$$

By combining both equations we obtain

$$b = b_0 \tan\left(\frac{\pi}{2} - \frac{\chi}{2}\right)$$

$$\left|\frac{db}{d\chi}\right| = \left|-\frac{b_0}{2}\frac{1}{\cos^2\left(\frac{\pi}{2} - \frac{\chi}{2}\right)}\right| = \frac{b_0}{2}\frac{1}{\cos^2\left(\frac{\pi}{2} - \frac{\chi}{2}\right)}. \tag{2.29}$$

Inserting the last two equations in Eq. (2.28) we obtain for the cross section

$$\sigma = \frac{b_0^2}{2}\frac{\tan\left(\frac{\pi}{2} - \frac{\chi}{2}\right)}{\sin\chi}\frac{1}{\cos^2\left(\frac{\pi}{2} - \frac{\chi}{2}\right)}. \tag{2.30}$$

By using $\tan x = \sin x / \cos x$ and

$$\cos\left(\frac{\pi}{2} - \frac{\chi}{2}\right) = \sin\left(\frac{\chi}{2}\right)$$

$$\sin\left(\frac{\pi}{2} - \frac{\chi}{2}\right) = \cos\left(\frac{\chi}{2}\right)$$

$$\sin\chi = 2\sin\left(\frac{\chi}{2}\right)\cos\left(\frac{\chi}{2}\right) \tag{2.31}$$

we find the Rutherford cross section

$$\sigma = \frac{b_0^2}{4}\frac{1}{\sin^4\left(\frac{\chi}{2}\right)}. \tag{2.32}$$

2.3 The Boltzmann Equation and the Fluid Equations

The equations describing the conservation of mass, momentum and energy are

$$\frac{\partial n}{\partial t} + \nabla \cdot (n\boldsymbol{u}) = \frac{\partial n}{\partial t} + \sum_i \frac{\partial}{\partial x_i}(nu_i) = 0 \tag{2.33}$$

$$mn\left(\frac{\partial u_i}{\partial t} + \sum_j u_j \frac{\partial u_i}{\partial x_j}\right) = -\sum_j \frac{\partial p_{ij}}{\partial x_j} \tag{2.34}$$

$$\frac{\partial}{\partial t}\left[mn\left(e + \frac{u^2}{2}\right)\right] + \sum_i \frac{\partial}{\partial x_i}\left[mnu_i\left(e + \frac{u^2}{2}\right) + \sum_j u_j p_{ij} + q_i\right] = 0. \tag{2.35}$$

The equations for conservation of mass and energy are scalars; the equation for conservation of momentum describes the i-th component of that vector.

The Maxwell-Boltzmann distribution is given by

$$f(\boldsymbol{x}, \boldsymbol{v}, t) = n\left(\frac{m}{2\pi k_B T}\right)^{3/2} \exp\left[-m\frac{(\boldsymbol{v} - \boldsymbol{u})^2}{2k_B T}\right], \tag{2.36}$$

where k_B is Boltzmann's constant, $\boldsymbol{u}$, n, and T are the bulk velocity, number density, and temperature of the gas, respectively.

Problem 2.7 By using the conservation equations for mass, momentum and energy, derive the evolution equation for p_{ij} assuming the flow is smooth, i.e., the flow has no discontinuities like shock waves or contact discontinuities.

Solution The idea is as follows: To derive an evolution equation for the pressure tensor we start from the equation for conservation of energy (2.35) and transform that equation in such a way that the conservation of mass and momentum can be used for simplification. By expanding Eq. (2.35) we obtain

$$m\frac{\partial ne}{\partial t} + \frac{m}{2}\frac{\partial nu^2}{\partial t} + m\sum_i \frac{\partial (nu_i e)}{\partial x_i}$$
$$+ \frac{m}{2}\sum_i \frac{\partial (nu_i u^2)}{\partial x_i} + \sum_i \sum_j \frac{\partial u_j p_{ij}}{\partial x_i} + \sum_i \frac{\partial q_i}{\partial x_i} = 0. \tag{2.37}$$

Here we used the fact that the mass m is constant, i.e., independent of time and space coordinates. Consider now the second and fourth term in that equation. By expanding we obtain

$$\begin{aligned}
&\frac{m}{2}\left(\frac{\partial nu^2}{\partial t}+\sum_i\frac{\partial(nu_iu^2)}{\partial x_i}\right)\\
&=\frac{m}{2}\left(u^2\frac{\partial n}{\partial t}+n\frac{\partial u^2}{\partial t}+u^2\sum_i\frac{\partial(nu_i)}{\partial x_i}+\sum_i nu_i\frac{\partial u^2}{\partial x_i}\right)\\
&=\frac{m}{2}\left(n\frac{\partial u^2}{\partial t}+\sum_i nu_i\frac{\partial u^2}{\partial x_i}\right),
\end{aligned}\tag{2.38}$$

where we used the continuity equation (2.33) for the summation of the first and third term in the second line. By using $u^2=u_x^2+u_y^2+u_z^2=\sum_j u_j^2$ for the partial derivatives with respect to t and x_i and by pulling out the number density n we can simplify Eq. (2.38) to obtain

$$\frac{m}{2}\left(\frac{\partial nu^2}{\partial t}+\sum_i\frac{\partial(nu_iu^2)}{\partial x_i}\right)=nm\left(\sum_j u_j\frac{\partial u_j}{\partial t}+\sum_i\sum_j u_iu_j\frac{\partial u_j}{\partial x_i}\right).$$

We change now the indices, i.e., $i\leftrightarrow j$. The change of indices has no influence on the summation. We then pull out the summation over index i multiplied by u_i and obtain

$$\begin{aligned}
\frac{m}{2}\left(\frac{\partial nu^2}{\partial t}+\sum_i\frac{\partial(nu_iu^2)}{\partial x_i}\right)&=nm\sum_i u_i\left(\frac{\partial u_i}{\partial t}+\sum_j u_j\frac{\partial u_i}{\partial x_j}\right)\\
&=-\sum_i\sum_j u_i\frac{\partial p_{ij}}{\partial x_j},
\end{aligned}$$

where we used the equation of momentum conservation (2.34). We substitute this result back into Eq. (2.37) and obtain

$$m\frac{\partial ne}{\partial t}+m\sum_i\frac{\partial(nu_ie)}{\partial x_i}+\sum_i\sum_j\frac{\partial u_jp_{ij}}{\partial x_i}-\sum_i\sum_j u_i\frac{\partial p_{ij}}{\partial x_j}+\sum_i\frac{\partial q_i}{\partial x_i}=0.\tag{2.39}$$

Consider now the third term

$$\sum_i \sum_j \frac{\partial u_j p_{ij}}{\partial x_i} = \sum_i \sum_j u_j \frac{\partial p_{ij}}{\partial x_i} + \sum_i \sum_j p_{ij} \frac{\partial u_j}{\partial x_i}$$
$$= \sum_i \sum_j u_i \frac{\partial p_{ij}}{\partial x_j} + \sum_i \sum_j p_{ij} \frac{\partial u_j}{\partial x_i}. \tag{2.40}$$

Note that we changed the summation index in the first term of the last line; this has no influence on the result. We also used the fact that the pressure tensor is symmetric, i.e., $p_{ij} = p_{ji}$, which can be seen from the definition of the pressure tensor (we refer to [5] for further details). With this result Eq. (2.39) can be simplified and we obtain

$$m\frac{\partial ne}{\partial t} + m\sum_i \frac{\partial (nu_i e)}{\partial x_i} + \sum_i \sum_j p_{ij} \frac{\partial u_j}{\partial x_i} + \sum_i \frac{\partial q_i}{\partial x_i} = 0. \tag{2.41}$$

The second term can be described by

$$m\sum_i \frac{\partial (nu_i e)}{\partial x_i} = m\sum_i u_i \frac{\partial ne}{\partial x_i} + m\sum_i ne \frac{\partial u_i}{\partial x_i} \tag{2.42}$$

giving

$$m\frac{\partial ne}{\partial t} + m\sum_i u_i \frac{\partial ne}{\partial x_i} + m\sum_i ne \frac{\partial u_i}{\partial x_i} + \sum_i \sum_j p_{ij} \frac{\partial u_j}{\partial x_i} + \sum_i \frac{\partial q_i}{\partial x_i} = 0.$$

By multiplying the last equation by 2 and substitute $2mne = \sum_j p_{jj}$ we obtain

$$\sum_j \frac{\partial p_{jj}}{\partial t} + \sum_i \sum_j u_i \frac{\partial p_{jj}}{\partial x_i} + \sum_i \sum_j p_{jj} \frac{\partial u_i}{\partial x_i}$$
$$+ 2\sum_i \sum_j p_{ij} \frac{\partial u_j}{\partial x_i} + 2\sum_i \frac{\partial q_i}{\partial x_i} = 0. \tag{2.43}$$

This equation can be interpreted as the evolution equation for the pressure tensor.

Problem 2.8 Use the Maxwell-Boltzmann distribution (2.36) to show that the definitions for (A) the number density n, (B) the bulk velocity $\boldsymbol{u}$, (C) the temperature T, and (D) the pressure tensor p_{ij} do indeed yield these quantities, and that the pressure tensor can be expressed as $p_{ij} = p(\boldsymbol{x}, t)\delta_{ij}$. Show too that (E) the heat flux $\boldsymbol{q}$ vanishes.

Solution The Maxwell-Boltzmann distribution is given by Eq. (2.36). In all following calculations we will make use of

$$\int_{-\infty}^{\infty} x^{\eta} e^{-\beta x^2} dx = 0 \qquad \text{for } x = 1, 3, 5, 7, \ldots, \tag{2.44}$$

since the Maxwell-Boltzmann distribution satisfies this relation for the expected value $E[X^{\eta}]$. For more details see also Problem 2.12 later in this chapter, where the expected values are calculated from the moment-generating function.

A. The number density is defined as

$$n = \int f(\boldsymbol{x}, \boldsymbol{v}, t) d^3 v.$$

With the substitution $\boldsymbol{c} = \boldsymbol{v} - \boldsymbol{u}$ and $\boldsymbol{c}^2 = c_x^2 + c_y^2 + c_z^2$ we find

$$\int f(\boldsymbol{x}, \boldsymbol{v}, t) d^3 v = n \left(\frac{m}{2\pi kT}\right)^{3/2} \int \exp\left[-m\frac{(\boldsymbol{v} - \boldsymbol{u})^2}{2kT}\right] d^3 v$$
$$= n \left(\frac{m}{2\pi kT}\right)^{3/2} \int \exp\left[-\frac{m\boldsymbol{c}^2}{2kT}\right] d^3 c = n.$$

B. The bulk velocity is defined by

$$n\boldsymbol{u} = \int \boldsymbol{v} f(\boldsymbol{x}, \boldsymbol{v}, t) d^3 v.$$

Consider the i-th component (i.e., nu_i) for which we have

$$\int v_i f(\boldsymbol{x}, \boldsymbol{v}, t) d^3 v = n \left(\frac{m}{2\pi kT}\right)^{3/2} \int v_i \exp\left[-m\frac{(\boldsymbol{v} - \boldsymbol{u})^2}{2kT}\right] d^3 v.$$

By using again the substitution $c_i = v_i - u_i$ we find

$$\int v_i f(\boldsymbol{x}, \boldsymbol{v}, t) d^3 v = n \left(\frac{m}{2\pi kT}\right)^{3/2} \int (c_i + u_i) \exp\left[-\frac{m\boldsymbol{c}^2}{2kT}\right] d^3 c$$
$$= n \left(\frac{m}{2\pi kT}\right)^{3/2} \left[\int c_i \exp\left[-\frac{m\boldsymbol{c}^2}{2kT}\right] d^3 c + \int u_i \exp\left[-\frac{m\boldsymbol{c}^2}{2kT}\right] d^3 c\right].$$

The first integral yields zero (see Eq. (2.44)) and the second integral can be solved with the help of part A. We find

$$\int v_i f(\boldsymbol{x}, \boldsymbol{v}, t) d^3 v = nu_i \left(\frac{m}{2\pi kT}\right)^{3/2} \int \exp\left[-\frac{m\boldsymbol{c}^2}{2kT}\right] d^3 c = nu_i.$$

C. The temperature T is defined by

$$\frac{3}{2}nkT = \frac{m}{2}\int (\boldsymbol{v}-\boldsymbol{u})^2 f(\boldsymbol{x},\boldsymbol{v},t)d^3v.$$

Again, by using $\boldsymbol{c} = \boldsymbol{v} - \boldsymbol{u}$ we find

$$\begin{aligned}\frac{m}{2}\int (\boldsymbol{v}-\boldsymbol{u})^2 f(\boldsymbol{x},\boldsymbol{v},t)d^3v &= nm\left(\frac{m}{2\pi kT}\right)^{3/2}\frac{1}{2}\int \boldsymbol{c}^2 \exp\left[-\frac{m\boldsymbol{c}^2}{2kT}\right]d^3c \\ &= \frac{1}{2}nm\left(\frac{m}{2\pi kT}\right)^{3/2} 3\frac{\pi^{3/2}}{2}\left(\frac{2kT}{m}\right)^{5/2} \\ &= \frac{3}{2}nkT,\end{aligned}$$

where the integration yields $(3/2)\pi^{3/2}\,(2kT/m)^{5/2}$. The factor 3 originates from the summation of $\boldsymbol{c}^2 = c_x^2 + c_y^2 + c_z^2$.

D. The pressure tensor is given by

$$p_{ij} = \int m(v_i - u_i)(v_j - u_j) f d^3v$$

With the substitution $c_i = v_i - u_i$ we obtain

$$p_{ij} = nm\left(\frac{m}{2\pi kT}\right)^{3/2}\int c_i c_j \exp\left[-\frac{m\boldsymbol{c}^2}{2kT}\right]d^3c.$$

With $d^3c = dc_i dc_j dc_k$ and $\boldsymbol{c}^2 = c_i^2 + c_j^2 + c_k^2$ we find that for $i \neq j$ the pressure tensor is zero (see part B). The integral only contributes for $i = j$ and we have

$$\begin{aligned}p_{ii} &= nm\left(\frac{m}{2\pi kT}\right)^{3/2}\int c_i^2 \exp\left[-\frac{m\boldsymbol{c}^2}{2kT}\right]d^3c \\ &= nm\left(\frac{m}{2\pi kT}\right)^{3/2}\frac{\pi^{3/2}}{2}\left(\frac{2kT}{m}\right)^{5/2} \\ &= nkT.\end{aligned}$$

The pressure tensor can therefore be written as $p_{ij} = p\delta_{ij}$, where $p = nkT$ for ideal gases.

E. The i-th component of the heat flux vector is defined by

$$q_i = \frac{m}{2}\int (v_i - u_i)(\boldsymbol{v}-\boldsymbol{u})^2 f d^3v.$$

Substitute again $\boldsymbol{c} = \boldsymbol{v} - \boldsymbol{u}$ and we obtain

$$q_i = n\left(\frac{m}{2\pi kT}\right)^{3/2} \frac{m}{2} \int c_i c^2 \exp\left[-\frac{mc^2}{2kT}\right] d^3c = 0.$$

The integral yields zero since we have integrals of the form (2.44) in each term.

Problem 2.9 Using the above results, derive the Euler equations.

Solution The *Euler equations* result from assuming that

$$p_{ij} = p(\boldsymbol{x}, t)\delta_{ij} \qquad \text{and} \qquad q_i = 0,$$

where $p(\boldsymbol{x}, t)$ is the scalar pressure. For the conservation of momentum we find from Eq. (2.34) for the i-th component

$$mn\left(\frac{\partial u_i}{\partial t} + \sum_j u_j \frac{\partial u_i}{\partial x_j}\right) = -\sum_j \frac{\partial p}{\partial x_j}\delta_{ij} = -\frac{\partial p}{\partial x_i}$$

or written as a vector

$$mn\left(\frac{\partial \boldsymbol{u}}{\partial t} + \boldsymbol{u} \cdot \nabla \boldsymbol{u}\right) = -\nabla p.$$

For the energy conservation we use Eq. (2.43) and $\sum_j p_{jj} = 3p$ and find

$$3\frac{\partial p}{\partial t} + 3\sum_i u_i \frac{\partial p}{\partial x_i} + 3p\sum_i \frac{\partial u_i}{\partial x_i} + 2\sum_i\sum_j p\delta_{ij}\frac{\partial u_j}{\partial x_i} = 0$$

$$3\frac{\partial p}{\partial t} + 3\sum_i u_i \frac{\partial p}{\partial x_i} + 3p\sum_i \frac{\partial u_i}{\partial x_i} + 2p\sum_i \frac{\partial u_i}{\partial x_i} = 0$$

$$3\frac{\partial p}{\partial t} + 3\sum_i u_i \frac{\partial p}{\partial x_i} + 5p\sum_i \frac{\partial u_i}{\partial x_i} = 0.$$

If we divide by 3 and write the summations with the help of vectors we eventually find

$$\frac{\partial p}{\partial t} + \boldsymbol{u} \cdot \nabla p + \frac{5}{3} p \nabla \cdot \boldsymbol{u} = 0.$$

Together with the continuity equation (2.33) we find the Euler equations,

$$\frac{\partial n}{\partial t} + \nabla \cdot (n\boldsymbol{u}) = 0 \tag{2.45a}$$

$$mn\left(\frac{\partial \boldsymbol{u}}{\partial t}+\boldsymbol{u}\cdot\nabla\boldsymbol{u}\right)=-\nabla p \tag{2.45b}$$

$$\frac{\partial p}{\partial t}+\boldsymbol{u}\cdot\nabla p+\frac{5}{3}p\nabla\cdot\boldsymbol{u}=0. \tag{2.45c}$$

Problem 2.10 Linearize the 1D Euler equations about the constant state $\Psi_0 = (n_0, u_0, p_0)$, i.e., consider perturbations $\delta\Psi$ such that $\Psi = \Psi_0 + \delta\Psi$. Derive a linear wave equation in terms of a single variable, say δn. Seek solutions to the linear wave equation in the form $\exp[i(\omega t - kx)]$, and show that the Euler equations admit a non-propagating zero-frequency wave and forward and backward propagating acoustic modes satisfying the dispersion relation $\omega' = \omega - u_0 k = \pm C_s k$, where C_s is a suitably defined sound speed.

Solution The 1D Euler equations are given by

$$\frac{\partial n}{\partial t}+\frac{\partial}{\partial x}(nu_x)=0 \tag{2.46a}$$

$$nm\left(\frac{\partial u_x}{\partial t}+u_x\frac{\partial u_x}{\partial x}\right)=-\frac{\partial p}{\partial x} \tag{2.46b}$$

$$\frac{\partial p}{\partial t}+u_x\frac{\partial p}{\partial x}+\frac{5}{3}p\frac{\partial u_x}{\partial x}=0. \tag{2.46c}$$

We linearize the 1D Euler equations about the constant state $\Psi_0 = (n_0, u_0, p_0)$, i.e., we consider a small perturbation so that $\Psi = \Psi_0 + \delta\Psi$. The number density n, the velocity u_x, and the pressure p are then given by

$$n = n_0 + \delta n \qquad u_x = u_0 + \delta u \qquad p = p_0 + \delta p.$$

We substitute these equations into the Euler equations and obtain

$$\frac{\partial \delta n}{\partial t}+u_0\frac{\partial \delta n}{\partial x}+n_0\frac{\partial \delta u}{\partial x}=0 \tag{2.47a}$$

$$mn_0\frac{\partial \delta u}{\partial t}+mn_0u_0\frac{\partial \delta u}{\partial x}=-\frac{\partial \delta p}{\partial x} \tag{2.47b}$$

$$\frac{\partial \delta p}{\partial t}+u_0\frac{\partial \delta p}{\partial x}+\frac{5}{3}p_0\frac{\partial \delta u}{\partial x}=0. \tag{2.47c}$$

Note that we neglected terms of order $\delta\Psi^2$ and $\delta\Psi\partial\delta\Psi/\partial t$ or $\delta\Psi\partial\delta\Psi/\partial x$ and that the derivatives of a constant are zero.

Alternative 1 We assume now that the solutions of the differential equations have the form $\delta\Psi = \delta\Psi_0 \exp[i(\omega t - kx)]$, where $\delta\Psi = \delta n, \delta u, \delta p$, i.e.,

$$\delta n = \delta n_0 e^{i(\omega t - kx)} \qquad \delta u = \delta u_0 e^{i(\omega t - kx)} \qquad \delta p = \delta p_0 e^{i(\omega t - kx)}.$$

For the partial derivatives with respect to time t and space coordinate x we find in general

$$\frac{\partial \delta \Psi}{\partial t} = i\omega \delta \Psi \qquad \frac{\partial \delta \Psi}{\partial x} = -ik\delta \Psi.$$

The linearized Euler equations can then be written as

$$\begin{aligned} i\omega\delta n - iku_0\delta n - ikn_0\delta u &= 0 & & (\omega - u_0 k)\,\delta n - kn_0\delta u = 0 \\ mn_0 i\omega\delta u - ikmn_0 u_0\delta u &= ik\delta p & \Rightarrow \quad & (\omega - u_0 k)\,\delta u - \frac{k}{mn_0}\delta p = 0 \\ i\omega\delta p - iku_0\delta p - ik\frac{5}{3}p_0\delta u &= 0 & & (\omega - u_0 k)\,\delta p - k\frac{5}{3}p_0\delta u = 0. \end{aligned}$$

By using $\omega' = \omega - u_0 k$ we introduce the matrix $\underline{\mathbf{A}}$ and the vector $\delta\boldsymbol{\Psi}$,

$$\underline{\mathbf{A}} = \begin{pmatrix} \omega' & -n_0 k & 0 \\ 0 & \omega' & -\frac{k}{mn_0} \\ 0 & -\frac{5}{3}p_0 k & \omega' \end{pmatrix} \qquad \delta\boldsymbol{\Psi} = \begin{pmatrix} \delta n \\ \delta u \\ \delta p \end{pmatrix},$$

so that the set of equations can be written as

$$\underline{\mathbf{A}} \cdot \delta\boldsymbol{\Psi} = 0.$$

The trivial solution is, of course, $\delta\boldsymbol{\Psi} = 0$. Non-trivial solutions are given by the eigenvalues of the matrix $\underline{\mathbf{A}}$. Therefore, we determine first the *characteristic equation* (also called the characteristic polynomial), $\det\underline{\mathbf{A}} = 0$,

$$\det\underline{\mathbf{A}} = \omega' \cdot \left[\omega'^2 - \frac{5}{3}\frac{k^2 p_0}{mn_0}\right] = 0.$$

The eigenvalues are given by the zeros of the characteristic equation, hence,

$$\omega_1' = 0 \qquad \omega_{2,3}' = \pm\sqrt{\frac{5}{3}\frac{k_B T}{m}}\,k,$$

where we used $p_0 = n_0 k_B T$ (ideal gas law). The sound speed is defined as $C_s = \sqrt{\gamma k_B T/m}$ with the adiabatic index γ (which is $\gamma = 5/3$ for ideal gases). We find

$$\omega_1' = 0 \qquad \omega_{2,3}' = \pm C_s k.$$

The Euler equations, indeed, admit a zero frequency (non-propagating) wave, and forward and backward propagating acoustic modes with sound speed C_s.

Alternative 2 Our starting point are again the linearized 1D Euler equations (2.47a)–(2.47c). By introducing the convective derivative

$$\frac{D}{Dt} = \frac{\partial}{\partial t} + u_0 \frac{\partial}{\partial x}, \tag{2.48}$$

we may write the linearized 1D Euler equations as

$$\frac{D\delta n}{Dt} + n_0 \frac{\partial \delta u}{\partial x} = 0$$

$$\frac{D\delta u}{Dt} + \frac{1}{n_0 m} \frac{\partial \delta p}{\partial x} = 0$$

$$\frac{D\delta p}{Dt} + \frac{5}{3} p_0 \frac{\partial \delta u}{\partial x} = 0.$$

This is a set of three differential equations with three *unknowns* (δn, δu, and δp). We want to solve that set of differential equation for δp. Therefore, we take the convective derivative of the first and third equation, and the partial derivative with respect to x of the second equation resulting in

$$\frac{D^2 \delta n}{Dt^2} + n_0 \frac{\partial}{\partial x} \frac{D\delta u}{Dt} = 0$$

$$\frac{\partial}{\partial x} \frac{D\delta u}{Dt} + \frac{1}{n_0 m} \frac{\partial^2 \delta p}{\partial x^2} = 0$$

$$\frac{D^2 \delta p}{Dt^2} + \frac{5}{3} p_0 \frac{\partial}{\partial x} \frac{D\delta u}{Dt} = 0.$$

We substitute now the term with δu in the second and third equation by using the first equation (continuity equation). We obtain a set of two differential equations,

$$-\frac{D^2 \delta n}{Dt^2} + \frac{1}{m} \frac{\partial^2 \delta p}{\partial x^2} = 0$$

$$\frac{D^2 \delta p}{Dt^2} - \frac{5}{3} \frac{p_0}{n_0} \frac{D^2 \delta n}{Dt^2} = 0.$$

Consider now an ideal gas with $p_0 = n_0 k_B T$. By substituting the term with δn in the second equation we obtain

$$\frac{D^2 \delta p}{Dt^2} - \frac{5}{3} \frac{k_B T}{m} \frac{\partial^2 \delta p}{\partial x^2} = \frac{D^2 \delta p}{Dt^2} - C_s^2 \frac{\partial^2 \delta p}{\partial x^2} = 0,$$

where we defined the sound speed C_s as above. This equation is a wave equation. By using the above definition of the convective derivative we obtain

$$\frac{D^2}{Dt^2} = \frac{\partial^2}{\partial t^2} + 2u_0 \frac{\partial^2}{\partial t \partial x} + u_0^2 \frac{\partial^2}{\partial x^2},$$

so that

$$\frac{\partial^2 \delta p}{\partial t^2} + 2u_0 \frac{\partial^2 \delta p}{\partial t \partial x} + u_0^2 \frac{\partial^2 \delta p}{\partial x^2} - C_s^2 \frac{\partial^2 \delta p}{\partial x^2} = 0. \tag{2.49}$$

We seek now solutions of the form $\delta p = \delta \bar{p} \exp\left[i(\omega t - kx)\right]$. We obviously find

$$\frac{\partial^2 \delta p}{\partial t^2} = -\omega^2 \delta p \qquad \frac{\partial^2 \delta p}{\partial x^2} = -k^2 \delta p \qquad \frac{\partial^2 \delta p}{\partial t \partial x} = \omega k \delta p,$$

so that Eq. (2.49) becomes

$$(-\omega^2 + 2u_0 \omega k - u_0^2 k^2) + C_s^2 k^2 = 0.$$

The term in brackets can be written as $(-\omega^2 + 2u_0\omega k - u_0^2 k^2) = -(\omega - u_0 k)^2$ so that by using ω' as defined above

$$\omega'^2 \equiv (\omega - u_0 k)^2 = C_s^2 k^2 \qquad \Longrightarrow \qquad \omega' \equiv \omega - u_0 k = \pm C_s k.$$

Note that ω' is the frequency of the wave seen by an observer who is co-moving with the background flow at speed u_0, and ω is the frequency seen by an observer outside (not co-moving) with the background flow.

A wave (seen from an observer outside the co-moving frame) that travels with the same velocity as the background medium u_0 and with dispersion $\omega = u_0 k$ has a zero frequency in the co-moving frame

$$\omega' = u_0 k - u_0 k = 0,$$

and is therefore, non-propagating in that co-moving frame.

2.4 The Chapman-Enskog Expansion

We consider an expansion of the distribution function f about the equilibrium or Maxwellian distribution f_0 in the form

$$f = f_0 + \epsilon f_1 + \epsilon^2 f_2 + \ldots,$$

(continued)

where $f_1, f_2, \ldots$ are successive corrections to f_0. We consider also the force-free Boltzmann equation,

$$\frac{\partial f}{\partial t} + v_k \frac{\partial f}{\partial x_k} = -\nu(f - f_0),$$

where the collision operator on the right side is approximated by a scattering frequency ν. By using the above expansion for the distribution function we find

$$\frac{\partial f_0}{\partial t} + v_k \frac{\partial f_0}{\partial x_k} = -\nu f_1. \tag{2.50}$$

Since f_0 is the Maxwell-Boltzmann distribution we can evaluate the left side and find an expression for the first correction f_1.

Problem 2.11 Complete the details for the derivation of the expressions above for $\partial f_0/\partial t$ and $\partial f_0/\partial x_k$. Use these results to complete the derivation of the expression for f_1.

Solution The first correction f_1 to the Maxwell-Boltzmann distribution f_0 can be calculated by using Eq. (2.50), where the Maxwell-Boltzmann distribution (2.36) is given by

$$f_0 = n \left(\frac{m}{2\pi k_B T}\right)^{3/2} \exp\left[-\frac{m(\boldsymbol{v} - \boldsymbol{u})^2}{2k_B T}\right]. \tag{2.51}$$

Note that the density n, velocity $\boldsymbol{u}$, and temperature T are functions of time and space, so that $n = n(\boldsymbol{x}, t)$, $\boldsymbol{u} = \boldsymbol{u}(\boldsymbol{x}, t)$, and $T = T(\boldsymbol{x}, t)$. By introducing these new variables we have to transform the derivatives according to the new time and spatial dependencies of the new variables (see also Problems 2.1 and 2.2).

The idea is as follows: First, we derive the transformations for the time derivative and the spatial derivative, respectively. Finally, these expressions will be substituted back into Eq. (2.50), which will lead to an expression for the first correction in terms of f_0.

A. The time derivative transforms as follows

$$\frac{\partial f_0}{\partial t} = \frac{\partial f_0}{\partial n}\frac{\partial n}{\partial t} + \frac{\partial f_0}{\partial \boldsymbol{u}}\frac{\partial \boldsymbol{u}}{\partial t} + \frac{\partial f_0}{\partial T}\frac{\partial T}{\partial t}, \tag{2.52}$$

where (using the definition of the Maxwell-Boltzmann distribution)

$$\frac{\partial f_0}{\partial n} = \frac{f_0}{n}, \tag{2.53a}$$

$$\frac{\partial f_0}{\partial \boldsymbol{u}} = f_0 \frac{m(\boldsymbol{v}-\boldsymbol{u})}{k_B T}, \tag{2.53b}$$

$$\frac{\partial f_0}{\partial T} = -\frac{3}{2}\frac{1}{T} f_0 + \frac{m(\boldsymbol{v}-\boldsymbol{u})^2}{2k_B T}\frac{1}{T} f_0. \tag{2.53c}$$

All derivatives can be expressed in terms of the Maxwell-Boltzmann distribution f_0. By substituting these results into Eq. (2.52) we obtain

$$\frac{\partial f_0}{\partial t} = f_0 \left[\frac{1}{n}\frac{\partial n}{\partial t} + \frac{m(\boldsymbol{v}-\boldsymbol{u})}{k_B T}\frac{\partial \boldsymbol{u}}{\partial t} - \frac{3}{2}\frac{1}{T}\frac{\partial T}{\partial t} + \frac{m(\boldsymbol{v}-\boldsymbol{u})^2}{2k_B T}\frac{1}{T}\frac{\partial T}{\partial t} \right].$$

By substituting $\boldsymbol{c} = \boldsymbol{v} - \boldsymbol{u}$ and by using a component description we find for the i-th component

$$\frac{\partial f_0}{\partial t} = f_0 \left[\frac{1}{n}\frac{\partial n}{\partial t} + \frac{m c_i}{k_B T}\frac{\partial u_i}{\partial t} - \frac{3}{2}\frac{1}{T}\frac{\partial T}{\partial t} + \frac{m c^2}{2k_B T}\frac{1}{T}\frac{\partial T}{\partial t} \right]. \tag{2.54}$$

Equation (2.54) includes time derivatives of the density n, velocity u_i, and temperature T. To replace these time derivatives we use the Euler equations (2.45a)–(2.45c) (with Einstein's summation convention),

$$\frac{\partial n}{\partial t} + \frac{\partial}{\partial x_k}(n u_k) = 0$$

$$mn\left(\frac{\partial u_i}{\partial t} + u_k \frac{\partial u_i}{\partial x_k}\right) = -\frac{\partial p}{\partial x_k}$$

$$\frac{\partial p}{\partial t} + u_k \frac{\partial p}{\partial x_k} + \frac{5}{3} p \frac{\partial u_k}{\partial x_k} = 0.$$

With $p = n k_B T$ we obtain

$$\frac{1}{n}\frac{\partial n}{\partial t} = -\frac{\partial u_k}{\partial x_k} - \frac{u_k}{n}\frac{\partial n}{\partial x_k}$$

$$\frac{\partial u_i}{\partial t} = -u_k \frac{\partial u_i}{\partial x_k} - \frac{k_B T}{mn}\frac{\partial n}{\partial x_i} - \frac{k_B}{m}\frac{\partial T}{\partial x_i}$$

$$\frac{1}{T}\frac{\partial T}{\partial t} = -\frac{2}{3}\frac{\partial u_k}{\partial x_k} - \frac{u_k}{T}\frac{\partial T}{\partial x_k}.$$

For the last equation we used the continuity equation to eliminate the terms proportional to $\partial n/\partial t$ and $\partial n/\partial x_k$ By substituting these results back into Eq. (2.54) we obtain

$$\frac{1}{f_0}\frac{\partial f_0}{\partial t} = -\frac{u_k}{n}\frac{\partial n}{\partial x_k} - u_k\frac{mc_i}{k_BT}\frac{\partial u_i}{\partial x_k} - \frac{c_i}{n}\frac{\partial n}{\partial x_i} - \frac{c_i}{T}\frac{\partial T}{\partial x_i} + \frac{3}{2}\frac{u_k}{T}\frac{\partial T}{\partial x_k} - \frac{2}{3}\frac{mc^2}{2k_BT}\frac{\partial u_k}{\partial x_k} - \frac{mc^2}{2k_BT}\frac{u_k}{T}\frac{\partial T}{\partial x_k}. \tag{2.55}$$

Equation (2.55) is just the transformation of the time derivative in Eq. (2.50).

B. We also need to calculate the spatial derivatives in Eq. (2.50), which is done similarly to the time derivative (2.52) and we obtain

$$\frac{\partial f_0}{\partial x_k} = \frac{\partial f_0}{\partial n}\frac{\partial n}{\partial x_k} + \frac{\partial f_0}{\partial \boldsymbol{u}}\frac{\partial \boldsymbol{u}}{\partial x_k} + \frac{\partial f_0}{\partial T}\frac{\partial T}{\partial x_k}$$
$$= f_0\left[\frac{1}{n}\frac{\partial n}{\partial x_k} + \frac{m(\boldsymbol{v}-\boldsymbol{u})}{k_BT}\frac{\partial \boldsymbol{u}}{\partial x_k} - \frac{3}{2}\frac{1}{T}\frac{\partial T}{\partial x_k} + \frac{m(\boldsymbol{v}-\boldsymbol{u})^2}{2k_BT}\frac{1}{T}\frac{\partial T}{\partial x_k}\right].$$

where we used Eqs. (2.53a)–(2.53c) to replace the derivatives of the Maxwell-Boltzmann distribution, as before. By multiplying this equation with v_k and dividing by f_0 we obtain

$$\frac{v_k}{f_0}\frac{\partial f_0}{\partial x_k} = \frac{v_k}{n}\frac{\partial n}{\partial x_k} + \frac{m}{k_BT}c_i v_k\frac{\partial u_i}{\partial x_k} - \frac{3}{2}\frac{1}{T}v_k\frac{\partial T}{\partial x_k} + \frac{mc^2}{2k_BT}\frac{1}{T}v_k\frac{\partial T}{\partial x_k}. \tag{2.56}$$

We have also used $\boldsymbol{c} = \boldsymbol{v} - \boldsymbol{u}$ and substituted $v_k = c_k + u_k$.

Now we substitute the results (2.55) and (2.56) back into Eq. (2.50) and obtain the somewhat lengthy expression

$$-\nu\frac{f_1}{f_0} = -\frac{u_k}{n}\frac{\partial n}{\partial x_k} - u_k\frac{mc_i}{k_BT}\frac{\partial u_i}{\partial x_k} - \frac{c_i}{n}\frac{\partial n}{\partial x_i} - \frac{c_i}{T}\frac{\partial T}{\partial x_i} + \frac{3}{2}\frac{u_k}{T}\frac{\partial T}{\partial x_k}$$
$$- \frac{2}{3}\frac{mc^2}{2k_BT}\frac{\partial u_k}{\partial x_k} - \frac{mc^2}{2k_BT}\frac{u_k}{T}\frac{\partial T}{\partial x_k} + \frac{c_k+u_k}{n}\frac{\partial n}{\partial x_k}$$
$$+ \frac{m}{k_BT}c_i(c_k+u_k)\frac{\partial u_i}{\partial x_k} - \frac{3}{2}\frac{1}{T}(c_k+u_k)\frac{\partial T}{\partial x_k} + \frac{mc^2}{2k_BT}\frac{1}{T}(c_k+u_k)\frac{\partial T}{\partial x_k}.$$

Note that $c_i\partial/\partial x_i = c_k\partial/\partial x_k$, since we use Einstein's summation convention. After some simplifications we obtain

$$-\nu\frac{f_1}{f_0} = -\frac{2}{3}\frac{mc^2}{2k_BT}\frac{\partial u_k}{\partial x_k} + \frac{m}{k_BT}c_ic_k\frac{\partial u_i}{\partial x_k} - \frac{5}{2}\frac{1}{T}c_k\frac{\partial T}{\partial x_k} + \frac{mc^2}{2k_BT}\frac{1}{T}c_k\frac{\partial T}{\partial x_k}.$$

From this equation it follows that

$$-\nu f_1 = f_0\left[\frac{m}{k_BT}\left(c_ic_k - \frac{1}{3}c^2\delta_{ik}\right)\frac{\partial u_i}{\partial x_k} + c_k\left(\frac{mc^2}{2k_BT} - \frac{5}{2}\right)\frac{1}{T}\frac{\partial T}{\partial x_k}\right]. \tag{2.57}$$

The first correction can be described through the Maxwell-Boltzmann distribution function.

Problem 2.12 Consider the 1D pdf

$$f(x) = \sqrt{\frac{\beta}{\pi}}e^{-\beta x^2} \qquad \text{for } -\infty < x < \infty. \tag{2.58}$$

(A) Show that the moment-generating function is given by $M(t) = \exp(t^2/4\beta)$. (B) Derive the expectations $E(X)$, $E(X^2)$, $E(X^3)$, $E(X^4)$, $E(X^5)$ and $E(X^6)$. (C) Hence show that one obtains the integrals

$$\frac{\sqrt{\pi}}{2\beta^{3/2}} = \int_{-\infty}^{\infty} x^2 e^{-\beta x^2}dx \tag{2.59a}$$

$$\frac{3\sqrt{\pi}}{4\beta^{5/2}} = \int_{-\infty}^{\infty} x^4 e^{-\beta x^2}dx \tag{2.59b}$$

$$\frac{15\sqrt{\pi}}{8\beta^{7/2}} = \int_{-\infty}^{\infty} x^6 e^{-\beta x^2}dx. \tag{2.59c}$$

Solution

A. The moment-generating function is defined by Eq. (1.15), so that

$$M(t) = \int_{-\infty}^{\infty} e^{tx}f(x)dx = \sqrt{\frac{\beta}{\pi}}\int_{-\infty}^{\infty} e^{-\beta x^2+tx}dx.$$

Consider the exponent. With the transformation

$$-\beta x^2 + tx = -\beta\left(x - \frac{t}{2\beta}\right)^2 + \frac{t^2}{4\beta} = -\beta y^2 + \frac{t^2}{4\beta},$$

where we substituted $y = x - t/2\beta$, we obtain

$$M(t) = \sqrt{\frac{\beta}{\pi}}\exp\left(\frac{t^2}{4\beta}\right)\int_{-\infty}^{\infty} e^{-\beta y^2}dy = \exp\left(\frac{t^2}{4\beta}\right),$$

since the integration yields $\sqrt{\pi/\beta}$.

B. The n-th moment (or the n-th derivative) of the moment-generating function is given by

$$M^n(t) = \sqrt{\frac{\beta}{\pi}} \int_{-\infty}^{\infty} x^n e^{-\beta x^2 + tx} dx = \frac{d^n}{dt^n} \left[\exp\left(\frac{t^2}{4\beta} \right) \right].$$

Since $E(X^n) = M^n(t = 0)$ we find with $M(0) = 1$ the following results

$$M'(t) = \frac{t}{2\beta} M(t) \qquad E(X) = 0$$

$$M''(t) = \left(\frac{1}{4} \frac{t^2}{\beta^2} + \frac{1}{2\beta} \right) M(t) \qquad E(X^2) = \frac{1}{2\beta}$$

$$M'''(t) = \left(\frac{3}{4} \frac{t}{\beta^2} + \frac{1}{8} \frac{t^3}{\beta^3} \right) M(t) \qquad E(X^3) = 0$$

$$M^4(t) = \left(\frac{3}{4} \frac{1}{\beta^2} + \frac{3}{4} \frac{t^2}{\beta^3} + \frac{1}{16} \frac{t^4}{\beta^4} \right) M(t) \qquad E(X^4) = \frac{3}{4} \frac{1}{\beta^2}$$

$$M^5(t) = \left(\frac{15}{8} \frac{t}{\beta^3} + \frac{5}{8} \frac{t^3}{\beta^4} + \frac{1}{32} \frac{t^5}{\beta^5} \right) M(t) \qquad E(X^5) = 0$$

$$M^6(t) = \left(\frac{15}{8} \frac{1}{\beta^3} + \frac{45}{16} \frac{t^2}{\beta^4} + \frac{15}{32} \frac{t^4}{\beta^5} + \frac{1}{64} \frac{t^6}{\beta^6} \right) M(t) \quad E(X^6) = \frac{15}{8} \frac{1}{\beta^3}.$$

C. With $E(X^n) = M^n(t = 0)$ it follows immediately that

$$E(X^n) = \sqrt{\frac{\beta}{\pi}} \int_{-\infty}^{\infty} x^n e^{-\beta x^2} dx$$

$$\Rightarrow \int_{-\infty}^{\infty} x^n e^{-\beta x^2} dx = \sqrt{\frac{\pi}{\beta}} E(X^n). \tag{2.60}$$

By combining the expectations derived in part (B) with Eq. (2.60) we find the equations given by (2.59a)–(2.59c).

Problem 2.13 Show that the Chapman-Enskog expression for f_1 satisfies the constraints

$$\int f_1 d^3 v = 0 \qquad \int c f_1 d^3 v = 0 \qquad \int c^2 f_1 d^3 v = 0 \tag{2.61}$$

Solution The first order correction term of the Chapman-Enskog expansion is given by

$$f_1 = -\frac{f_0}{\nu} \left[\frac{m}{k_B T} \left(c_i c_k - \frac{1}{3} c^2 \delta_{ik} \right) \frac{\partial u_i}{\partial x_k} + c_k \left(\frac{mc^2}{2k_B T} - \frac{5}{2} \right) \frac{1}{T} \frac{\partial T}{\partial x_k} \right], \tag{2.62}$$

with the scattering frequency ν and the Maxwell-Boltzmann distribution f_0 (compare with Eq. (2.57)). For the sake of brevity we write the Maxwell-Boltzmann distribution function as

$$f_0 = n\left(\frac{\beta}{\pi}\right)^{3/2} e^{-\beta c^2}, \tag{2.63}$$

where $\beta = m/(2kT)$ and $\boldsymbol{c} = \boldsymbol{v} - \boldsymbol{u}$. This has the advantage that we can use the results of the previous Problem 2.12; compare with the probability density function given by Eq. (2.51). With $d^3v = d^3c$ the constraints can also be written as

$$\begin{aligned} 0 &= \int c^\alpha f_1 d^3c \qquad (2.64)\\ &= \frac{1}{\nu}\int c^\alpha f_0 \left[\frac{m}{k_B T}\frac{\partial u_i}{\partial x_k}\left(c_i c_k - \frac{1}{3}c^2\delta_{ik}\right) + \frac{1}{T}\frac{\partial T}{\partial x_k}c_k\left(\frac{mc^2}{2kT} - \frac{5}{2}\right)\right] d^3c. \end{aligned}$$

where $\alpha = 0, 1, 2$ for the zeroth, first, and second constraint respectively. In Eq. (2.64) the constants k_B, T, m and the partial derivatives with respect to x_k are independent of d^3c and, thus, the integral operates only on the terms in rounded brackets (which include the velocity c). Each of the integrals on the right hand side has to vanish independently, therefore, we consider both integrals separately. For simplicity we substitute f_0 by Eq. (2.63) and neglect all constant factors and obtain the following two conditions

$$A: \qquad \int c^\alpha e^{-\beta c^2}\left(c_i c_k - \frac{1}{3}c^2\delta_{ik}\right) d^3c = 0 \tag{2.65a}$$

$$B: \qquad \int c^\alpha e^{-\beta c^2} c_k \left(\beta c^2 - \frac{5}{2}\right) d^3c = 0. \tag{2.65b}$$

Basically, the constraints (2.61) reduce to the two conditions A and B.

First Constraint: $\alpha = 0$

A. In this case the expression given by Eq. (2.65a) becomes

$$\int_{-\infty}^{\infty} e^{-\beta c^2}\left(c_i c_k - \frac{1}{3}c^2\delta_{ik}\right) d^3c. \tag{2.66}$$

Here we have to distinguish between the two cases $i = k$ and $i \neq k$.

Case $i = k$: We find for Eq. (2.66)

$$\int_{-\infty}^{\infty}\int_{-\infty}^{\infty}\int_{-\infty}^{\infty} e^{-\beta(c_i^2+c_j^2+c_k^2)}\left[c_i^2 - \frac{1}{3}(c_i^2 + c_j^2 + c_k^2)\right] dc_i dc_j dc_k.$$

Note that the integral of each component within the squared brackets yields the same result, i.e., the integrals with c_j^2 and c_k^2 yield the same result as c_i^2, hence we set $(c_i^2 + c_j^2 + c_k^2) = 3c_i^2$ and obtain

$$\int_{-\infty}^{\infty}\int_{-\infty}^{\infty}\int_{-\infty}^{\infty} e^{-\beta(c_i^2+c_j^2+c_k^2)}\left[c_i^2 - c_i^2\right] dc_i dc_j dc_k = 0.$$

Case $i \neq k$: In this case Eq. (2.66) becomes

$$\int_{-\infty}^{\infty}\int_{-\infty}^{\infty}\int_{-\infty}^{\infty} e^{-\beta(c_i^2+c_j^2+c_k^2)} c_i c_k dc_i dc_j dc_k = 0.$$

The result is zero due to odd orders of c under the integral, compare with Eq. (2.44).

B. Let us now consider the second integral given by Eq. (2.65b). For $\alpha = 0$ we obtain

$$\int_{-\infty}^{\infty} e^{-\beta c^2} c_k \left[\beta c^2 - \frac{5}{2}\right] d^3c. \tag{2.67}$$

By expanding $c^2 = c_i^2 + c_j^2 + c_k^2$ we find

$$\int_{-\infty}^{\infty}\int_{-\infty}^{\infty}\int_{-\infty}^{\infty} e^{-\beta(c_i^2+c_j^2+c_k^2)}\left[\beta(c_k c_i^2 + c_k c_j^2 + c_k^3) - \frac{5}{2}c_k\right] dc_i dc_j dc_k = 0.$$

This integral is zero, where we used again Eq. (2.44).

Obviously, the first constraint is fulfilled.

Second Constraint: $\boldsymbol{\alpha} = 1$ The result of this constraint is a vector, but for convenience we consider only the j-th component of the vector $\boldsymbol{c}$.

A. For the first integral we find

$$\int e^{-\beta c^2}\left(c_i c_j c_k - \frac{1}{3}c_j c^2 \delta_{ik}\right) d^3c. \tag{2.68}$$

Here we have to distinguish the following cases:

Case $i = k = j$: In this case Eq. (2.68) becomes

$$\int_{-\infty}^{\infty}\int_{-\infty}^{\infty}\int_{-\infty}^{\infty} e^{-\beta(c_i^2+c_l^2+c_m^2)}\left[c_i^3 - \frac{1}{3}c_i(c_i^2 + c_l^2 + c_m^2)\right] dc_i dc_l dc_m = 0,$$

where we used the indices i, l, m to avoid confusion with the indices j, k. One can see immediately that the integral vanishes according to Eq. (2.44), since we have always an odd order of c_i under the integral.

Case $i = k \neq j$: In this case Eq. (2.68) becomes

$$\int_{-\infty}^{\infty}\int_{-\infty}^{\infty}\int_{-\infty}^{\infty} e^{-\beta(c_i^2+c_j^2+c_l^2)} \left[c_i^2 c_j - \frac{1}{3}c_j(c_i^2 + c_j^2 + c_l^2)\right] dc_i dc_j dc_l = 0,$$

which vanishes according to Eq. (2.44), since we have always odd orders of c_j under the integral.

Case $i \neq k = j$: In this case Eq. (2.68) becomes

$$\int e^{-\beta(c_i^2+c_j^2+c_l^2)} c_i c_j^2 \, dc_i dc_j dc_l = 0,$$

which vanishes also according to Eq. (2.44).

Case $i \neq k \neq j$ and $i \neq j$: In this last case, where all indices are mutually distinct, Eq. (2.68) becomes

$$\int e^{-\beta(c_i^2+c_j^2+c_k^2)} c_i c_j c_k \, dc_i dc_j dc_k = 0,$$

which vanishes also, because we have an odd order of c for all indices i, j, and k.

B. For the second integral we obtain (by using the j-th component of the vector)

$$\int e^{-\beta c^2} \left(\beta c_j c_k c^2 - \frac{5}{2} c_j c_k\right) d^3c, \tag{2.69}$$

where we pulled c_j and c_k into the brackets. We have to distinguish between the cases $j = k$ and $j \neq k$.

Case $j = k$: In this case Eq. (2.69) becomes

$$\int_{-\infty}^{\infty}\int_{-\infty}^{\infty}\int_{-\infty}^{\infty} e^{-\beta(c_j^2+c_l^2+c_m^2)} \left[\beta(c_j^2 + c_l^2 + c_m^2)c_j^2 - \frac{5}{2}c_j^2\right] dc_j dc_l dc_m.$$

The first part of that integral becomes then

$$\beta \int_{-\infty}^{\infty}\int_{-\infty}^{\infty}\int_{-\infty}^{\infty} e^{-\beta(c_j^2+c_l^2+c_m^2)} (c_j^4 + c_j^2 c_l^2 + c_j^2 c_m^2) dc_j dc_l dc_m = \frac{5}{4}\frac{\pi^{3/2}}{\beta^{5/2}}.$$

Note that while evaluating the integral we can set $c_j^2 c_l^2 + c_j^2 c_m^2 = 2c_j^2 c_l^2$, since both integrals provide the same result. The second part of the integral yields

$$-\frac{5}{2}\int_{-\infty}^{\infty}\int_{-\infty}^{\infty}\int_{-\infty}^{\infty} c_j^2 e^{-\beta(c_j^2+c_l^2+c_m^2)} dc_j dc_l dc_m = -\frac{5}{4}\frac{\pi^{3/2}}{\beta^{5/2}}.$$

By adding up both parts we obtain

$$\int e^{-\beta c^2}\left(\beta c_j c_k c^2 - \frac{5}{2} c_j c_k\right) d^3c = 0.$$

Case $j \neq k$: In this case Eq. (2.69) becomes

$$\int_{-\infty}^{\infty}\int_{-\infty}^{\infty}\int_{-\infty}^{\infty} e^{-\beta(c_j^2+c_k^2+c_l^2)}\left[\beta(c_j^2 + c_k^2 + c_l^2)c_j c_k - \frac{5}{2}c_j c_k\right] dc_j dc_k dc_l.$$

Obviously, this integral yields zero according to Eq. (2.44), since we have odd orders of c_j and c_k under the integral.

Obviously, the second constraint is also fulfilled.

Third Constraint: $\alpha = 2$

A. In this case Eq. (2.65a) becomes

$$\int e^{-\beta c^2}\left(c^2 c_i c_k - \frac{1}{3}c^4 \delta_{ik}\right) d^3c, \tag{2.70}$$

where we have pulled the c^2 into the brackets. Here we have to distinguish between the cases $i = k$ and $i \neq k$.

Case $i = k$: In this case Eq. (2.70) becomes

$$\int e^{-\beta c^2}\left(c^2 c_i^2 - \frac{1}{3}c^4\right) d^3c = \Lambda_1 + \Lambda_2.$$

Let us consider both terms separately and we obtain for the first part

$$\Lambda_1 = \int_{-\infty}^{\infty}\int_{-\infty}^{\infty}\int_{-\infty}^{\infty} e^{-\beta(c_i^2+c_l^2+c_m^2)}(c_i^4 + c_i^2 c_l^2 + c_i^2 c_m^2) dc_i dc_l dc_m.$$

Note that the integration of the terms containing $c_i^2 c_l^2$ and $c_i^2 c_m^2$ yield the same result. Therefore we simplify the equation and obtain for the first term

$$\Lambda_1 = \int_{-\infty}^{\infty}\int_{-\infty}^{\infty}\int_{-\infty}^{\infty} e^{-\beta(c_i^2+c_l^2+c_m^2)}(c_i^4 + 2c_i^2 c_l^2) dc_i dc_l dc_m.$$

For the second part of the integral we consider first the expression c^4 which can be written as

$$(c_i^2 + c_l^2 + c_m^2)^2 = c_i^4 + c_l^4 + c_m^4 + 2c_i^2c_l^2 + 2c_i^2c_m^2 + 2c_l^2c_m^2$$
$$= 3c_i^4 + 6c_i^2c_l^2,$$

where we used the fact, that the integrations over c_i^4, c_l^4, and c_m^4 yield the same result as well as the integrations over $c_i^2c_l^2, c_i^2c_m^2$, and $c_l^2c_m^2$. The second part of the integral can then be written as

$$\Lambda_2 = -\int_{-\infty}^{\infty}\int_{-\infty}^{\infty}\int_{-\infty}^{\infty} e^{-\beta(c_i^2+c_l^2+c_m^2)}\frac{1}{3}(3c_i^4 + 6c_i^2c_l^2)dc_i dc_l dc_m.$$

Since $\Lambda_1 = -\Lambda_2$ we find that $\Lambda_1 + \Lambda_2 = 0$ and therefore the integral vanishes for $i = k$.

Case $i \neq k$: In this case Eq. (2.70) becomes

$$\int_{-\infty}^{\infty}\int_{-\infty}^{\infty}\int_{-\infty}^{\infty} e^{-\beta(c_i^2+c_k^2+c_l^2)}(c_i^2 + c_k^2 + c_l^2)c_ic_k\, dc_i dc_k dc_l = 0.$$

This integral vanishes also, since we have odd orders of c_i and c_k under the integral.

With both results we find

$$\int c^2 e^{-\beta c^2}\left(c_ic_k - \frac{1}{3}c^2\delta_{ik}\right)d^3c = 0.$$

B. For the second integral we find

$$\int e^{-\beta c^2} c_k \left(\beta c^4 - \frac{5}{2}c^2\right)d^3c = 0,$$

where we pulled c^2 into the brackets. One can see immediately that this integral vanishes, since we have always an odd order of c_k under the integral.

By adding both results up we find that the third constraint is also fullfilled. All results show that the first correction f_1 to the Maxwell-Boltzmann distribution f_0 indeed satisfies the above mentioned conditions (2.61).

Problem 2.14 Show that the terms $\propto$ $(1/T)\partial T/\partial x_k$ in the pressure term of Eq. (2.57) vanish identically.

Solution The first correction to the pressure term is given by

$$p_{ij}^1 = m\int c_ic_jf_1d^3c.$$

According to Eq. (2.57) the term of f_1 that is proportional to $\propto (1/T)\partial T/\partial x_k$ is given by

$$\tilde{f}_1 = \frac{f_0}{\nu} c_k \left(\beta c^2 - \frac{5}{2}\right) \frac{1}{T} \frac{\partial T}{\partial x_k}.$$

Let's consider both terms separately. Since the temperature and the scattering frequency are independent of c we have

$$A: \qquad \int_{-\infty}^{\infty} c_i c_j c_k c^2 e^{-\beta c^2} d^3c = 0 \tag{2.71a}$$

$$B: \qquad \int_{-\infty}^{\infty} c_i c_j c_k e^{-\beta c^2} d^3c = 0, \tag{2.71b}$$

where we used the Maxwell-Boltzmann distribution function f_0 from Eq. (2.63). We consider first the integral of Eq. (2.71a).

A. Here we consider Eq. (2.71a). We have to distinguish between the following cases: (a) $i = j = k$, (b) $i = j \neq k$ (it is an easy matter to show that $i \neq j = k$ yields the same results) and (c) $i \neq j \neq k$ and $i \neq k$.

Case $i = j = k$: In this case Eq. (2.71a) becomes

$$\int_{-\infty}^{\infty}\int_{-\infty}^{\infty}\int_{-\infty}^{\infty} c_i^3 (c_i^2 + c_l^2 + c_m^2) e^{-\beta(c_i^2+c_l^2+c_m^2)} dc_i dc_l dc_m = 0.$$

This integral vanishes, since we have an odd order of c_i under the integral in each term, see Eq. (2.44).

Case $i \neq j = k$: In this case Eq. (2.71a) becomes

$$\int_{-\infty}^{\infty}\int_{-\infty}^{\infty}\int_{-\infty}^{\infty} c_i c_j^2 (c_i^2 + c_j^2 + c_l^2) e^{-\beta(c_i^2+c_j^2+c_l^2)} dc_i dc_j dc_l.$$

As before we have an odd order of c_i under the integral in each term. Therefore, the integral vanishes.

Case $i \neq j \neq k$ and $i \neq k$: In this case Eq. (2.71a) becomes

$$\int_{-\infty}^{\infty}\int_{-\infty}^{\infty}\int_{-\infty}^{\infty} c_i c_j c_k (c_i^2 + c_j^2 + c_k^2) e^{-\beta(c_i^2+c_j^2+c_k^2)} dc_i dc_j dc_k.$$

This integral vanishes also, since we have odd orders of c_i, c_j, and c_k under the integral in each term.

B. For the second integral of Eq. (2.71b) we have to distinguish the same three cases:

Case $i = j = k$: In this case Eq. (2.71b) becomes

$$\int_{-\infty}^{\infty}\int_{-\infty}^{\infty}\int_{-\infty}^{\infty} c_i^3 e^{-\beta(c_i^2+c_l^2+c_m^2)} dc_i dc_l dc_m = 0,$$

since we have an odd order of c_i under the integral.

Case $i \neq j = k$: In this case Eq. (2.71b) becomes

$$\int_{-\infty}^{\infty}\int_{-\infty}^{\infty}\int_{-\infty}^{\infty} c_i c_j^2 e^{-\beta(c_i^2+c_j^2+c_l^2)} dc_i dc_j dc_l,$$

which will also vanish due to odd orders of c_i under the integral.

Case $i \neq j \neq k$ and $i \neq k$: In this case Eq. (2.71a) becomes

$$\int_{-\infty}^{\infty}\int_{-\infty}^{\infty}\int_{-\infty}^{\infty} c_i c_j c_k e^{-\beta(c_i^2+c_j^2+c_k^2)} dc_i dc_j dc_k,$$

which also vanishes due to odd orders of c under the integral.

By combining the results from part A and B we find indeed that the terms proportional to $(1/T)\partial T/\partial x_k$ in the pressure term vanish identically.

Problem 2.15 Show that the heat flux vector is given by

$$q_i = -\lambda\frac{\partial T}{\partial x_i} \qquad \text{with} \qquad \lambda = \frac{5}{2}\frac{nk^2T}{m\nu}.$$

Solution The heat flux vector can be calculated by

$$q_i = \frac{m}{2}\int c_i c^2 f_1 d^3c,$$

with the first order correction f_1 given by Eq. (2.62), and the Maxwell-Boltzmann distribution function f_0 given by Eq. (2.63), with $\beta = m/(2kT)$ and $\boldsymbol{c} = \boldsymbol{v} - \boldsymbol{u}$. Again, the integral operates only on the terms in rounded brackets

$$q_i = -\frac{m}{2\nu}\left\{\frac{m}{kT}\int f_0 c_i c^2\left(c_j c_k - \frac{1}{3}c^2\delta_{jk}\right)d^3c\frac{\partial u_i}{\partial x_k}\right.$$
$$\left.+\int c_i c^2 f_0 c_k\left(\beta c^2 - \frac{5}{2}\right)d^3c\frac{1}{T}\frac{\partial T}{\partial x_k}\right\}.$$

It can easily be seen that the first integral is

$$\int f_0 c_i c^2\left(c_j c_k - \frac{1}{3}c^2\delta_{jk}\right)d^3c = 0,$$

since we find in each term odd orders of c, see Eq. (2.44). The heat flux is therefore, given by

$$q_i = -\frac{m}{2\nu}\frac{1}{T}\frac{\partial T}{\partial x_k}\int c_i c_k c^2 f_0 \left(\beta c^2 - \frac{5}{2}\right) d^3c.$$

As before, for $i \neq k$ this integral is zero, therefore we consider only the case $i = k$ and obtain

$$q_i = -n\left(\frac{\beta}{\pi}\right)^{3/2}\frac{m}{2\nu}\frac{1}{T}\frac{\partial T}{\partial x_i}\int c_i^2 c^2 e^{-\beta c^2}\left(\beta c^2 - \frac{5}{2}\right) d^3c,$$

where we substituted f_0 by Eq. (2.63). Let us consider the first term of the sum. We obtain

$$n\left(\frac{\beta}{\pi}\right)^{3/2}\beta\frac{m}{2\nu}\frac{1}{T}\frac{\partial T}{\partial x_i}\int c_i^2 c^4 e^{-\beta c^2} d^3c = n\frac{35}{16}\frac{1}{\beta^2}\frac{m}{\nu T}\frac{\partial T}{\partial x_i}. \tag{2.72}$$

For the second term we find

$$n\frac{5}{2}\left(\frac{\beta}{\pi}\right)^{3/2}\frac{m}{2\nu}\frac{1}{T}\frac{\partial T}{\partial x_i}\int c_i^2 c^2 e^{-\beta c^2} d^3c = n\frac{25}{16}\frac{1}{\beta^2}\frac{m}{\nu T}\frac{\partial T}{\partial x_i}. \tag{2.73}$$

By subtracting Eq. (2.73) from Eq. (2.72) we find for the heat flux

$$q_i = -n\frac{35}{16}\frac{1}{\beta^2}\frac{m}{\nu T}\frac{\partial T}{\partial x_i} - n\frac{25}{16}\frac{1}{\beta^2}\frac{m}{\nu T}\frac{\partial T}{\partial x_i} = -n\frac{5}{8}\frac{1}{\beta^2}\frac{m}{\nu T}\frac{\partial T}{\partial x_i} = -\frac{5}{2}\frac{nk^2T}{m\nu}\frac{\partial T}{\partial x_i},$$

where we used $\beta = m/2kT$.

2.5 Application 1: Structure of Weak Shock Waves

The one-dimensional *Rankine-Hugoniot* conditions are given by

$$s\,[\rho] = [\rho u] \equiv [m]$$
$$s\,[\rho u] = \left[\rho u^2 + p\right]$$
$$s\,[e] = [(e+p)u]\,,$$

(continued)

where $s = dx/dt$ is the speed of the discontinuity and e is the total energy with

$$e \equiv \frac{1}{2}\rho u^2 + \frac{p}{\gamma - 1} = \frac{1}{2}\rho u^2 + \rho\epsilon. \tag{2.74}$$

Here, $\epsilon = p/\rho(\gamma-1)$ is the expression for the internal energy. Since the Euler equations are Galilean invariant, we may transform the Rankine-Hugoniot conditions into a coordinate system moving with a uniform velocity such that the speed of the discontinuity is zero, $s = 0$. The steady-state Rankine-Hugoniot conditions can then be written as

$$\rho_0 u_0 = \rho_1 u_1 \tag{2.75a}$$

$$\rho_0 u_0^2 + p_0 = \rho_1 u_1^2 + p_1 \tag{2.75b}$$

$$(e_0 + p_0)u_0 = (e_1 + p_1)u_1. \tag{2.75c}$$

If we let $m = \rho_0 u_0 = \rho_1 u_1$, we can distinguish between two classes of discontinuities. If $m = 0$, the discontinuity is called a *contact discontinuity* or *slip line*. Since $u_0 = u_1 = 0$, these discontinuities convect with the fluid. From (2.75b) we observe that $p_0 = p_1$ across a contact discontinuity but in general $\rho_0 \neq \rho_1$. By contrast, if $m \neq 0$, then the discontinuity is called a *shock wave*. Since $u_0 \neq 0$ and $u_1 \neq 0$, the gas crosses the shock, or equivalently, the shock propagates through the fluid. The side of the shock that comprises gas that has *not* been shocked is the *front* or *upstream* of the shock, while the shocked gas is the *back* of *downstream* of the shock.

Problem 2.16 Explicitly, derive the $O(\varepsilon)$ and $O(\varepsilon^2)$ expansions of the Euler equations (e.g., Eqs. (2.46a)–(2.46c)).

Solution The 1D Euler equations can be rewritten as

$$\frac{\partial \bar{\rho}}{\partial \bar{t}} + \frac{\partial}{\partial \bar{x}}(\bar{\rho}\bar{u}) = 0$$

$$\bar{\rho}\left(\frac{\partial \bar{u}}{\partial \bar{t}} + \bar{u}\frac{\partial \bar{u}}{\partial \bar{x}}\right) = -\frac{a_{c0}^2}{\gamma V_p^2}\frac{\partial \bar{p}}{\partial \bar{x}}$$

$$\frac{\partial \bar{p}}{\partial \bar{t}} + \bar{u}\frac{\partial \bar{p}}{\partial \bar{x}} + \gamma\bar{p}\frac{\partial \bar{u}}{\partial \bar{x}} = 0,$$

where we introduced the dimensionless variables $\bar{t} = t/T, \bar{x} = x/L, \bar{\rho} = \rho/\rho_0, \bar{p} = p/p_0$, and $\bar{u} = u/V_p$. Here, T and L are a characteristic time and length scale respectively, and V_p is a characteristic phase velocity. Also ρ_0 and p_0 are equilibrium values for the density and pressure far upstream of any shock transition.

By introducing fast and slow variables $\xi = \bar{x} - \bar{t}$ and $\tau = \epsilon\bar{t}$ we find

$$\frac{\partial}{\partial \bar{x}} = \frac{\partial \xi}{\partial \bar{x}}\frac{\partial}{\partial \xi} = \frac{\partial}{\partial \xi} \qquad \frac{\partial}{\partial \bar{t}} = \frac{\partial \tau}{\partial \bar{t}}\frac{\partial}{\partial \tau} + \frac{\partial \xi}{\partial \bar{t}}\frac{\partial}{\partial \xi} = \varepsilon\frac{\partial}{\partial \tau} - \frac{\partial}{\partial \xi}. \tag{2.76}$$

If we expand the flow variables about a uniform far-upstream background, we obtain

$$\bar{\rho} = 1 + \varepsilon\rho_1 + \varepsilon^2\rho_2 + \ldots$$
$$\bar{u} = \varepsilon u_1 + \varepsilon^2 u_2 + \ldots$$
$$\bar{p} = 1 + \varepsilon p_1 + \varepsilon^2 p_2 + \ldots$$

A. **Continuity Equation**
For the continuity equation we find the general expression

$$\left(\varepsilon\frac{\partial}{\partial \tau} - \frac{\partial}{\partial \xi}\right)\left(1 + \varepsilon\rho_1 + \varepsilon^2\rho_2 + \ldots\right)$$
$$+ \frac{\partial}{\partial \xi}\left[\left(1 + \varepsilon\rho_1 + \varepsilon^2\rho_2 + \ldots\right)\left(\varepsilon u_1 + \varepsilon^2 u_2 + \ldots\right)\right] = 0.$$

- Terms of order $O(\varepsilon)$:

$$-\frac{\partial \rho_1}{\partial \xi} + \frac{\partial u_1}{\partial \xi} = 0 \qquad \Rightarrow \qquad \rho_1 = u_1. \tag{2.77}$$

- Terms of order $O(\varepsilon^2)$:

$$\frac{\partial \rho_1}{\partial \tau} - \frac{\partial \rho_2}{\partial \xi} + \frac{\partial u_2}{\partial \xi} + \frac{\partial \rho_1 u_1}{\partial \xi} = 0.$$

B. **Momentum Equation**
The general form of the momentum equation is given by

$$\varepsilon\left(1 + \varepsilon\rho_1 + \varepsilon^2\rho_2 + \ldots\right)\frac{\partial}{\partial \tau}\left(\varepsilon u_1 + \varepsilon^2 u_2 + \ldots\right)$$
$$-\left(1 + \varepsilon\rho_1 + \varepsilon^2\rho_2 + \ldots\right)\frac{\partial}{\partial \xi}\left(\varepsilon u_1 + \varepsilon^2 u_2 + \ldots\right)$$
$$+\left(1 + \varepsilon\rho_1 + \varepsilon^2\rho_2 + \ldots\right)\left(\varepsilon u_1 + \varepsilon^2 u_2 + \ldots\right)\frac{\partial}{\partial \xi}\left(\varepsilon u_1 + \varepsilon^2 u_2 + \ldots\right)$$
$$= -\frac{\bar{a}_{c0}^2}{\gamma}\frac{\partial}{\partial \xi}\left(1 + \varepsilon p_1 + \varepsilon^2 p_2 + \ldots\right),$$

where we used $\bar{a}_{c0}^2 = a_{c0}^2/V_p^2$.

- Terms of order $O(\varepsilon)$:

$$\frac{\partial u_1}{\partial \xi} = \frac{\bar{a}_{c0}^2}{\gamma}\frac{\partial p_1}{\partial \xi} \quad \Rightarrow \quad u_1 = \frac{\bar{a}_{c0}^2}{\gamma} p_1. \tag{2.78}$$

- Terms of order $O(\varepsilon^2)$:

$$\frac{\partial u_1}{\partial \tau} - \frac{\partial u_2}{\partial \xi} - \rho_1 \frac{\partial u_1}{\partial \xi} + u_1 \frac{\partial u_1}{\partial \xi} = -\frac{\bar{a}_{c0}^2}{\gamma}\frac{\partial p_2}{\partial \xi}.$$

C. **Energy Equation**
The general form of the energy equation is given by

$$\varepsilon \frac{\partial}{\partial \tau}\left(1 + \varepsilon p_1 + \varepsilon^2 p_2 + \ldots\right) - \frac{\partial}{\partial \xi}\left(1 + \varepsilon p_1 + \varepsilon^2 p_2 + \ldots\right)$$
$$+ \left(\varepsilon u_1 + \varepsilon^2 u_2 + \ldots\right) \frac{\partial}{\partial \xi}\left(1 + \varepsilon p_1 + \varepsilon^2 p_2 + \ldots\right)$$
$$+ \gamma \left(1 + \varepsilon p_1 + \varepsilon^2 p_2 + \ldots\right) \frac{\partial}{\partial \xi}\left(\varepsilon u_1 + \varepsilon^2 u_2 + \ldots\right) = 0.$$

- Terms of order $O(\varepsilon)$:

$$-\frac{\partial p_1}{\partial \xi} + \gamma \frac{\partial u_1}{\partial \xi} = 0 \quad \Rightarrow \quad p_1 = \gamma u_1. \tag{2.79}$$

- Terms of order $O(\varepsilon^2)$:

$$\frac{\partial p_1}{\partial \tau} - \frac{\partial p_2}{\partial \xi} + u_1 \frac{\partial p_1}{\partial \xi} + \gamma \frac{\partial u_2}{\partial \xi} + \gamma p_1 \frac{\partial u_1}{\partial \xi} = 0.$$

Problem 2.17 Derive the nonlinear wave equation

$$\frac{\partial u_1}{\partial \tau} + \frac{\gamma + 1}{2} u_1 \frac{\partial u_1}{\partial \xi} = 0, \tag{2.80}$$

which is called the *inviscid* form of Burgers' equation.

Solution From the first order expansions of the Euler equations (see previous Problem 2.16), we find the relations

$$\rho_1 = u_1 \qquad u_1 = \frac{\bar{a}_{c0}^2}{\gamma} p_1 \qquad p_1 = \gamma u_1. \tag{2.81}$$

From the last two relations it follows that $\bar{a}_{c0}^2 = 1$. For the second order equations, $O(\varepsilon^2)$, we find

$$-\frac{\partial \rho_2}{\partial \xi} + \frac{\partial u_2}{\partial \xi} = -\frac{\partial}{\partial \xi}(\rho_1 u_1) - \frac{\partial \rho_1}{\partial \tau} \tag{2.82a}$$

$$-\frac{\partial u_2}{\partial \xi} + \frac{\bar{a}_{c0}^2}{\gamma}\frac{\partial p_2}{\partial \xi} = \rho_1 \frac{\partial u_1}{\partial \xi} - \frac{\partial u_1}{\partial \tau} - u_1 \frac{\partial u_1}{\partial \xi} \tag{2.82b}$$

$$-\frac{\partial p_2}{\partial \xi} + \gamma \frac{\partial u_2}{\partial \xi} = -\frac{\partial p_1}{\partial \tau} - \gamma p_1 \frac{\partial u_1}{\partial \xi} - u_1 \frac{\partial p_1}{\partial \xi}. \tag{2.82c}$$

Equation (2.82c) can be rewritten as

$$\frac{\partial p_2}{\partial \xi} = \frac{\partial p_1}{\partial \tau} + \gamma p_1 \frac{\partial u_1}{\partial \xi} + u_1 \frac{\partial p_1}{\partial \xi} + \gamma \frac{\partial u_2}{\partial \xi}.$$

Substituting this result into Eq. (2.82b) and setting $\bar{a}_{c0}^2 = 1$ we obtain

$$-\frac{\partial u_2}{\partial \xi} + \frac{1}{\gamma}\left(\frac{\partial p_1}{\partial \tau} + \gamma p_1 \frac{\partial u_1}{\partial \xi} + u_1 \frac{\partial p_1}{\partial \xi} + \gamma \frac{\partial u_2}{\partial \xi}\right) = \rho_1 \frac{\partial u_1}{\partial \xi} - \frac{\partial u_1}{\partial \tau} - u_1 \frac{\partial u_1}{\partial \xi},$$

which can be simplified to

$$\frac{1}{\gamma}\frac{\partial p_1}{\partial \tau} + p_1 \frac{\partial u_1}{\partial \xi} + \frac{1}{\gamma} u_1 \frac{\partial p_1}{\partial \xi} = \rho_1 \frac{\partial u_1}{\partial \xi} - \frac{\partial u_1}{\partial \tau} - u_1 \frac{\partial u_1}{\partial \xi}.$$

Now we substitute $p_1 = \gamma u_1$ and $\rho_1 = u_1$ from Eq. (2.81) and obtain

$$\frac{\partial u_1}{\partial \tau} + \gamma u_1 \frac{\partial u_1}{\partial \xi} + u_1 \frac{\partial u_1}{\partial \xi} = u_1 \frac{\partial u_1}{\partial \xi} - \frac{\partial u_1}{\partial \tau} - u_1 \frac{\partial u_1}{\partial \xi}.$$

Finally, this can be simplified to

$$\frac{\partial u_1}{\partial \tau} + \frac{\gamma + 1}{2} u_1 \frac{\partial u_1}{\partial \xi} = 0,$$

which is called the *inviscid* form of Burgers' equation.

Problem 2.18 Solve the linear wave equation

$$\frac{\partial u}{\partial t} + c\frac{\partial u}{\partial x} = 0, \tag{2.83}$$

where c is a constant and the initial condition $u(x, t = 0) = f(x)$. Write down the solution if $f(x) = \sin(kx)$.

Solution The initial curve at time $t = 0$ can be parameterized through

$$t = 0 \qquad x = x_0 \qquad u(x, t = 0) = f(x_0). \tag{2.84}$$

The set of characteristic equations (see also Problem 2.3) and their solutions are given by

$$\frac{dt}{d\tau} = 1 \qquad t = \tau + const_1 \qquad t = \tau$$

$$\frac{dx}{d\tau} = c \qquad x = c\tau + const_2 \qquad x = c\tau + x_0$$

$$\frac{du}{d\tau} = 0 \qquad u = const_3 \qquad u = f(x_0).$$

From the last equations it follows that u is constant along the characteristic curve $x_0 = x - ct$. In particular, $x_0 = x_0(x, t)$. The characteristics can be inverted and we find $u = f(x_0) = f(x - ct)$. Together with $f(x) = \sin(kx)$ the solution is given by

$$u = \sin(kx - ckt) = \sin(kx - \omega t)\,, \tag{2.85}$$

where we used the dispersion $\omega = ck$.

Problem 2.19 Consider the initial data

$$U(x, 0) = \begin{cases} 0 & x \geq 0 \\ 1 & x < 0 \end{cases} \tag{2.86}$$

for the partial differential equation written in conservative form

$$\frac{\partial U}{\partial t} + \frac{1}{2}\frac{\partial U^2}{\partial x} = 0.$$

Sketch the characteristics. What is the shock propagation speed necessary to prevent the characteristics from crossing?

Solution First, we rewrite Burgers' equation in the form

$$\frac{\partial U}{\partial t} + U\frac{\partial U}{\partial x} = 0.$$

The set of characteristic equations is then given by

$$\frac{dt}{ds} = 1 \qquad \frac{dx}{ds} = U \qquad \frac{dU}{ds} = 0.$$

From the last equation, $dU/ds = 0$, we find that $U = const.$ along the characteristics. In fact we have $U = U_0(x_0)$, where $U_0(x_0) = U(x, 0)$ is given by the initial data, Eq. (2.86). Since U is constant it follows that

$$\frac{dx}{ds} = U = U_0(x_0) \quad \Longrightarrow \quad x(s) = U_0(x_0)s + x_0.$$

And since $dt/ds = 1$, so that $s = t$, we find for the characteristic curve

$$x(t) = U_0(x_0)t + x_0 \quad \Longrightarrow \quad x(t) = \begin{cases} x_0 & \text{for } x_0 \geq 0 \\ x_0 + t & \text{for } x_0 < 0 \end{cases},$$

where we used the initial conditions given in Eq. (2.86). The curves are shown in Fig. 2.2. To prevent the characteristics from crossing, we introduce a shock with propagating speed (valid for the inviscid Burgers' equation)

$$s = \frac{\left[\frac{1}{2}U^2\right]}{[U]} = \frac{U_0 + U_1}{2} = \frac{1}{2},$$

with $U_0 = 0$ and $U_1 = 1$, and where $s = (U_0 + U_1)/2$ is the shock jump relation for the inviscid Burgers' equation, connecting the speed of propagation s of the discontinuity with the amounts by which the velocity U jumps. Figure 2.3 shows this case.

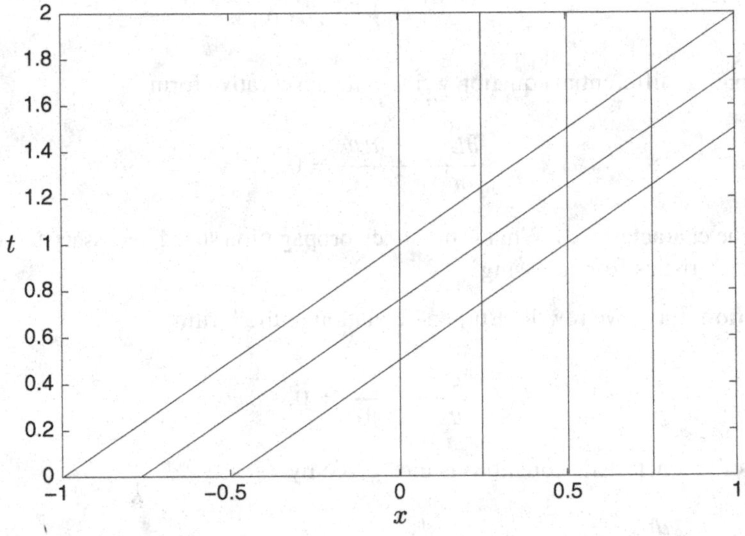

Fig. 2.2 Shown are the characteristic curves. Apparently, the curves intersect

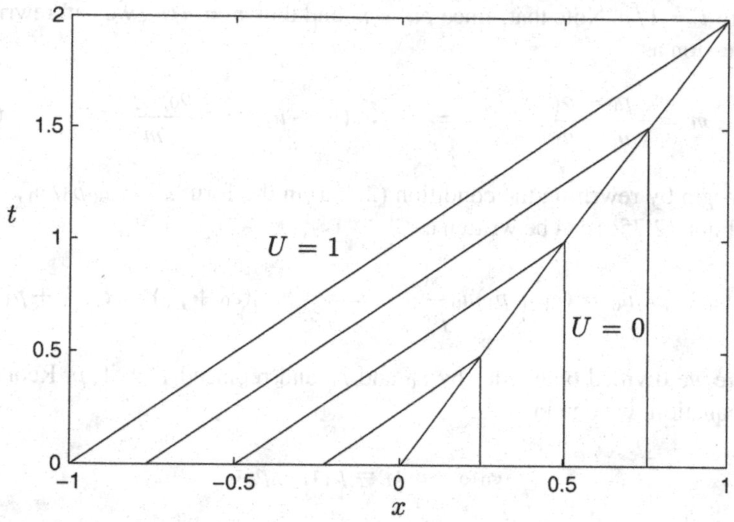

Fig. 2.3 Shown are the characteristic curves with the shock speed $s = 0.5$

Problem 2.20 Starting from the stationary Rankine-Hugoniot conditions (2.75a)–(2.75c), show that

$$m^2 = -\frac{p_0 - p_1}{\tau_0 - \tau_1},$$

where $\tau \equiv 1/\rho$. Show also that $e_0\tau_0 - e_1\tau_1 = p_1\tau_1 - p_0\tau_0$ and hence that

$$\epsilon_1 - \epsilon_0 + \frac{p_0 + p_1}{2}(\tau_1 - \tau_0) = 0,$$

(the Hugoniot equation for the shock) where $\epsilon \equiv p\tau/(\gamma - 1)$.

Solution

A. We begin with Eq. (2.75b) and rewrite this equation as

$$\rho_1 u_1^2 - \rho_0 u_0^2 = p_0 - p_1 \qquad \Longrightarrow \qquad \frac{\rho_1^2 u_1^2}{\rho_1} - \frac{\rho_0^2 u_0^2}{\rho_0} = p_0 - p_1,$$

where we multiplied each term on the left-hand side with $1 = \rho_1/\rho_1 = \rho_0/\rho_0$. Using now Eq. (2.75a) and letting $\rho_0 u_0 = \rho_1 u_1 = m$, we find immediately

$$m^2\left[\frac{1}{\rho_1} - \frac{1}{\rho_0}\right] = p_0 - p_1 \qquad \Longrightarrow \qquad m^2 = -\frac{p_0 - p_1}{\tau_0 - \tau_1},$$

where $\tau = 1/\rho$. Note that, since $\rho u = m$ and thus $\tau = u/m$, we can rewrite the expression as

$$m = -\frac{p_0 - p_1}{u_0 - u_1} \qquad \Longrightarrow \qquad (u_0 - u_1) = -\frac{p_0 - p_1}{m}. \tag{2.87}$$

B. We begin by rewriting the condition (2.75a) in the form $u_1 = u_0\rho_0/\rho_1$, so that condition (2.75c) can be written as

$$(e_0 + p_0)u_0 = (e_1 + p_1)u_0\frac{\rho_0}{\rho_1} \qquad \Longrightarrow \qquad \tau_0(e_0 + p_0) = \tau_1(e_1 + p_1),$$

where we divided both sides by u_0 and ρ_0 and replaced $\tau = 1/\rho$. Reordering the equation, we obtain

$$e_0\tau_0 - e_1\tau_1 = p_1\tau_1 - p_0\tau_0. \tag{2.88}$$

C. We begin with the left-hand side of Eq. (2.88) and substitute

$$e = \rho\epsilon + \frac{1}{2}\rho u^2 = \left(\epsilon + \frac{1}{2}u^2\right)\frac{1}{\tau},$$

where $\tau = 1/\rho$, ϵ is the internal energy, and e is the specific total energy (see the introduction of this section). We obtain

$$\left(\epsilon_0 + \frac{1}{2}u_0^2\right) - \left(\epsilon_1 + \frac{1}{2}u_1^2\right) = p_1\tau_1 - p_0\tau_0$$

$$\epsilon_0 - \epsilon_1 + \frac{1}{2}\left(u_0^2 - u_1^2\right) = p_1\tau_1 - p_0\tau_0$$

$$\epsilon_0 - \epsilon_1 + \frac{1}{2}\left(u_0 - u_1\right)\left(u_0 + u_1\right) = p_1\tau_1 - p_0\tau_0.$$

We substitute now $(u_0 - u_1)$ by Eq. (2.87) and use $u = m\tau$ for the term $(u_0 + u_1)$, and obtain

$$\epsilon_0 - \epsilon_1 - \frac{1}{2}\frac{p_0 - p_1}{m}(m\tau_0 + m\tau_1) = p_1\tau_1 - p_0\tau_0$$

$$\epsilon_0 - \epsilon_1 - \frac{p_0 - p_1}{2}(\tau_0 + \tau_1) = p_1\tau_1 - p_0\tau_0$$

$$\epsilon_0 - \epsilon_1 - \frac{1}{2}p_0\tau_0 - \frac{1}{2}p_0\tau_1 + \frac{1}{2}p_1\tau_0 + \frac{1}{2}p_1\tau_1 = p_1\tau_1 - p_0\tau_0$$

$$\epsilon_0 - \epsilon_1 + \frac{1}{2}p_0\tau_0 - \frac{1}{2}p_0\tau_1 + \frac{1}{2}p_1\tau_0 - \frac{1}{2}p_1\tau_1 = 0$$

$$\epsilon_0 - \epsilon_1 + \frac{1}{2}(p_0 + p_1)(\tau_0 - \tau_1) = 0,$$

which is the Hugoniot equation for a shock.

Problem 2.21 Show that the Cole-Hopf transformation

$$u = -2\kappa\frac{\phi_x}{\phi} \tag{2.89}$$

removes the nonlinear term in the Burgers' equation

$$u_t + uu_x = \kappa u_{xx},$$

and yields the heat equation as the transformed equation. For the initial problem $u(x, t = 0) = F(x)$, show that this transforms to the initial problem

$$\Phi = \Phi(x) = \exp\left[-\frac{1}{2\kappa}\int_0^x F(\eta)d\eta\right], \quad t = 0,$$

for the heat equation. Show that the solution for u is

$$u(x,t) = \frac{\int_{-\infty}^{\infty}\frac{x-\eta}{t}e^{-G/2\kappa}d\eta}{\int_{-\infty}^{\infty}e^{-G/2\kappa}d\eta},$$

where

$$G(\eta; x, t) = \int_0^\eta F(\eta')d\eta' + \frac{(x-\eta)^2}{2t}.$$

Solution Note that the velocity u is a one-dimensional function of time and space, thus $u = u(x,t)$. The Cole-Hopf transformation (2.89) introduces a function ϕ which is also a function of time and space, $\phi = \phi(x,t)$. To avoid any ambiguities we define the initial condition at time $t = 0$ as $\Phi = \Phi(x) = \phi(x, 0)$.

A. With the Cole-Hopf transformation the derivatives of u with respect to t and x are given by

$$u_x = -2\kappa\frac{\phi\phi_{xx} - \phi_x^2}{\phi^2}, \qquad u_t = -2\kappa\frac{\phi\phi_{xt} - \phi_x\phi_t}{\phi^2},$$

$$u_{xx} = -2\kappa\frac{\phi_{xxx}}{\phi} + 6\kappa\frac{\phi_x\phi_{xx}}{\phi^2} - 4\kappa\frac{\phi_x^3}{\phi^3}.$$

Burgers' equation becomes then

$$\begin{aligned} u_t + uu_x - \kappa u_{xx} &= -\frac{\phi_{xt}}{\phi} + \frac{\phi_x \phi_t}{\phi^2} + \kappa \frac{\phi_{xxx}}{\phi} - \kappa \frac{\phi_x \phi_{xx}}{\phi^2} \\ &= \phi \left(\kappa \phi_{xx} - \phi_t \right)_x - \phi_x \left(\kappa \phi_{xx} - \phi_t \right) = 0. \end{aligned}$$

Note that the expressions in both brackets are identical. For any non-trivial solution of ϕ, i.e., $\phi \neq 0$, this equation is fulfilled if the term in brackets equals zero, and thus

$$\kappa \phi_{xx} = \phi_t. \tag{2.90}$$

Equation (2.90) is called the *heat equation* and we note that the Cole-Hopf transformation reduces Burgers' equation to the problem of solving the heat equation.

B. By using Eq. (2.89) we find for the initial problem (setting $t = 0$, so that $\phi(x, 0) = \Phi(x)$) the following ordinary differential equation

$$F(x) = -2\kappa \frac{\phi_x(x, 0)}{\phi(x, 0)} \qquad \Longrightarrow \qquad \frac{d\Phi(x)}{\Phi(x)} = -\frac{F(x)}{2\kappa} dx,$$

where $u(x, t = 0) = F(x)$. The general solution is given by

$$\begin{aligned} \ln \Phi(x) - \ln \Phi(0) &= -\frac{1}{2\kappa} \int_0^x F(\eta) d\eta \\ \Phi(x) &= C \exp\left[-\frac{1}{2\kappa} \int_0^x F(\eta) d\eta \right], \end{aligned} \tag{2.91}$$

where we used $C = \Phi(0)$.

C. In order to solve Burgers' equation we essentially need to solve the heat equation (2.90) and transform the solution back according to the Cole-Hopf transformation (2.89). Let us begin by rewriting the heat equation as

$$\begin{aligned} \frac{\partial \phi(x, t)}{\partial t} &= \kappa \frac{\partial^2 \phi(x, t)}{\partial x^2} & -\infty < x < \infty, \quad 0 < t \\ \phi(x, t = 0) &= \Phi(x) & -\infty < x < \infty. \end{aligned} \tag{2.92}$$

with the initial profile $\Phi(x)$ at time $t = 0$, given by Eq. (2.91). A very common approach is to transform the function $\phi(x, t)$ into Fourier space

$$\phi(x, t) = \frac{1}{2\pi} \int_{-\infty}^{\infty} dk\, \hat{\phi}(k, t) e^{ikx}, \quad \hat{\phi}(k, t) = \int_{-\infty}^{\infty} dx\, \phi(x, t) e^{-ikx} \tag{2.93a}$$

$$\Phi(x) = \frac{1}{2\pi} \int_{-\infty}^{\infty} dk\, \hat{\Phi}(k) e^{ikx}, \qquad \hat{\Phi}(k) = \int_{-\infty}^{\infty} dx\, \Phi(x) e^{-ikx}. \tag{2.93b}$$

where, $\hat{\phi}(k,t)$ and $\hat{\Phi}(k)$ are the Fourier transforms of $\phi(x,t)$ and $\Phi(x)$. By substituting the Fourier transform of $\phi(x,t)$ into the heat equation (2.92) we obtain

$$\frac{\partial \hat{\phi}(k,t)}{\partial t} = -\kappa k^2 \hat{\phi}(k,t) \qquad k \in \mathbb{R}, \quad 0 \le t \tag{2.94}$$
$$\hat{\phi}(k,0) = \hat{\Phi}(k) \qquad k \in \mathbb{R}.$$

For each constant wave mode k, the function $\hat{\phi}(k,t)$ fullfils the initial problem with the initial condition $\hat{\phi}(k,0) = \hat{\Phi}(k)$. The ordinary differential equation (2.94) can easily be solved,

$$\frac{\partial \hat{\phi}(k,t)}{\hat{\phi}(k,t)} = -\kappa k^2 \partial t \qquad \Rightarrow \qquad \ln \hat{\phi}(k,t) - \ln \hat{\phi}(k,0) = -\kappa k^2 t,$$

where we integrated from $t = 0$ to t. Since $\hat{\phi}(k,0) = \hat{\Phi}(k)$ we find our solution

$$\hat{\phi}(k,t) = \hat{\Phi}(k) e^{-\kappa k^2 t}. \tag{2.95}$$

Now we transform this solution back into real space, using Eq. (2.93a),

$$\phi(x,t) = \frac{1}{2\pi} \int_{-\infty}^{\infty} dk\, \hat{\Phi}(k) e^{-\kappa k^2 t} e^{ikx}.$$

Substituting now the initial condition $\hat{\Phi}(k)$ by Eq. (2.93b) we find

$$\phi(x,t) = \frac{1}{2\pi} \int_{-\infty}^{\infty} dx'\, \Phi(x') \int_{-\infty}^{\infty} dk\, e^{-ikx'} e^{-\kappa k^2 t} e^{ikx}.$$

Note that we use the x'-coordinate for the back transformation of the initial condition. The integral with respect to k is readily solved by (see Problem 2.27 for a detailed analysis of the integration)

$$K(x - x') = \int_{-\infty}^{\infty} e^{-ikx'} e^{-\kappa k^2 t} e^{ikx} dk = \sqrt{\frac{\pi}{\kappa t}} e^{-\frac{(x-x')^2}{4\kappa t}}, \tag{2.96}$$

so that

$$\phi(x,t) = \frac{1}{2\pi} \int_{-\infty}^{\infty} dx'\, \Phi(x') K(x - x'), \tag{2.97}$$

where $K(x - x')$ is the so-called *heat kernel*. Note that this integration is only convergent for $\kappa > 0$ and $t > 0$. Thus, we find the *general solution*

$$\phi(x,t) = \frac{1}{\sqrt{4\pi\kappa t}} \int_{-\infty}^{\infty} \Phi(x') e^{-\frac{(x-x')^2}{4\kappa t}} \, dx'.$$

This solution is a convolution of the *fundamental solution*

$$\psi(x,t) = \frac{1}{\sqrt{4\pi\kappa t}} \exp\left[-\frac{x^2}{4\kappa t}\right]$$

and the function $\Phi(x')$. Substituting $\Phi(x')$ by Eq. (2.91) we obtain

$$\phi(x,t) = \frac{C}{\sqrt{4\pi\kappa t}} \int_{-\infty}^{\infty} \exp\left[-\frac{1}{2\kappa}\int_0^{x'} F(\eta)d\eta - \frac{(x-x')^2}{4\kappa t}\right] dx'.$$

The spatial derivative is then

$$\phi(x,t)_x = -\frac{C}{\sqrt{4\pi\kappa t}} \int_{-\infty}^{\infty} \frac{x-x'}{2\kappa t} \exp\left[-\frac{1}{2\kappa}\int_0^{x'} F(\eta)d\eta - \frac{(x-x')^2}{4\kappa t}\right] dx'.$$

According to Eq. (2.89) the solution of Burgers' equation is then

$$u(x,t) = \frac{\int_{-\infty}^{\infty} \frac{x-x'}{t} \exp\left[-\frac{1}{2\kappa}\int_0^{x'} F(\eta)d\eta - \frac{(x-x')^2}{4\kappa t}\right] dx'}{\int_{-\infty}^{\infty} \exp\left[-\frac{1}{2\kappa}\int_0^{x'} F(\eta)d\eta - \frac{(x-x')^2}{4\kappa t}\right] dx'}. \tag{2.98}$$

Problem 2.22 Show that the exponential solution of the characteristic form of the steady Burgers' equation admits a solution that can be expressed as a hyperbolic tan (tanh) profile, given $u(-\infty) = u_0$ and $u(\infty) = u_1$.

Solution Burgers' equation is given by

$$\frac{\partial u}{\partial t} + u\frac{\partial u}{\partial x} = \kappa \frac{\partial^2 u}{\partial x^2}$$

The steady state Burger's equation can also be written as

$$\frac{d}{dx}\left[\frac{u^2}{2} - \frac{d}{dx}(\kappa u)\right] = 0 \qquad \Longrightarrow \qquad \frac{d}{dx}\left[u^2 - 2\kappa\frac{du}{dx}\right] = 0.$$

where we multiplied by 2 and assumed that κ is independent of x. Obviously, the term in brackets has to be constant, and the dimensions of that constant are of the order velocity squared. Thus, we introduce the constant u_c and have

$$u^2 - 2\kappa\frac{du}{dx} = u_c^2,$$

where u_c is a constant velocity that has to be determined by the boundary conditions. We can rewrite that equation and obtain

$$\int dx = \int du\frac{2\kappa}{u^2 - u_c^2} = -\frac{2\kappa}{u_c^2}\int du\frac{1}{1-\frac{u^2}{u_c^2}} = -\frac{2\kappa}{u_c}\int dy\frac{1}{1-y^2},$$

where we used the substitution $y = u/u_c$. According to Gradshteyn and Ryzhik [2], Eq. (2.01–16), the integral can be solved by

$$x = -\frac{2\kappa}{u_0}\operatorname{arctanh}\left(\frac{u(x)}{u_c}\right) + x_0,$$

where x_0 is an integration constant and $u(x)$ obeys the condition $-u_c < u(x) < u_c$. Solving this equation for $u(x)$ gives

$$u(x) = -u_c \tanh\left[(x - x_0)\frac{u_c}{2\kappa}\right],$$

where we used $\tanh(-a) = -\tanh(a)$. Now it is clear that the integration constant x_0 shifts the $\tanh(x)$ by a distance x_0 in the positive or negative x direction, depending on the sign of x_0. The constant $2\kappa/u_c$ has the dimension of a length. For convenience, we introduce a characteristic length scale $l = 2\kappa/u_c$ so that our solution is

$$u(x) = -u_c \tanh\left(\frac{x - x_0}{l}\right). \tag{2.99}$$

Example: Shock. Here we consider the example of a shock located at x_0 and a background flow with the upstream velocity $u_0 = u(x = -\infty)$ and the downstream velocity $u_1 = u(x = +\infty)$.

A. By using $\tanh(\pm\infty) = \pm 1$ and Eq. (2.99) we find the relations $u_0 = u_c$ and $u_1 = -u_c$, and therefore $u_0 = -u_1$. This also implies $u(x = x_0) = 0$, because $\tanh(0) = 0$, which means, that the shock speed at the position of the shock x_0 is zero, thus $u(x_0) = 0$. However, in some cases one is rather interested in $u_0 \neq -u_1$. It is obvious that, in this case, we have to add a constant $C \neq 0$ to Eq. (2.99),

$$u(x) = -u_c \tanh\left(\frac{x - x_0}{l}\right) + C, \tag{2.100}$$

to guarantee that $u(x_0) \neq 0$. We find immediately from Eq. (2.100) that

$$u(x = -\infty) = u_c + C \tag{2.101a}$$

$$u(x = x_0) = C \tag{2.101b}$$

$$u(x = +\infty) = -u_c + C. \tag{2.101c}$$

To specify the constant C we impose another boundary condition,

$$u(x = x_0) = \frac{u_0 + u_1}{2} \equiv C, \tag{2.102}$$

which means, that at the shock position x_0 the shock speed has decreased to the constant arithmetic mean $C = (u_0 + u_1)/2$, which is zero only for $u_0 = -u_1$. The constant u_c is determined by Eqs. (2.101a) and (2.101c). Subtracting the second equation from the first equation and dividing by 2 gives

$$u_c = \frac{u_0 - u_1}{2}.$$

Substituting C and u_c back into Eq. (2.100) gives

$$u(x) = \frac{u_0 + u_1}{2} - \frac{u_0 - u_1}{2} \tanh\left(\frac{x - x_0}{l}\right), \tag{2.103}$$

which provides the correct results for the boundary conditions. In the case that $C = 0$ and, thus, $u_0 = -u_1 \equiv u_e$ we find the result as given by Eq. (2.99). As an example, Fig. 2.4 shows the velocity profile (2.103) across a shock for $u_0 = 1$ and $u_1 = 0.25$ and the shock position $x_0 = 5$.

B. The exponential solution of the characteristic form of Burgers' equation is given by

$$u = u_0 + \frac{u_1 - u_0}{e^{2z} + 1}, \quad \text{where} \quad z = \frac{u_1 - u_0}{4\kappa}\left(x - \frac{u_0 + u_1}{2}t\right). \tag{2.104}$$

Note that the hyperbolic tangent can be expressed by

$$\tanh z = 1 - \frac{2}{e^{2z} + 1}.$$

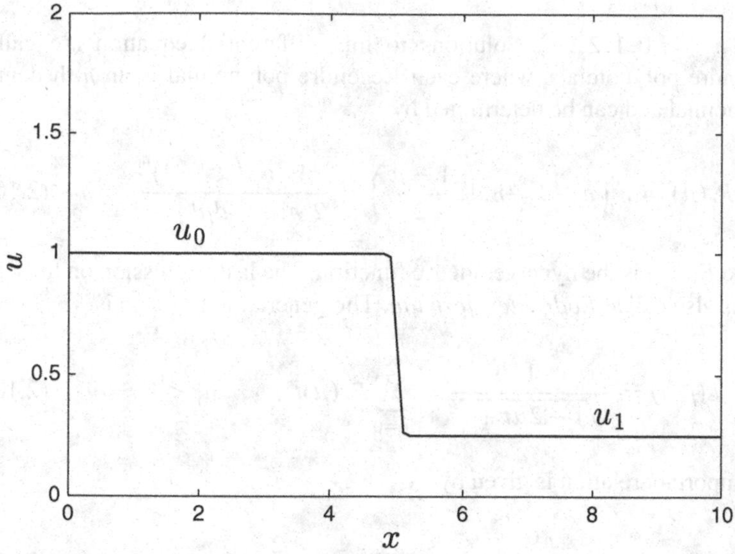

Fig. 2.4 Shown is the velocity profile for $x_0 = 5$, $u_0 = 1$ and $u_1 = 0.25$

By expressing $u_0 = u_0/2 + u_0/2$ in Eq. (2.104) and adding a 'zero' in the form of $0 = u_1/2 - u_1/2$, we obtain

$$u = \frac{u_0}{2} + \frac{u_1}{2} + \frac{u_0}{2} - \frac{u_1}{2} + \frac{u_1 - u_1}{e^{2z}+1} = \frac{u_0 + u_1}{2} + \frac{u_0 - u_1}{2} - \frac{u_0 - u_1}{e^{2z}+1}$$

$$= \frac{u_0 + u_1}{2} + \frac{u_0 - u_1}{2}\left[1 - \frac{2}{e^{2z}+1}\right] = \frac{u_0 + u_1}{2} + \frac{u_0 - u_1}{2}\tanh(z).$$

This result is identical to Eq. (2.103).

2.6 Application 2: The Diffusion and Telegrapher Equations

Legendre's differential equation is an ordinary differential equation, given by

$$0 = (1-\mu^2)\frac{dP_n^2(\mu)}{d\mu^2} - 2\mu\frac{dP_n(\mu)}{d\mu} + n(n+1)P_n(\mu) \tag{2.105a}$$

$$= \frac{d}{d\mu}\left[(1-\mu^2)\frac{dP_n}{d\mu}\right] + n(n+1)P_n, \tag{2.105b}$$

(continued)

with $n = 0, 1, 2, \ldots$. Solutions to this differential equation are called Legendre polynomials, where each Legendre polynomial is an n-th degree polynomial and can be determined by

$$P_n(\mu) = F\left(n+1, -n, 1; \frac{1-\mu}{2}\right) = \frac{1}{2^n n!}\frac{d^n\left(\mu^2-1\right)^n}{d\mu^n}, \tag{2.106}$$

where $F(\ldots)$ is the hypergeometric function. The last expression on the right side is also called *Rodrigues' formula*. The generating function is

$$L(\mu, t) = \frac{1}{\sqrt{1-2\mu t+t^2}} = \sum_{n=0}^{\infty} P_n(\mu)t^n, \qquad |t| < 1. \tag{2.107}$$

An important relation is given by

$$\int_{-1}^{1} P_n(\mu)P_m(\mu)d\mu = \frac{2}{2m+1}\delta_{nm}. \tag{2.108}$$

Problem 2.23 Legendre polynomials $P_n(\mu)$ and $P_m(\mu)$ satisfy Legendre's differential equation (2.105a). Show that for $n \neq m$, the following orthogonality condition holds,

$$\int_{-1}^{1} P_n(\mu)P_m(\mu)d\mu = 0 \qquad \text{for } n \neq m. \tag{2.109}$$

Solution Using Eq. (2.105a), Legendre's differential equation (for two Legendre polynomials P_n and P_m) is described by

$$\frac{d}{d\mu}\left[(1-\mu^2)\frac{dP_n}{d\mu}\right] + n(n+1)P_n = 0$$
$$\frac{d}{d\mu}\left[(1-\mu^2)\frac{dP_m}{d\mu}\right] + m(m+1)P_m = 0,$$

where $n \neq m$. Multiply now the first equation with P_m and the second with P_n, then subtract the second from the first. We obtain

$$\begin{aligned}
&[m(m+1) - n(n+1)]\, P_m P_n \\
&= P_m \frac{d}{d\mu}\left[(1-\mu^2)\frac{dP_n}{d\mu}\right] - P_n \frac{d}{d\mu}\left[(1-\mu^2)\frac{dP_m}{d\mu}\right] \\
&= \frac{d}{d\mu}\left[(1-\mu^2)P_m\frac{dP_n}{d\mu} - (1-\mu^2)P_n\frac{dP_m}{d\mu}\right].
\end{aligned}$$

It is easy to show that the last line equals the second. Now we integrate both sides with respect to μ and obtain

$$
\begin{aligned}
&[m(m+1) - n(n+1)] \int_{-1}^{1} P_m P_n d\mu \\
&= \int_{-1}^{+1} \frac{d}{d\mu}\left[(1-\mu^2)\left(P_m \frac{dP_n}{d\mu} - P_n \frac{dP_m}{d\mu}\right)\right] d\mu \\
&= \left[(1-\mu^2)\left(P_m \frac{dP_n}{d\mu} - P_n \frac{dP_m}{d\mu}\right)\right]_{-1}^{+1} = 0,
\end{aligned}
$$

because the term $(1-\mu^2)$ vanishes for $\mu = \pm 1$. For $n \neq m$ we have the orthogonality condition given by Eq. (2.109).

Problem 2.24 The generating function for the Legendre polynomials is given by Eq. (2.107). By differentiating the generating function with respect to t and equating coefficients, derive the recursion relation

$$
(n+1)P_{n+1} + nP_{n-1} = (2n+1)\mu P_n, \qquad n = 1, 2, 3, \ldots .
$$

Solution We consider the partial derivative of Eq. (2.107) with respect to t and obtain

$$
\frac{\partial L}{\partial t} = \frac{\mu - t}{(1 - 2\mu t + t^2)^{3/2}} = \sum_{n=0}^{\infty} n P_n t^{n-1}.
$$

By multiplying both sides with $\left(1 - 2\mu t + t^2\right)$ we find

$$
\frac{\mu - t}{\sqrt{1 - 2\mu t + t^2}} \equiv (\mu - t) \sum_{n=0}^{\infty} P_n t^n = \left(1 - 2\mu t + t^2\right) \sum_{n=0}^{\infty} n P_n t^{n-1},
$$

where we used the definition of the generating function (2.107) for the first equality. By expanding the parenthesis in front of both sums we find

$$
\sum_{n=0}^{\infty} \mu P_n t^n - \sum_{n=0}^{\infty} P_n t^{n+1} = \sum_{n=0}^{\infty} n P_n t^{n-1} - \sum_{n=0}^{\infty} 2\mu n P_n t^n + \sum_{n=0}^{\infty} n P_n t^{n+1}.
$$

Compare now each coefficient with the same power of t. We find

$$
\mu P_n - P_{n-1} = (n+1)P_{n+1} - 2\mu n P_n + (n-1)P_{n-1}.
$$

Rearranging leads to the recursion relation

$$
(n+1)P_{n+1} + nP_{n-1} = (2n+1)\mu P_n.
$$

Problem 2.25 By using the generating function and Problem 2.23 above, show that

$$\int_{-1}^{1} P_n^2(\mu)d\mu = \frac{2}{2n+1}. \tag{2.110}$$

Solution By multiplying the generating function (2.107) with itself we obtain

$$\frac{1}{1-2\mu t+t^2} = \left(\sum_{n=0}^{\infty} P_n t^n\right)\left(\sum_{m=0}^{\infty} P_m t^m\right) = \sum_{n=0}^{\infty}\sum_{m=0}^{\infty} P_n P_m t^{n+m},$$

where $|t| < 1$. By integrating with respect to μ we find

$$\int_{-1}^{1} \frac{1}{1-2\mu t+t^2} d\mu = \sum_{n=0}^{\infty}\sum_{m=0}^{\infty} \int_{-1}^{1} P_n P_m d\mu \; t^{n+m} = \sum_{n=0}^{\infty} \int_{-1}^{1} P_n^2 d\mu \; t^{2n},$$

since we know from Problem 2.23 that $\int_{-1}^{1} P_n P_m d\mu = 0$ for $m \neq n$. Let us consider now the left-hand side of that equation. By using $\alpha = 1+t^2$ and $x = -2\mu t$ we have

$$\begin{aligned}
\int_{-1}^{1} \frac{1}{1-2\mu t+t^2} d\mu &= \frac{1}{2t}\int_{-2t}^{2t} \frac{1}{\alpha+x} dx = \frac{1}{2t}\left[\ln(\alpha+x)\right]_{-2t}^{2t} \\
&= \frac{1}{2t}\ln\left(\frac{\alpha+2t}{\alpha-2t}\right) = \frac{1}{2t}\ln\left(\frac{1+2t+t^2}{1-2t+t^2}\right) \\
&= \frac{1}{2t}\ln\left(\left(\frac{1+t}{1-t}\right)^2\right) = \frac{1}{t}\ln\left(\frac{1+t}{1-t}\right).
\end{aligned}$$

According to Gradshteyn and Rhyzhik [2], Eq. (1.513–1) the logarithm can be written as an infinite sum

$$\ln\left(\frac{1+t}{1-t}\right) = 2\sum_{n=0}^{\infty} \frac{1}{2n+1} t^{2n+1} \qquad \text{with} \qquad |t| \leq 1.$$

Together with the factor $1/t$ we find for the left-hand side

$$\int_{-1}^{1} \frac{1}{1-2\mu t+t^2} d\mu = \sum_{n=0}^{\infty} \frac{2}{2n+1} t^{2n} = \sum_{n=0}^{\infty} \int_{-1}^{1} P_n^2 d\mu \; t^{2n}.$$

By comparing the coefficients for each n we find

$$\int_{-1}^{1} P_n^2 d\mu = \frac{2}{2n+1}.$$

A basic problem in space physics and astrophysics is the transport or charged particles in the presence of a magnetic field that is ordered on some large scale and highly random and temporal on the other scales. We discuss a simplified form of the Fokker-Planck transport equation that describes particle transport via particle scattering in pitch angle in a magnetically turbulent medium since it resembles closely the basic Boltzmann equation. In the absence of both focusing and adiabatic energy changes, the BGK form of the Boltzmann equation reduces to the simplest possible integro-differential equation

$$\frac{\partial f}{\partial t} + \mu v \frac{\partial f}{\partial r} = \frac{\langle f \rangle - f}{\tau}, \tag{2.111}$$

where $f(r, t, v, \mu)$ is a gyrophase averaged velocity distribution function at position r and time t for particles of speed v and pitch-angle cosine $\mu = \cos\theta$ with $\mu \in [-1, 1]$ and where

$$\langle f \rangle = \frac{1}{2} \int_{-1}^{1} f d\mu$$

is the mean or isotropic distribution function averaged over μ.

Problem 2.26 Starting from Eq. (2.111) derive the infinite set of partial differential equations

$$(2n+1)\frac{\partial f_n}{\partial t} + (n+1)v\frac{\partial f_{n+1}}{\partial r} + nv\frac{\partial f_{n-1}}{\partial r} + (2n+1)\frac{f_n}{\tau} = \frac{f_0}{\tau}\delta_{n0} \tag{2.112}$$

with $n = 0, 1, 2, \ldots$.

Solution By expanding $f = f(r, t, v, \mu)$ in an infinite series of Legendre polynomials $P_n(\mu)$,

$$f = \frac{1}{4\pi} \sum_{n=0}^{\infty} (2n+1) P_n(\mu) f_n(r, t, v),$$

we can rewrite Eq. (2.111) as (neglecting the factor $1/4\pi$)

$$\sum_{n=0}^{\infty}(2n+1)P_n\frac{\partial f_n}{\partial t} + \mu v \sum_{n=0}^{\infty}(2n+1)P_n\frac{\partial f_n}{\partial r}$$
$$+ \frac{1}{\tau}\sum_{n=0}^{\infty}(2n+1)P_n f_n = \frac{1}{\tau}\frac{1}{2}\int_{-1}^{1}\sum_{n=0}^{\infty}(2n+1)P_n f_n d\mu.$$

The right-hand side of that equation can be written as

$$\frac{1}{\tau}\frac{1}{2}\int_{-1}^{1}\sum_{n=0}^{\infty}(2n+1)P_n f_n\, d\mu = \frac{1}{\tau}\frac{1}{2}\sum_{n=0}^{\infty}(2n+1)f_n\int_{-1}^{1}P_n P_0 d\mu,$$

where we included $P_0(\mu) = 1$ in the integral. With the orthogonality relation (2.108),

$$\int_{-1}^{1}P_n P_0\, d\mu = \delta_{n0}\frac{2}{2n+1},$$

we find

$$\frac{1}{\tau}\frac{1}{2}\int_{-1}^{1}\sum_{n=0}^{\infty}(2n+1)P_n f_n d\mu = \frac{1}{\tau}\sum_{n=0}^{\infty}f_n(r,t,v)\delta_{n0} = \frac{f_0}{\tau},$$

since the delta function contributes only for $n = 0$. The differential equation reduces to

$$\sum_{n=0}^{\infty}(2n+1)P_n\frac{\partial f_n}{\partial t} + \mu v\sum_{n=0}^{\infty}(2n+1)P_n\frac{\partial f_n}{\partial r} + \frac{1}{\tau}\sum_{n=0}^{\infty}(2n+1)P_n f_n = \frac{f_0}{\tau}.$$

With the recurrence relation $(2n+1)\mu P_n = (n+1)P_{n+1} + nP_{n-1}$ we can rewrite the second term and find

$$\sum_{n=0}^{\infty}(2n+1)P_n\frac{\partial f_n}{\partial t} + v\sum_{n=0}^{\infty}(n+1)P_{n+1}\frac{\partial f_n}{\partial r}$$
$$+ v\sum_{n=0}^{\infty}nP_{n-1}\frac{\partial f_n}{\partial r} + \frac{1}{\tau}\sum_{n=0}^{\infty}(2n+1)P_n f_n = \frac{f_0}{\tau}.$$

Now we multiply the equation by P_m and integrate with respect to μ,

$$\sum_{n=0}^{\infty}(2n+1)\int_{-1}^{1}P_m P_n\, d\mu\frac{\partial f_n}{\partial t} + v\sum_{n=0}^{\infty}(n+1)\int_{-1}^{1}P_m P_{n+1} d\mu\frac{\partial f_n}{\partial r}$$
$$+ v\sum_{n=0}^{\infty}n\int_{-1}^{1}P_m P_{n-1}\, d\mu\frac{\partial f_n}{\partial r} + \frac{1}{\tau}\sum_{n=0}^{\infty}(2n+1)\int_{-1}^{1}P_m P_n\, d\mu f_n$$
$$= \frac{f_0}{\tau}\int_{-1}^{1}P_m\, d\mu = \frac{f_0}{\tau}\int_{-1}^{1}P_0 P_m\, d\mu.$$

With the orthogonality relation Eq. (2.108) we find

$$\sum_{n=0}^{\infty}(2n+1)\frac{2}{2m+1}\delta_{mn}\frac{\partial f_n}{\partial t}+v\sum_{n=0}^{\infty}(n+1)\frac{2}{2m+1}\delta_{mn+1}\frac{\partial f_n}{\partial r}$$
$$+v\sum_{n=0}^{\infty}n\frac{2}{2m+1}\delta_{mn-1}\frac{\partial f_n}{\partial r}+\frac{1}{\tau}\sum_{n=0}^{\infty}(2n+1)\frac{2}{2m+1}\delta_{mn}f_n$$
$$=\frac{f_0}{\tau}\frac{2}{2m+1}\delta_{m0}.$$

Note that the second term contributes only for $n = m-1$, while the third term contributes only for $n = m+1$. All other terms on the left side contribute only for $n = m$, so that

$$\frac{\partial f_m}{\partial t}+v\frac{2m}{2m+1}\frac{\partial f_{m-1}}{\partial r}+v\frac{2(m+1)}{2m+1}\frac{\partial f_{m+1}}{\partial r}+\frac{1}{\tau}f_m=\frac{f_0}{\tau}\frac{2}{2m+1}\delta_{m0}.$$

By multiplying this equation with $(2m+1)/2$ and swapping the indices $m \leftrightarrow n$ we obtain Eq. (2.112).

Problem 2.27 Show that the integral

$$\int_{-\infty}^{\infty}\exp\left(-\beta\omega^2 t-i\alpha\omega t-i\omega r\right)d\omega=\sqrt{\frac{\pi}{\beta t}}\exp\left(-\frac{(r+\alpha t)^2}{4\beta t}\right). \tag{2.113}$$

Solution The exponential function can be rewritten as

$$\exp\left(-\beta\omega^2 t-i\alpha\omega t-i\omega r\right)=\exp\left[-\beta t\left(\omega^2+i\omega\frac{\alpha t+r}{\beta t}\right)\right]$$
$$=\exp\left[-\beta t\left(\omega+i\frac{\alpha t+r}{2\beta t}\right)^2-\frac{(\alpha t+r)^2}{4\beta t}\right].$$

By substituting $x=\omega+i\left(\alpha t+r\right)/2\beta t$ we obtain

$$\int_{-\infty}^{\infty}\exp\left(-\beta\omega^2 t-i\alpha\omega t-i\omega r\right)d\omega=\exp\left[-\frac{(\alpha t+r)^2}{4\beta t}\right]\int_{-\infty}^{\infty}e^{-\beta t x^2}dx$$
$$=\sqrt{\frac{\pi}{\beta t}}\exp\left[-\frac{(\alpha t+r)^2}{4\beta t}\right].$$

Chapter 3
Collisional Charged Particle Transport in a Magnetized Plasma

3.1 The Kinetic Equation and Moments for a Magnetized Plasma

The non-relativistic Boltzmann equation is given by Eq. (2.1), where the particle-particle collisions are represented by the collision term $(\delta f/\delta t)_{coll}$ on the right-hand side. Consider now a collisionless magnetized plasma, where the collision term is zero. A prominent example for such a plasma is the solar wind, which consists of protons, electron, ions, ... Charged particles are primarily affected by electromagnetic fields, so that the collisionless Boltzmann equation for a particle species a can be written as:

$$\frac{\partial f_a}{\partial t} + \boldsymbol{v} \cdot \nabla f_a + \frac{q_a}{m_a} (\boldsymbol{E} + \boldsymbol{v} \times \boldsymbol{B}) \cdot \nabla_v f_a = 0, \tag{3.1}$$

where $\boldsymbol{F} = q (\boldsymbol{E} + \boldsymbol{v} \times \boldsymbol{B})$ is the Lorentz force. In the neighborhood of each discrete charged particle, the fields can be large and dominate the macroscopic large-scale fields. thus, $\boldsymbol{E}$ and $\boldsymbol{B}$ fluctuate strongly on short length scales compared to the Debye length (which is the distance over which charged carriers are screened). We take $\boldsymbol{E}$ and $\boldsymbol{B}$ to be the average of the actual electric and mange fields over many Debye lengths, and the effects of the short-range electromagnetic fluctuations or collisions will be included through a collision operator

$$C_a(f_a) = \left.\frac{\delta f_a}{\delta t}\right|_{coll},$$

(continued)

A. Dosch, G.P. Zank, *Transport Processes in Space Physics and Astrophysics*, Lecture Notes in Physics 918, DOI 10.1007/978-3-319-24880-6_3

also called the *Fokker-Planck operator*, so that the Boltzmann equation (2.1) can be written as

$$\frac{\partial f_a}{\partial t} + \boldsymbol{v} \cdot \nabla f_a + \frac{q_a}{m_a}\left(\boldsymbol{E} + \boldsymbol{v} \times \boldsymbol{B}\right) \cdot \nabla_v f_a = C_a(f_a). \tag{3.2}$$

The collision operator

$$C_a = \sum_b C_{ab}(f_a, f_b) \tag{3.3}$$

is a sum of the contributions from collisions with each particle species b, including self-collisions $a = b$. Like the Boltzmann collision operator, the number density, momentum, and energy moments of the Fokker-Planck collisional operator must satisfy

$$\int C_{ab}(f_a, f_b) d^3v = 0 \tag{3.4a}$$

$$\int m_a \boldsymbol{v} C_{ab}(f_a, f_b) d^3v = -\int m_b \boldsymbol{v} C_{ba}(f_b, f_a) d^3v \tag{3.4b}$$

$$\int \frac{1}{2} m_a v^2 C_{ab}(f_a, f_b) d^3v = -\int \frac{1}{2} m_b v^2 C_{ba}(f_b, f_a) d^3v, \tag{3.4c}$$

since the force a species a exerts on a species b must be equal and opposite to that which species b exerts on species a, so that no net momentum or energy change results from collisions. For $b = a$ we have

$$\int C_{aa}(f_a) d^3v = 0 \tag{3.5a}$$

$$\int m_a \boldsymbol{v} C_{aa}(f_a) d^3v = 0 \tag{3.5b}$$

$$\int \frac{1}{2} m_a v^2 C_{aa}(f_a) d^3v = 0. \tag{3.5c}$$

Any model collision operator has to satisfy these properties!

By taking moments of the kinetic equation (3.2), on can derive the fluid equations (see Problem 3.1)

$$\frac{\partial n}{\partial t} + \nabla \cdot (n\boldsymbol{u}) = 0 \tag{3.6a}$$

(continued)

$$\frac{\partial}{\partial t}(mn\boldsymbol{u}) + \nabla \cdot \boldsymbol{P} = qn\left(\boldsymbol{E} + \boldsymbol{u} \times \boldsymbol{B}\right) + \int m\boldsymbol{v}C(f)d^3v \tag{3.6b}$$

$$\frac{\partial}{\partial t}\left[\frac{1}{2}mnu^2 + \frac{3}{2}nkT\right] + \nabla \cdot \boldsymbol{Q} = qn\boldsymbol{E} \cdot \boldsymbol{u} + \int \frac{1}{2}mv^2C(f)d^3v. \tag{3.6c}$$

The right-hand sides of the fluid equations (3.6a)–(3.6c) contain the rate of change of momentum and energy due to the electromagnetic fields and the collisional transfer of momentum and energy via collisions to and from other species, and may be expressed as

$$\int m\boldsymbol{v}C(f)d^3v = \boldsymbol{R} \tag{3.7}$$

$$\int \frac{1}{2}mv^2C(f)d^3v = Q + \boldsymbol{R} \cdot \boldsymbol{u} \tag{3.8}$$

Note that the *scalar* Q describes the rate of thermal energy transfer and *not* the heat flux vector $\boldsymbol{Q}$ (see below). $\boldsymbol{R}$ is the rate of transfer of momentum to the particle species of interest due to collisions with other species in the plasma.

Problem 3.1 By taking moments of the kinetic equation (3.2), derive the fluid equations (3.6a)–(3.6c).

Solution We begin with Eq. (3.2) and neglect the subscript a for convenience in all following calculations. We calculate the zeroth, first, and second moment for the continuity, momentum-, and energy conservation equations. In order to calculate the moments we define

$$n\langle A\rangle = \int Af\,d^3v.$$

If we substitute A by $1, \boldsymbol{v}, \boldsymbol{v}^2, \boldsymbol{v}^3, \ldots$ we find for the first three moments

$$n = \int f\,d^3v \qquad n\langle \boldsymbol{v}\rangle = n\boldsymbol{u} = \int \boldsymbol{v}f\,d^3v \qquad n\langle \boldsymbol{v}^2\rangle = \int \boldsymbol{v}^2f\,d^3v. \tag{3.9}$$

A. **Zeroth Moment.** We integrate Eq. (3.2) with respect to velocity d^3v and obtain

$$\int \frac{\partial f}{\partial t}d^3v + \int \boldsymbol{v} \cdot \nabla f d^3v + \int \frac{q}{m}\left(\boldsymbol{E} + \boldsymbol{v} \times \boldsymbol{B}\right) \cdot \nabla_v f d^3v = \int C(f)d^3v.$$

The time derivative in the first term can be pulled in front of the integral. The nabla operator in the second term pertains to the *spatial* derivative and can also be pulled in front of the integral. However, the integral in the third term includes derivatives in *velocity space*, thus, the nabla operator *cannot* be pulled in front of the integral. Instead, we have to use integration by parts,

$$\frac{q}{m}\int (\boldsymbol{E}+\boldsymbol{v}\times\boldsymbol{B})\cdot\nabla_v f d^3v$$
$$= [f(\boldsymbol{E}+\boldsymbol{v}\times\boldsymbol{B})]_{v=-\infty}^{v=\infty} - \int f\nabla_v\cdot(\boldsymbol{E}+\boldsymbol{v}\times\boldsymbol{B})\,d^3v.$$

The first term on the right-hand side is zero since any physical distribution function vanishes for $\boldsymbol{v}\to\pm\infty$. The electric and magnetic fields are functions of $\boldsymbol{x}$ and t, but not of $\boldsymbol{v}$. Thus, the nabla operator operates only on the $(\boldsymbol{v}\times\boldsymbol{B})$ term. It is an easy matter to show that $\nabla_v\cdot(\boldsymbol{v}\times\boldsymbol{B})=0$, since the i-th component of the remaining vector (after the cross-product) is independent of the velocity in i-direction. Thus, the derivative with respect to velocity is zero. We obtain

$$\frac{\partial}{\partial t}\int f d^3v + \nabla\cdot\int \boldsymbol{v} f d^3v = \int C(f) d^3v.$$

With the moments expressed in Eqs. (3.9) and with $\int C(f)d^3v = 0$ we derive the continuity equation

$$\frac{\partial n}{\partial t} + \nabla\cdot(n\boldsymbol{u}) = 0.$$

B. **First Moment.** Here we multiply Eq. (3.2) with $m\boldsymbol{v}$ and integrate with respect to d^3v. For simplicity we consider the j-th component of the velocity vector. We find

$$\int m v_j \frac{\partial f}{\partial t} d^3v + \int m v_j \boldsymbol{v}\cdot\nabla f d^3v$$
$$+ \int q v_j (\boldsymbol{E}+\boldsymbol{v}\times\boldsymbol{B})\,\nabla_v f d^3v = \int m v_i C(f) d^3v.$$

For convenience let us consider each term separately, beginning with the first term on the left-hand side. As before, we can pull the time derivative in front of the integral

$$\frac{\partial}{\partial t}\int m v_j f d^3v = \frac{\partial}{\partial t}(m n u_j)\,, \tag{3.10}$$

where we used the expressions in (3.9). For the second term we rewrite $\boldsymbol{v} \cdot \nabla f = v_k \partial f / \partial x_k$, where we used Einstein's summation convention. The derivative is independent of the integration and we can write

$$\frac{\partial}{\partial x_k} \int m v_j v_k f d^3 v = \frac{\partial}{\partial x_k} \left[mn \langle v_j v_k \rangle \right] = \frac{\partial P_{jk}}{\partial x_k} = (\nabla \cdot \boldsymbol{P})_j, \tag{3.11}$$

where we used the momentum flux tensor $P_{jk} \equiv mn \langle v_j v_k \rangle$ (see [5] for more information). Again, for the third term we use integration by parts and obtain

$$\int q v_j (\boldsymbol{E} + \boldsymbol{v} \times \boldsymbol{B}) \cdot \nabla_v f d^3 v$$
$$= q \left[f v_j (\boldsymbol{E} + \boldsymbol{v} \times \boldsymbol{B}) \right]_{-\infty}^{\infty} - q \int f \nabla_v \cdot \left[v_j (\boldsymbol{E} + \boldsymbol{v} \times \boldsymbol{B}) \right] d^3 v.$$

The first term on the right-hand side is zero since the distribution function vanishes as $\boldsymbol{v} \to \pm\infty$. For the second term on the right-hand side we find

$$\nabla_v \cdot \left[v_j (\boldsymbol{E} + \boldsymbol{v} \times \boldsymbol{B}) \right] = \frac{\partial}{\partial v_i} \left[v_j (\boldsymbol{E} + \boldsymbol{v} \times \boldsymbol{B})_i \right] = (\boldsymbol{E} + \boldsymbol{v} \times \boldsymbol{B})_i \frac{\partial v_j}{\partial v_i}$$
$$= (\boldsymbol{E} + \boldsymbol{v} \times \boldsymbol{B})_i \delta_{ij} = (\boldsymbol{E} + \boldsymbol{v} \times \boldsymbol{B})_j$$

since the i-th component of $(\boldsymbol{E} + \boldsymbol{v} \times \boldsymbol{B})$ is independent of v_i and, therefore, the derivative vanishes. We obtain

$$\int q v_j (\boldsymbol{E} + \boldsymbol{v} \times \boldsymbol{B}) \cdot \nabla_v f d^3 v = -q \int f \left[\boldsymbol{E} + \boldsymbol{v} \times \boldsymbol{B} \right]_j d^3 v$$
$$= -q E_j \int f d^3 v - q \left[\int (f \boldsymbol{v}) \, d^3 v \times \boldsymbol{B} \right]_j$$
$$= -nq (\boldsymbol{E} + \boldsymbol{u} \times \boldsymbol{B})_j, \tag{3.12}$$

where the subscript j denotes the j-th component of that vector. Note that $\boldsymbol{E}$ and $\boldsymbol{B}$ are independent of velocity $\boldsymbol{v}$. By adding up Eqs. (3.10)–(3.12) and by using the vector description instead of the j-th component, we find

$$\frac{\partial}{\partial t} (mn\boldsymbol{u}) + \nabla \cdot \boldsymbol{P} = nq (\boldsymbol{E} + \boldsymbol{u} \times \boldsymbol{B}) + \int m \boldsymbol{v} C(f) d^3 v,$$

the conservation of momentum.

C. **Second Moment.** Here we multiply Eq. (3.2) with $mv^2/2$ and integrate with respect to d^3v. We have

$$\frac{m}{2}\int v^2\frac{\partial f}{\partial t}d^3v + \frac{m}{2}\int v^2 \boldsymbol{v}\cdot\nabla f d^3v$$
$$+\frac{q}{2}\int v^2\left(\boldsymbol{E}+\boldsymbol{v}\times\boldsymbol{B}\right)\cdot\nabla_v f d^3v = \int \frac{1}{2}mv^2 C(f) d^3v.$$

Again, let us consider each term separately. The first term on the left-hand side can be rewritten as

$$\frac{\partial}{\partial t}\frac{m}{2}\int v^2 f d^3v = \frac{\partial}{\partial t}\frac{m}{2}n\left\langle v^2\right\rangle = \frac{\partial}{\partial t}\left[\frac{1}{2}mnu^2 + \frac{3}{2}nkT\right], \tag{3.13}$$

where we used the fact, that the total energy is the sum of the kinetic energy associated with the mean flow (the term proportional to u^2) and the thermal energy (the term proportional to T, see [5] for further information). In the second term we can pull the nabla operator in front of the integral and find

$$\nabla\cdot\frac{m}{2}\int v^2\boldsymbol{v} f d^3v = \nabla\cdot\frac{1}{2}mn\left\langle v^2\boldsymbol{v}\right\rangle = \nabla\cdot\boldsymbol{Q} \tag{3.14}$$

with the heat flux vector $\boldsymbol{Q} = mn\left\langle v^2\boldsymbol{v}\right\rangle/2$. For the last term on the left-hand side we have to apply integration by parts and obtain

$$\frac{q}{2}\int v^2\left(\boldsymbol{E}+\boldsymbol{v}\times\boldsymbol{B}\right)\cdot\nabla_v f d^3v$$
$$= \frac{q}{2}\left[f v^2\left(\boldsymbol{E}+\boldsymbol{v}\times\boldsymbol{B}\right)\right]_{-\infty}^{\infty} - \frac{q}{2}\int f\nabla_v\left[v^2\left(\boldsymbol{E}+\boldsymbol{v}\times\boldsymbol{B}\right)\right]d^3v.$$

The first term on the right-hand side is zero since the distribution function vanishes as $\boldsymbol{v}\to\pm\infty$. For the second term on the right-hand side we use

$$\nabla_v\left[v^2\left(\boldsymbol{E}+\boldsymbol{v}\times\boldsymbol{B}\right)\right] = \nabla_v\left[v^2\right]\cdot\left(\boldsymbol{E}+\boldsymbol{v}\times\boldsymbol{B}\right) + v^2\nabla_v\cdot\left[\left(\boldsymbol{E}+\boldsymbol{v}\times\boldsymbol{B}\right)\right]$$
$$= 2\boldsymbol{v}\cdot\boldsymbol{E}.$$

As before, the second term on the right-hand side is zero. With $\nabla_v v^2 = 2\boldsymbol{v}$ and $\boldsymbol{v}\cdot(\boldsymbol{v}\times\boldsymbol{B}) = 0$, because $\boldsymbol{v}\perp(\boldsymbol{v}\times\boldsymbol{B})$. We find

$$\frac{q}{2}\int v^2\left(\boldsymbol{E}+\boldsymbol{v}\times\boldsymbol{B}\right)\cdot\nabla_v f d^3v = -q\boldsymbol{E}\cdot\int \boldsymbol{v} f d^3v = -qn\boldsymbol{E}\cdot\boldsymbol{u}. \tag{3.15}$$

By adding up Eqs. (3.13)–(3.15) we find

$$\frac{\partial}{\partial t}\left[\frac{1}{2}mnu^2 + \frac{3}{2}nkT\right] + \nabla \cdot \boldsymbol{Q} = qn\boldsymbol{E} \cdot \boldsymbol{u} + \int \frac{1}{2}mv^2 C(f) d^3v,$$

the conservation of energy.

Problem 3.2 Derive the momentum equation (3.16) and the energy equation (3.17) from the conservation laws (3.6a)–(3.6c).

Solution We derive first the momentum equation and then the energy equation.

1. Momentum Equation. Let us start with Eq. (3.6b). With $\int m\boldsymbol{v}C(f)d^3v = \boldsymbol{R}$, where $\boldsymbol{R}$ is the total force exerted on the particle of interest due to collisions with other species in the plasma, we find for the j-th component of the momentum conservation equation

$$\frac{\partial}{\partial t}(mnu_j) + \frac{\partial P_{jk}}{\partial x_k} = qn\left(\boldsymbol{E} + \boldsymbol{u} \times \boldsymbol{B}\right)_j + R_j,$$

where we used $(\nabla \cdot \boldsymbol{P})_j = \partial P_{jk}/\partial x_k$; compare also with part B of the previous Problem 3.1. For the momentum flux tensor we use

$$P_{jk} = p\delta_{jk} + \pi_{jk} + mnu_j u_k,$$

so that the momentum conservation equation becomes

$$\frac{\partial}{\partial t}(mnu_j) + \frac{\partial p}{\partial x_j} + \frac{\partial \pi_{jk}}{\partial x_k} + \frac{\partial}{\partial x_k}\left[mnu_j u_k\right] = qn\left(\boldsymbol{E} + \boldsymbol{u} \times \boldsymbol{B}\right)_j + R_j.$$

In the next step we pull the two terms that include the pressure p and the viscosity tensor π_{jk} to the right-hand side. The mass m is independent of t and x. Since we want to use the continuity equation for the remaining terms on the left-hand side we derive

$$\begin{aligned} mu_j\frac{\partial n}{\partial t} + mn\frac{\partial u_j}{\partial t} + mu_j\frac{\partial}{\partial x_k}(nu_k) + mnu_k\frac{\partial u_j}{\partial x_k} \\ = -\frac{\partial p}{\partial x_j} - \frac{\partial \pi_{jk}}{\partial x_k} + qn\left(\boldsymbol{E} + \boldsymbol{u} \times \boldsymbol{B}\right)_j + \boldsymbol{R}_j. \end{aligned}$$

Using the continuity equation, the first and the third term on the left-hand side vanish and we obtain

$$mn\left(\frac{\partial u_j}{\partial t} + u_k\frac{\partial u_j}{\partial x_k}\right) = -\frac{\partial p}{\partial x_j} - \frac{\partial \pi_{jk}}{\partial x_k} + qn\left(\boldsymbol{E} + \boldsymbol{u} \times \boldsymbol{B}\right)_j + \boldsymbol{R}_j.$$

In rewriting this equation as a vector we obtain the momentum equation in terms of the flow velocity,

$$mn\left(\frac{\partial \boldsymbol{u}}{\partial t} + (\boldsymbol{u}\cdot\nabla)\boldsymbol{u}\right) = -\nabla p - \nabla\cdot\pi + qn\left(\boldsymbol{E} + \boldsymbol{u}\times\boldsymbol{B}\right) + \boldsymbol{R}. \tag{3.16}$$

2. Energy Equation. In the following we calculate the energy equation from the conservation of energy (3.6c),

$$\frac{\partial}{\partial t}\left[\frac{1}{2}mnu^2 + \frac{3}{2}nkT\right] + \nabla\cdot\boldsymbol{Q} = qn\boldsymbol{E}\cdot\boldsymbol{u} + Q + \boldsymbol{R}\cdot\boldsymbol{u},$$

where we substituted $\int \frac{1}{2}mv^2C(f)d^3v = Q + \boldsymbol{R}\cdot\boldsymbol{u}$. Note that Q describes the rate of thermal energy transfer, whereas $\boldsymbol{Q}$ is the energy flux (see below). Let us consider now each term separately and note that we use Einstein's summation convention for simplicity.

A. The first term on the left-hand side can be written as

$$\frac{\partial}{\partial t}\left[\frac{1}{2}mnu^2 + \frac{3}{2}nkT\right] = mnu_j\frac{\partial u_j}{\partial t} + \frac{m}{2}u^2\frac{\partial n}{\partial t} + \frac{3}{2}n\frac{\partial kT}{\partial t} + \frac{3}{2}kT\frac{\partial n}{\partial t},$$

where the particle mass m is time-independent and $\partial u^2/\partial t = 2\boldsymbol{u}\cdot\partial\boldsymbol{u}/\partial t$.

B. For the second term on the left-hand side we find

$$\nabla\cdot\boldsymbol{Q} = \frac{\partial q_j}{\partial x_j} + \frac{5}{2}\frac{\partial}{\partial x_j}\left(u_jp\right) + \frac{\partial}{\partial x_j}\left(\pi_{jk}u_k\right) + \frac{m}{2}\frac{\partial}{\partial x_j}\left(nu^2u_j\right),$$

where we used the energy flux

$$Q_i = q_i + \frac{5}{2}u_ip + \pi_{ij}u_j + \frac{1}{2}mnu^2u_i.$$

Recall that the pressure $p = nkT$, so that the second term on the right-hand side can be written as

$$\begin{aligned}\frac{5}{2}\frac{\partial}{\partial x_j}\left(u_jp\right) &= \frac{3}{2}\frac{\partial}{\partial x_j}\left(u_jnkT\right) + \frac{\partial}{\partial x_j}\left(u_jp\right)\\ &= \frac{3}{2}kT\frac{\partial nu_j}{\partial x_j} + \frac{3}{2}u_jn\frac{\partial kT}{\partial x_j} + u_j\frac{\partial p}{\partial x_j} + p\frac{\partial u_j}{\partial x_j},\end{aligned}$$

where we used $5/2 = 3/2 + 1$. We obtain finally

$$\nabla \cdot \boldsymbol{Q} = \frac{\partial q_j}{\partial x_j} + \frac{3}{2}kT\frac{\partial n u_j}{\partial x_j} + \frac{3}{2}u_j n\frac{\partial kT}{\partial x_j} + u_j\frac{\partial p}{\partial x_j} + p\frac{\partial u_j}{\partial x_j}$$
$$+ u_k\frac{\partial \pi_{jk}}{\partial x_j} + \pi_{jk}\frac{\partial u_k}{\partial x_j} + \frac{m}{2}u^2\frac{\partial n u_j}{\partial x_j} + m n u_j u_k\frac{\partial u_k}{\partial x_j},$$

where we used $u_j \partial u^2/\partial x_j = 2u_j u_k \partial u_k/\partial x_j$.

C. For the last term on the right-hand side we multiply Eq. (3.16) by $\boldsymbol{u}$ and rearrange for $\boldsymbol{R}\cdot\boldsymbol{u}$. We then obtain

$$\boldsymbol{R}\cdot\boldsymbol{u} = m n u_j\left(\frac{\partial u_j}{\partial t} + u_k\frac{\partial u_j}{\partial x_k}\right) + u_j\frac{\partial p}{\partial x_j} + u_j\frac{\partial \pi_{jk}}{\partial x_k} - q n u_j E_j.$$

Note that the term with the magnetic field vanishes, since $\boldsymbol{u}\cdot(\boldsymbol{u}\times\boldsymbol{B}) = 0$, because $\boldsymbol{u} \perp (\boldsymbol{u}\times\boldsymbol{B}) = 0$.

The equation for energy conservation (3.6c) becomes then

$$\underbrace{m n u_j\frac{\partial u_j}{\partial t}}_{2} + \underbrace{\frac{m}{2}u^2\frac{\partial n}{\partial t}}_{4} + \frac{3}{2}n\frac{\partial kT}{\partial t} + \underbrace{\frac{3}{2}kT\frac{\partial n}{\partial t}}_{5} + \frac{\partial q_j}{\partial x_j} + \underbrace{\frac{3}{2}kT\frac{\partial n u_j}{\partial x_j}}_{5} + \frac{3}{2}u_j n\frac{\partial kT}{\partial x_j}$$
$$+ \underbrace{u_j\frac{\partial p}{\partial x_j}}_{3} + p\frac{\partial u_j}{\partial x_j} + \underbrace{u_k\frac{\partial \pi_{jk}}{\partial x_j}}_{6} + \pi_{jk}\frac{\partial u_k}{\partial x_j} + \underbrace{\frac{m}{2}u^2\frac{\partial n u_j}{\partial x_j}}_{4} + \underbrace{m n u_j u_k\frac{\partial u_k}{\partial x_j}}_{2}$$
$$= \underbrace{q n u_j E_j}_{1} + Q + \underbrace{m n u_j\left(\frac{\partial u_j}{\partial t} + u_k\frac{\partial u_j}{\partial x_k}\right)}_{2} + \underbrace{u_j\frac{\partial p}{\partial x_j}}_{3} + \underbrace{u_j\frac{\partial \pi_{jk}}{\partial x_k}}_{6} \underbrace{- q n u_j E_j}_{1}.$$

Note that the viscosity tensor π is symmetric, so that $u_j \partial \pi_{jk}/\partial x_k = u_k \partial \pi_{jk}/\partial x_j$ under the summation. We also find $u_j u_k \partial u_k/\partial x_j = u_k u_j \partial u_j/\partial x_k$. For convenience we marked each term that cancels out. By using the continuity equation we find

$$\frac{3}{2}n\frac{\partial (kT)}{\partial t} + \frac{\partial q_j}{\partial x_j} + \frac{3}{2}u_j n\frac{\partial (kT)}{\partial x_j} + p\frac{\partial u_j}{\partial x_j} + \pi_{jk}\frac{\partial u_k}{\partial x_j} = Q$$

or written as a vector

$$\frac{3}{2}n\left(\frac{\partial (kT)}{\partial t} + \boldsymbol{u}\cdot\nabla(kT)\right) + p\nabla\cdot\boldsymbol{u} = -\nabla\cdot\boldsymbol{q} - \pi : \nabla\boldsymbol{u} + Q. \qquad (3.17)$$

3.2 Markov Processes, the Chapman-Kolmogorov Equation, and the Fokker-Planck Equation

Problem 3.3 Consider a coin tossing event. Suppose

$$X(t_n) \equiv X_n = \sum_{i=1}^{n} x_i, \tag{3.18}$$

where x_i are given by $+1$ for a head and -1 for a tail.

A. Determine the sample spaces $\mathcal{C}_n$ of the random variable X_n for the first 5 tosses, i.e., $n = 1, 2, 3, 4, 5$.
B. Determine the probability of all outcomes for each sample space! Interpret your findings! (Hint: a table might be helpful.)
C. Explain what distribution function should be used and calculate the pdf of the random variable X_n for the general case of n tosses.

Solution

A. The sample spaces of the random variable X_n for the first 5 tosses is given by

$$\begin{aligned}
\mathcal{C}_1 &= \{c : c = -1, 1\} \\
\mathcal{C}_2 &= \{c : c = -2, 0, 2\} \\
\mathcal{C}_3 &= \{c : c = -3, 1, 1, 3\} \\
\mathcal{C}_4 &= \{c : c = -4, -2, 0, 2, 4\} \\
\mathcal{C}_5 &= \{c : c = -5, -3, -1, 1, 3, 5\}.
\end{aligned}$$

B. Let us consider the first toss with $n = 1$. The sum X_1 can be either -1 (for tail) or $+1$ (for head). Both results have the same probability, namely 0.5. After 2 tosses the sum has the possible values $X_2 = -2$ (TT), $X_2 = 0$ (TH or HT), and $X_2 = +2$ (HH). Thus, the number of all possible outcomes is 4. There is only one way to reach $X_2 = -2$ or $X_2 = +2$, so both results have the probability 0.25, but there are two ways to reach $X_2 = 0$, thus the probability to get $X_2 = 0$ is 0.5. Similar considerations can be made for $n = 3, 4, 5, \ldots$ and we obtain the probabilities as shown in Table 3.1. Obviously, the form of this table resembles Pascal's triangle, where each entry can be described by the binomial coefficient

$$\binom{n}{a} = \frac{n!}{a!(n-a)!}$$

Table 3.1 The schematic shows the probabilities for each outcome after 1–5 coin tosses

Steps n / Sum X_n	-5	-4	-3	-2	-1	0	1	2	3	4	5
1					$\frac{1}{2}$		$\frac{1}{2}$				
2				$\frac{1}{4}$		$\frac{2}{4}$		$\frac{1}{4}$			
3			$\frac{1}{8}$		$\frac{3}{8}$		$\frac{3}{8}$		$\frac{1}{8}$		
4		$\frac{1}{16}$		$\frac{4}{16}$		$\frac{6}{16}$		$\frac{4}{16}$		$\frac{1}{16}$	
5	$\frac{1}{32}$		$\frac{5}{32}$		$\frac{10}{32}$		$\frac{10}{32}$		$\frac{5}{32}$		$\frac{1}{32}$

The table resembles Pascal's triangle

C. Since there are only two possible outcomes for tossing a coin it is straightforward to use the binomial distribution function to describe this experiment,

$$f(a) = \binom{n}{a} p^a (1-p)^{n-a},$$

where n is the number of tosses, $p = 0.5$ is the probability of obtaining head (or tail), and $a = (n + X_n)/2$ is the number of heads (or tails). Note that $a \neq X_n$, since the entry for the binomial coefficient in Pascal's triangle always starts with 0 and not with $X_n = -1, -2, \ldots$. Note also that the outcome of Eq. (3.18) is restricted by the number of tosses n in such a way that the sum $X_n = n - 2i$, where $i = 0, 1, 2, \ldots, n$. That means, after 3 tosses the sum cannot yield zero. The pdf is then given by

$$f(X_n) = \binom{n}{\frac{n+X_n}{2}} \frac{1}{2^n},$$

for $X_n = n - 2i$, where $i = 0, 1, 2, \ldots, n$.

3.3 Collision Dynamics, the Rosenbluth Potentials, and the Landau Collision Operator

Problem 3.4 Show that

$$\frac{\langle (\Delta v_y)^2 \rangle^{ab}}{\Delta t} = \frac{\langle (\Delta v_z)^2 \rangle^{ab}}{\Delta t} = \frac{L^{ab}}{4\pi} \int \frac{1}{V_{rel}} f_b(\boldsymbol{v}') d^3 v'$$

and that

$$\frac{\langle (\Delta v_x)^2 \rangle^{ab}}{\Delta t} = \frac{\pi}{4} \left(1 + \frac{m_a}{m_b}\right)^2 \left(\frac{q_a q_b}{2\pi \varepsilon m_a}\right)^4 \left[\frac{1}{r_{min}^2} - \frac{1}{\lambda_D^2}\right] \int \frac{f_b(\boldsymbol{v}')}{V_{rel}^5} d^3 v'.$$

Solution We recall that the velocity changes in x, y, and z direction are given by

$$\Delta v_x = \left(1+\frac{m_a}{m_b}\right)\left(\frac{q_a q_b}{2\pi\varepsilon m_a}\right)^2 \frac{1}{2r^2 V_{rel}^3}$$

$$\Delta v_y = \frac{q_a q_b}{2\pi\varepsilon m_a}\frac{\cos\theta}{V_{rel} r}$$

$$\Delta v_z = \frac{q_a q_b}{2\pi\varepsilon m_a}\frac{\sin\theta}{V_{rel} r}.$$

The characteristic length L^{ab} is given by Eq. (3.31), see Sect. 3.4 below. For the average we apply the operator

$$\frac{\left\langle (\Delta v_i)^2 \right\rangle^{ab}}{\Delta t} = \int_{r_{min}}^{\lambda_D}\int_0^{2\pi}\int (\Delta v_i)^2 r f_b(\boldsymbol{v}') V_{rel} dr d\theta d^3 v'.$$

With that we find

$$\begin{aligned}\frac{\left\langle (\Delta v_y)^2 \right\rangle^{ab}}{\Delta t} &= \left(\frac{q_a q_b}{2\pi\varepsilon m_a}\right)^2 \int_{r_{min}}^{\lambda_D}\int_0^{2\pi}\int \frac{\cos^2\theta}{V_{rel}^2 r^2} r f_b(\boldsymbol{v}') V_{rel} dr d\theta d^3 v' \\ &= \frac{1}{4\pi}\left(\frac{q_a q_b}{\varepsilon m_a}\right)^2 \int_{r_{min}}^{\lambda_D}\frac{dr}{r}\int \frac{f_b(\boldsymbol{v}')}{V_{rel}} d^3 v' \\ &= \frac{L^{ab}}{4\pi}\int \frac{f_b(\boldsymbol{v}')}{V_{rel}} d^3 v',\end{aligned}$$

where we used $\int_0^{2\pi}\cos^2\theta d\theta = \pi$. Similarly we find for the z component

$$\begin{aligned}\frac{\left\langle (\Delta v_z)^2 \right\rangle^{ab}}{\Delta t} &= \left(\frac{q_a q_b}{2\pi\varepsilon m_a}\right)^2 \int_{r_{min}}^{\lambda_D}\int_0^{2\pi}\int \frac{\sin^2\theta}{V_{rel}^2 r^2} r f_b(\boldsymbol{v}') V_{rel} dr d\theta d^3 v' \\ &= \frac{1}{4\pi}\left(\frac{q_a q_b}{\varepsilon m_a}\right)^2 \int_{r_{min}}^{\lambda_D}\frac{dr}{r}\int \frac{f_b(\boldsymbol{v}')}{V_{rel}} d^3 v' \\ &= \frac{L^{ab}}{4\pi}\int \frac{f_b(\boldsymbol{v}')}{V_{rel}} d^3 v',\end{aligned}$$

where we used $\int_0^{2\pi}\sin^2\theta d\theta = \pi$. The velocity mean square displacements in y- and z-direction are identical. For the x-component we find

$$\frac{\left\langle (\Delta v_x)^2 \right\rangle^{ab}}{\Delta t} = \left(1+\frac{m_a}{m_b}\right)^2\left(\frac{q_a q_b}{2\pi\varepsilon m_a}\right)^4 \int_{r_{min}}^{\lambda_D}\int_0^{2\pi}\int \frac{r f_b(\boldsymbol{v}') V_{rel}}{4r^4 V_{rel}^6} dr d\theta d^3 v'$$

$$= \frac{\pi}{2}\left(1+\frac{m_a}{m_b}\right)^2\left(\frac{q_a q_b}{2\pi\varepsilon m_a}\right)^4 \int_{r_{min}}^{\lambda_D}\frac{dr}{r^3}\int\frac{f_b(\boldsymbol{v}')}{V_{rel}^5}d^3v'$$

$$= \frac{\pi}{4}\left(1+\frac{m_a}{m_b}\right)^2\left(\frac{q_a q_b}{2\pi\varepsilon m_a}\right)^4 \left[\frac{1}{r_{min}^2}-\frac{1}{\lambda_D^2}\right]\int\frac{f_b(\boldsymbol{v}')}{V_{rel}^5}d^3v'.$$

Problem 3.5 By direct substitution, show that the Landau collision operator (and hence the other forms) satisfy the conservation laws

$$\int C_{ab}(f_a,f_b)d^3v = 0 \tag{3.19a}$$

$$\int m_a\boldsymbol{v}C_{ab}(f_a,f_b)d^3v = -\int m_b\boldsymbol{v}C_{ba}(f_b,f_a)d^3v \tag{3.19b}$$

$$\int \frac{1}{2}m_a v^2 C_{ab}(f_a,f_b)d^3v = -\int \frac{1}{2}m_b v^2 C_{ba}(f_b,f_a)d^3v. \tag{3.19c}$$

Solution The Landau form of the collision operator is given by

$$C_{ab}(f_a,f_b) = \frac{K}{m_a}\frac{\partial}{\partial v_i}\int V_{ij}\left[\frac{f_a(\boldsymbol{v})}{m_b}\frac{\partial f_b(\boldsymbol{v}')}{\partial v_j'} - \frac{f_b(\boldsymbol{v}')}{m_a}\frac{\partial f_a(\boldsymbol{v})}{\partial v_j}\right]d^3v', \tag{3.20}$$

where we used the constant $K = \ln \Lambda/8\pi\,(q_a q_b/\varepsilon_0)^2$ for simplicity and the tensor

$$V_{ij} = \frac{V_{rel}^2\delta_{ij} - V_{rel,i}V_{rel,j}}{V_{rel}^3}, \tag{3.21}$$

with the vector $V_{rel} = \boldsymbol{v} - \boldsymbol{v}'$ and $V_{rel}^3 = |\boldsymbol{v} - \boldsymbol{v}'|^3$ in the denominator. Keep in mind that the tensor V_{ij} depends on $\boldsymbol{v}$ *and* $\boldsymbol{v}'$. In the following calculations we will frequently use the substitution

$$T(\boldsymbol{v},\boldsymbol{v}') = V_{ij}\left[\frac{f_a(\boldsymbol{v})}{m_b}\frac{\partial f_b(\boldsymbol{v}')}{\partial v_j'} - \frac{f_b(\boldsymbol{v}')}{m_a}\frac{\partial f_a(\boldsymbol{v})}{\partial v_j}\right], \tag{3.22}$$

so that the Landau collision operator can be written as

$$C_{ab}(f_a,f_b) = \frac{K}{m_a}\frac{\partial}{\partial v_i}\int T(\boldsymbol{v},\boldsymbol{v}')d^3v'.$$

Note that the expression $T(\boldsymbol{v},\boldsymbol{v}')$ vanishes, if $v_i \to \pm\infty$. This can easily be seen in Eq. (3.22); the first term on the right-hand side vanishes because $f_a(v_i = \pm\infty, v_j, v_k) = 0$, since any physical distribution function vanishes as $\boldsymbol{v} \to \pm\infty$. The second term in Eq. (3.22) also vanishes, since $\partial f_a(v_i = \pm\infty, v_j, v_k)/\partial v_j = 0$, for the same reason. Note that this is the derivative of the v_j-component.

A. For the first conservation law (3.19a) we have

$$\int C_{ab}(f_a, f_b) d^3 v = \frac{K}{m_a} \int \frac{\partial}{\partial v_i} \int T(\boldsymbol{v}, \boldsymbol{v}') d^3 v' d^3 v.$$

First, we swap the integrations with respect to $\boldsymbol{v}$ and $\boldsymbol{v}'$, so that

$$\int C_{ab}(f_a, f_b) d^3 v = \frac{K}{m_a} \int d^3 v' \int dv_i dv_j dv_k \frac{\partial}{\partial v_i} T(\boldsymbol{v}, \boldsymbol{v}').$$

In the next step we execute the integration with respect to v_i and obtain

$$\int C_{ab}(f_a, f_b) d^3 v = \frac{K}{m_a} \int d^3 v' \int dv_j dv_k \left[T(\boldsymbol{v}, \boldsymbol{v}')\right]_{v_i=-\infty}^{v_i=\infty} = 0.$$

Since $f_a(v_i = \pm\infty, v_j, v_k) = 0$ for any physical distribution function, we find immediately that the zeroth order vanishes.

B. For the first moment (3.19b) we consider only the l-th component of the vector, $\boldsymbol{v} \to v_l$. The left hand side of Eq. (3.19b) can then be written as

$$\int m_a v_l C_{ab}(f_a, f_b) d^3 v = K \int d^3 v \int d^3 v' \, v_l \frac{\partial}{\partial v_i} T(\boldsymbol{v}, \boldsymbol{v}').$$

By swapping the integrations with respect to $\boldsymbol{v}$ and $\boldsymbol{v}'$ we obtain

$$\int m_a v_l C_{ab}(f_a, f_b) d^3 v = K \int d^3 v' \int d^3 v \, v_l \frac{\partial}{\partial v_i} T(\boldsymbol{v}, \boldsymbol{v}').$$

Considering only the integral with respect to $\boldsymbol{v}$ we have to distinguish the following two cases:

$l \neq i$ In this cases we can set either $l = j$ or $l = k$. We choose here $l = j$ so that

$$\int dv_i dv_j dv_k \, v_j \frac{\partial}{\partial v_i} T(\boldsymbol{v}, \boldsymbol{v}') = \int dv_j dv_k \, v_j \left[T(\boldsymbol{v}, \boldsymbol{v}')\right]_{v_i=-\infty}^{v_i=+\infty} = 0.$$

$l = i$ In this cases we find use integration by parts, so that

$$\int d^3 v \, v_i \frac{\partial}{\partial v_i} T(\boldsymbol{v}, \boldsymbol{v}') = \left[v_i T(\boldsymbol{v}, \boldsymbol{v}')\right]_{v_i=-\infty}^{v_i=+\infty} - \int d^3 v \, T(\boldsymbol{v}, \boldsymbol{v}').$$

The first term on the right-hand side vanishes (see above), but the second term remains.

With that, the left side of the conservation law (3.19b) can be written as

$$\int m_a v_l C_{ab}(f_a, f_b) d^3 v$$

$$= -K \int d^3 v' \int d^3 v \; V_{ij} \left[\frac{f_a(\boldsymbol{v})}{m_b} \frac{\partial f_b(\boldsymbol{v}')}{\partial v_j'} - \frac{f_b(\boldsymbol{v}')}{m_a} \frac{\partial f_a(\boldsymbol{v})}{\partial v_j} \right], \tag{3.23}$$

where we substituted $T(\boldsymbol{v}, \boldsymbol{v}')$ back into the equation. Similarly, we obtain for the particle species b (just by exchanging $a \leftrightarrow b$ *and (!)* $\boldsymbol{v} \leftrightarrow \boldsymbol{v}'$)

$$\int m_b v_l C_{ba}(f_b, f_a) d^3 v$$

$$= -K \int d^3 v' \int d^3 v \; V_{ij} \left[\frac{f_b(\boldsymbol{v}')}{m_a} \frac{\partial f_a(\boldsymbol{v})}{\partial v_j} - \frac{f_a(\boldsymbol{v})}{m_b} \frac{\partial f_b(\boldsymbol{v}')}{\partial v_j'} \right]. \tag{3.24}$$

Note that V_{ij} is unchanged under the transformation $v \leftrightarrow v'$. By swapping the terms in square brackets in Eq. (3.24) and by comparing with Eq. (3.23) we find immediately the relation (3.19b).

C. The second moment is given by Eq. (3.19c) and can be written as

$$\int \frac{1}{2} m_a v^2 C_{ab}(f_a, f_b) d^3 v = K \int d^3 v \int d^3 v' \; \frac{1}{2} v^2 \frac{\partial}{\partial v_i} T(\boldsymbol{v}, \boldsymbol{v}').$$

Again, by swapping the integrations with respect to $\boldsymbol{v}$ and $\boldsymbol{v}'$ we obtain

$$\int \frac{1}{2} m_a v^2 C_{ab}(f_a, f_b) d^3 v = K \int d^3 v' \int d^3 v \; \frac{1}{2} v^2 \frac{\partial}{\partial v_i} T(\boldsymbol{v}, \boldsymbol{v}').$$

Now we want to investigate only the integration with respect to v, bearing in mind that $v^2 = v_x^2 + v_y^2 + v_z^2$. The last integral becomes then

$$\int d^3 v \; \frac{1}{2} v^2 \frac{\partial}{\partial v_i} T(\boldsymbol{v}, \boldsymbol{v}') = \sum_{l=i,j,k} \int d^3 v \; \frac{1}{2} v_l^2 \frac{\partial}{\partial v_i} T(\boldsymbol{v}, \boldsymbol{v}')$$

$$= \sum_{l=i,j,k} \left[\frac{1}{2} v_l^2 T(v, v') \right]_{v_i=-\infty}^{v_i=+\infty} - \int d^3 v \; v_i \; T(\boldsymbol{v}, \boldsymbol{v}').$$

Note that the last term has no sum, since $\sum_l \partial v_l^2 / \partial v_i = 2 v_i$. That means that only the i-th component remains from that sum. Obviously, the first term on the right hand side vanishes as before and we are left with

$$\int \frac{1}{2} m_a v^2 C_{ab}(f_a, f_b) d^3 v = -K \int d^3 v' \int d^3 v \; v_i \; V_{ij} \left[\frac{f_a(\boldsymbol{v})}{m_b} \frac{\partial f_b(\boldsymbol{v}')}{\partial v'_j} - \frac{f_b(\boldsymbol{v}')}{m_a} \frac{\partial f_a(\boldsymbol{v})}{\partial v_j} \right], \tag{3.25}$$

where we substituted the integral and the function $T(\boldsymbol{v}, \boldsymbol{v}')$ back into the expression. A similar expression can be found for particle species b by exchanging $a \leftrightarrow b$ and $v \leftrightarrow v'$ and we obtain

$$\begin{aligned}
\int \frac{1}{2} m_b v^2 C_{ba}(f_b, f_a) d^3 v &= -K \int d^3 v' \int d^3 v \; v'_i \; V_{ij} \left[\frac{f_b(\boldsymbol{v}')}{m_a} \frac{\partial f_a(\boldsymbol{v})}{\partial v_j} - \frac{f_a(\boldsymbol{v})}{m_b} \frac{\partial f_b(\boldsymbol{v}')}{\partial v'_j} \right] \\
&= K \int d^3 v' \int d^3 v \; v'_i \; V_{ij} \left[\frac{f_a(\boldsymbol{v})}{m_b} \frac{\partial f_b(\boldsymbol{v}')}{\partial v'_j} - \frac{f_b(\boldsymbol{v}')}{m_a} \frac{\partial f_a(\boldsymbol{v})}{\partial v_j} \right].
\end{aligned} \tag{3.26}$$

Note that v_i was also changed to v'_i. In the last step we simply changed the terms in brackets to get rid of the minus sign in front of the expression and to find a similar expression for particle species a. Now we simply add up Eqs. (3.25) and (3.26) and obtain

$$\int \frac{1}{2} m_a v^2 C_{ab}(f_a, f_b) d^3 v + \int \frac{1}{2} m_b v^2 C_{ba}(f_b, f_a) d^3 v = -K \int d^3 v' \int d^3 v \; (v_i - v'_i) \; V_{ij} \left[\frac{f_a(\boldsymbol{v})}{m_b} \frac{\partial f_b(\boldsymbol{v}')}{\partial v'_j} - \frac{f_b(\boldsymbol{v}')}{m_a} \frac{\partial f_a(\boldsymbol{v})}{\partial v_j} \right].$$

Note that $v_i - v'_i = V_{rel,i}$. Remember that we use Einstein's summation convention and that we therefore implicitly sum over all indices that appear more than once. Obviously, the term $(v_i - v'_i) \; V_{ij}$ is the only term that includes the index i more than once. Therefore, we may sum this term over all spatial coordinates, leading to (recall that $V_{rel,i} = v_i - v'_i$ and V_{ij} is given by Eq. (3.21))

$$\begin{aligned}
\sum_i V_{rel,i} V_{ij} &= \frac{1}{V_{rel}^3} \left(\sum_i V_{rel}^2 V_{rel,i} \delta_{ij} - V_{rel,i}^2 V_{rel,j} \right) \\
&= \frac{V_{rel}^2}{V_{rel}^3} \left[\left(\sum_i V_{rel,i} \delta_{ij} \right) - V_{rel,j} \right] \\
&= 0,
\end{aligned} \tag{3.27}$$

where we used $\sum_i V_{rel,i}\delta_{ij} = V_{rel,j}$ in the first term on the right hand side and $\sum_i V_{rel,i}^2 = V_{rel}^2$ in the second term. It follows immediately that

$$\int \frac{1}{2} m_a v^2 C_{ab}(f_a, f_b) d^3v + \int \frac{1}{2} m_b v^2 C_{ba}(f_b, f_a) d^3v = 0$$

and, thus, the conservation law given by Eq. (3.19c).

Problem 3.6 By using the Landau form of the collision operator, show that $C_{ab}(f_a, f_b) = 0$, if f_a and f_b are Maxwellian distributions with equal temperatures $T_a = T_b = T$.

Solution The Landau form of the collision operator is given by Eq. (3.20). Consider now f_a and f_b as Maxwellian distributions with

$$f_a = n_a \left(\frac{m_a}{2\pi k_B T_a} \right)^{3/2} \exp\left[-\frac{m_a v^2}{2k_B T_a} \right],$$

$$f_b = n_b \left(\frac{m_b}{2\pi k_B T_b} \right)^{3/2} \exp\left[-\frac{m_b v'^2}{2k_B T_b} \right].$$

Note that the primed velocities pertain to f_b. The derivations are then

$$\frac{\partial f_a}{\partial v_j} = -\frac{m_a v_j}{k_B T_a} n_a \left(\frac{m_a}{2\pi k_B T_a} \right)^{3/2} \exp\left[-\frac{m_a v^2}{2k_B T_a} \right]$$

$$\frac{\partial f_b}{\partial v'_j} = -\frac{m_b v'_j}{k_B T_b} n_b \left(\frac{m_b}{2\pi k_B T_b} \right)^{3/2} \exp\left[-\frac{m_b v'^2}{2k_B T_b} \right].$$

For the expression in square brackets in Eq. (3.20) we find

$$\left[\frac{f_a(\boldsymbol{v})}{m_b} \frac{\partial f_b(\boldsymbol{v}')}{\partial v'_j} - \frac{f_b(\boldsymbol{v}')}{m_a} \frac{\partial f_a(\boldsymbol{v})}{\partial v'_j} \right]$$

$$= \frac{n_a n_b}{k_B} \left(\frac{m_a m_b}{4\pi^2 k_B^2 T_a T_b} \right)^{3/2} \exp\left[-\frac{m_a v^2}{2k_B T_a} - \frac{m_b v'^2}{2k_B T_b} \right] \left(\frac{v_j}{T_a} - \frac{v'_j}{T_b} \right)$$

$$= \frac{n_a n_b}{k_B T} \left(\frac{m_a m_b}{4\pi^2 k_B^2 T^2} \right)^{3/2} \exp\left[-\frac{m_a v^2}{2k_B} - \frac{m_b v'^2}{2k_B T} \right] (v_j - v'_j),$$

where we have set $T_a = T_b = T$ in the first step. Note that $V_{rel,j} = \left(v_j - v'_j \right)$.

Substituting this expression back into Eq. (3.20) we obtain

$$C_{ab}(f_a, f_b) = \frac{\ln \Lambda}{8\pi m_a} \left(\frac{q_a q_b}{\varepsilon_0}\right)^2 \frac{n_a n_b}{k_B T} \left(\frac{m_a m_b}{4\pi^2 k_B^2 T^2}\right)^{3/2} \times \frac{\partial}{\partial v_i} \int \exp\left[-\frac{m_a v^2}{2k_B} - \frac{m_b v'^2}{2k_B T}\right] V_{ij} V_{rel,j} d^3 v'. \quad (3.28)$$

Note that the expression under the integral is a sum over index j (using Einstein's summation convention). According to Eq. (3.27) we find that $V_{ij} V_{rel,j} = 0$, thus, the collision operator vanishes.

3.4 Electron-Proton Collisions

The Coulomb collision operator can be simplified if the colliding particles move at very different speeds, such as electrons colliding with protons moving at some average speed $\boldsymbol{u}_p$. The scattering operator for electron-proton scattering can therefore be expressed as the sum of the Lorentz scattering operator $\mathcal{L}(f_e)$ and C_{ep}^1 (see [5]),

$$C_{ep}(f_e) = \nu_{ep}(v) \left(\mathcal{L}(f_e) - \frac{m_e \boldsymbol{v} \cdot \boldsymbol{u}_p}{kT_e} f_{Me}\right), \quad (3.29)$$

where

$$\nu_{ep}(v) \equiv \frac{n_p L^{ep}}{4\pi v^3} = \frac{n_p e^4}{4\pi m_e^2 \epsilon_0^2 v^3} \ln \Lambda \quad (3.30)$$

is a velocity-dependent electron-proton collision frequency and the general form

$$L^{ab} = \left(\frac{q_a q_b}{\varepsilon m_a}\right)^2 \int_{r_{min}}^{\lambda_D} \frac{dr}{r} = \left(\frac{q_a q_b}{\varepsilon m_a}\right)^2 \ln \Lambda. \quad (3.31)$$

where $\ln \Lambda$ is the Coulomb logarithm.

The first part of the collision operator (3.29) describes the collisions of electrons with *infinitely heavy* stationary protons, implying that only the electron direction and not the velocity changes in a collision. Consequently, there is only diffusion in velocity space on a sphere of constant radius

(continued)

$v = constant$, and the collision operator is spherically symmetric. Finally, note that the proton mass is completely absent from the collision operator, depending only on charge e. This makes it straightforward to model electron collisions in a plasma comprising several different ion species.

Problem 3.7 How would the result (3.29) change if the electrons scattered off a background of (A) α particles (He—nuclei), and (B) a mixture of protons p and α particles, as found in the solar wind emitted by the sun?

Solution

A. Since the collision operator depends only on the charge of the scattering centers we find

$$C_{e\alpha}(f_e) = \nu_{e\alpha}(v)\left(\mathcal{L}(f_e) - \frac{m_e \boldsymbol{v}\cdot\boldsymbol{u}_\alpha}{kT_e} f_{Me}\right), \tag{3.32}$$

with the collision frequency

$$\nu_{e\alpha}(v) = \frac{4n_\alpha e^4}{4\pi m_e^2 \epsilon_0^2 v^3} \ln\Lambda, \tag{3.33}$$

since $q_e = e$ and $q_\alpha = 2e$.

B. The general collision operator for multiple ion species is the sum over each individual collision operator

$$C_{ei} = \sum_j C_{ej} = C_{ep} + C_{e\alpha}.$$

Substituting Eqs. (3.29) and (3.32) we obtain

$$C_{ei} = \nu_{ep}(v)\left(\mathcal{L}(f_e) - \frac{m_e \boldsymbol{v}\cdot\boldsymbol{u}_p}{kT_e} f_{Me}\right) + \nu_{e\alpha}(v)\left(\mathcal{L}(f_e) - \frac{m_e \boldsymbol{v}\cdot\boldsymbol{u}_\alpha}{kT_e} f_{Me}\right).$$

Under the assumption that the protons and α-particles move with the same background velocity $\boldsymbol{u}_{p,\alpha}$ we can simplify this expression,

$$C_{ei} = \nu_{ei}(v)\left(\mathcal{L}(f_e) - \frac{m_e \boldsymbol{v}\cdot\boldsymbol{u}_{p,\alpha}}{kT_e} f_{Me}\right),$$

where $\nu_{ei}(v) = \left[\nu_{ep}(v) + \nu_{e\alpha}(v)\right]$. By collecting the collision frequencies for protons and α-particles we find

$$\nu_{ei}(v) = \frac{\left(n_p + 4n_\alpha\right) e^4}{4\pi m_e^2 \epsilon_0^2 v^3} \ln \Lambda .$$

By introducing an *effective* charge Z_{eff} with

$$Z_{eff} = \frac{\sum_j n_j Z_j^2}{\sum_j n_j Z_j} = \frac{\sum_j n_j Z_j^2}{n_e} = \frac{n_p + 4n_\alpha}{n_e},$$

where we used the fact that for a quasineutral plasma $\sum_j n_j Z_j = n_e$, and that $Z_p = 1$ and $Z_\alpha = 2$, we find $n_e Z_{eff} = n_p + 4n_\alpha$, so that the collision frequency can be written as

$$\nu_{ei}(v) = \frac{n_e Z_{eff} e^4}{4\pi m_e^2 \epsilon_0^2 v^3} \ln \Lambda . \tag{3.34}$$

3.5 Collisions with a Maxwellian Background

The Chandrasekhar function (Fig. 3.1) is defined as

$$G(x) = \frac{f(x) - xf'(x)}{2x^2}, \tag{3.35}$$

where $f(x)$ is the familiar error function,

$$f(x) = \frac{2}{\sqrt{\pi}} \int_0^x e^{-z^2} dz; \quad f'(x) = \frac{2}{\sqrt{\pi}} e^{-x^2}; \quad f''(x) = -2xf'(x). \tag{3.36}$$

Problem 3.8 Show that the assumption of a Maxwell-Boltzmann distribution function $f_b(v)$ yields the solution to the partial differential equation

$$\frac{1}{v^2}\frac{\partial}{\partial v}\left(v^2 \frac{\partial \phi_b}{\partial v}\right) = f_b(v) \quad \text{as} \quad \phi_b'(v) = \frac{m_b n_b}{4\pi T_b} G\left(\frac{v}{v_T}\right), \tag{3.37}$$

where $G(x)$ is the Chandrasekhar function (3.35) and ϕ_b' is the Rosenbluth potential.

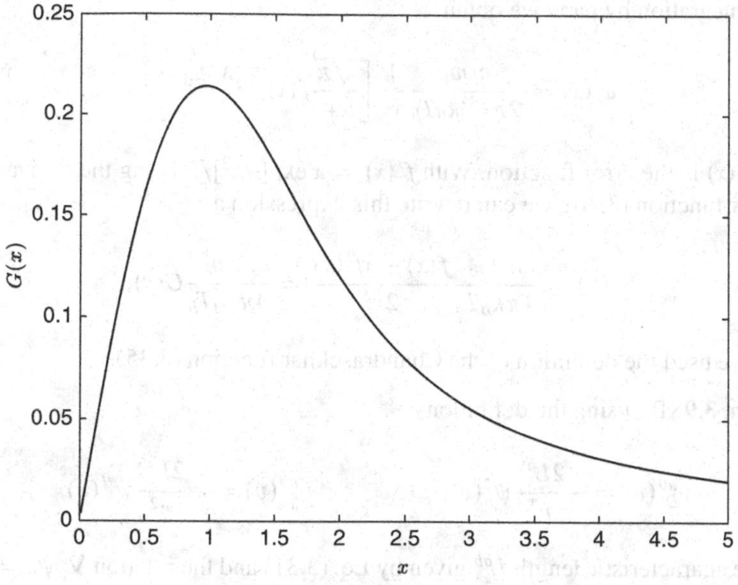

Fig. 3.1 Shown is the Chandrasekhar function (3.35) in the interval [0, 5]

Solution The Maxwell-Boltzmann distribution function is given by

$$f_b(v) = n_b \left(\frac{m_b}{2\pi k_B T_b} \right)^{3/2} \exp\left[-\frac{m_b v^2}{2k_B T_b} \right] = \frac{n_b}{\pi^{3/2} v_T^3} \exp\left[-\frac{v^2}{v_T^2} \right], \quad (3.38)$$

where we used $v_T = \sqrt{2k_B T_b / m_b}$. The starting point is the differential equation (3.37). We multiply this equation by v^2 and integrate with respect to v,

$$\int \frac{\partial}{\partial v} \left(v^2 \frac{\partial \phi_b}{\partial v} \right) dv = \int v^2 f_b(v) dv. \quad (3.39)$$

Substituting f_b on the right-hand side by the Maxwell-Boltzmann distribution function (3.38) and executing the integral on the left-hand side we find (after dividing by v^2)

$$\frac{\partial \phi_b}{\partial v} \equiv \phi'(v) = \frac{n_b}{\pi^{3/2} v_T^3} \frac{1}{v^2} \int v^2 \exp\left[-\frac{v^2}{v_T^2} \right] dv.$$

Now, we substitute $v = x v_T$ and obtain

$$\phi'(x) = \frac{n_b}{\pi^{3/2} v_T^2} \frac{1}{x^2} \int x^2 \exp\left[-x^2 \right] dx.$$

Using integration by parts we obtain

$$\phi'(x) = \frac{n_b m_b}{2\pi^{3/2} k_B T_b} \frac{1}{x^2} \left[\frac{\sqrt{\pi}}{4} f(x) - \frac{x}{2} e^{-x^2} \right],$$

where $f(x)$ is the error function. With $f'(x) = x \exp[-x^2]/2$ being the derivative of the error function (3.36), we can rewrite this expression as

$$\phi'(v) = \frac{n_b m_b}{4\pi k_B T_b} \frac{f(x) - x f'(x)}{2x^2} = \frac{n_b m_b}{4\pi k_B T_b} G(x),$$

where we used the definition of the Chandrasekhar function (3.35).

Problem 3.9 By using the definitions

$$\nu_D^{ab}(v) = -\frac{2L^{ab}}{v^3} \psi_b'(v), \qquad \nu_\parallel^{ab}(v) = -\frac{2L^{ab}}{v^2} \psi_b''(v)$$

with the characteristic length L^{ab} given by Eq. (3.31) and the relation $\nabla_v^2 \psi_b = \phi_b$ in spherical velocity-space coordinates, derive the collision frequencies

$$\nu_D^{ab}(v) = \bar{\nu}_{ab} \frac{f(x_b) - G(x_b)}{x_a^3} \quad \text{and} \quad \nu_\parallel^{ab}(v) = 2\bar{\nu}_{ab} \frac{G(x_b)}{x_a^3},$$

where $G(x_b)$ is the Chandrasekhar function (3.35) and

$$\bar{\nu}_{ab} \equiv \frac{n_b q_a^2 q_b^2 \ln \Lambda}{4\pi \varepsilon^2 m_a^2 v_{Ta}^3}.$$

Solution

A. We start with the definition for $\nu_\parallel^{ab}(v)$ and obtain

$$\nu_\parallel^{ab}(v) = -\frac{2L^{ab}}{v^2} \psi_b''(v) = \frac{n_b}{8\pi} \frac{2L^{ab}}{v^2} \frac{d}{dv} \left[f(x_b) - G(x_b) \right],$$

where we used the definition $\psi_b'(v) = -n_b/8\pi \left[f(x_b) - G(x_b) \right]$, see [5]. By using $v = x_b v_{Tb}$ we find $d/dv = (1/v_{Tb}) d/dx_b$ and we can substitute

$$\begin{aligned}
\nu_\parallel^{ab}(v) &= \frac{n_b}{4\pi} \frac{L^{ab}}{v^2} \frac{1}{v_{Tb}} \frac{d}{dx_b} \left[f(x_b) - \frac{f(x_b)}{2x_b^2} + \frac{f'(x_b)}{2x_b} \right] \\
&= \frac{n_b}{4\pi} \frac{L^{ab}}{v^2} \frac{1}{v_{Tb}} \\
&\quad \times \left[f'(x_b) - \frac{2x_b^2 f'(x_b) - 4x_b f(x_b)}{4x_b^4} + \frac{2x_b f''(x_b) - 2f'(x_b)}{4x_b^2} \right].
\end{aligned}$$

In the first step we substituted the Chandrasekhar function $G(x_b)$ by Eq. (3.35). In the last step we calculated the derivative of the expression in squared brackets. By simplifying we obtain

$$\nu_{\parallel}^{ab}(v) = \frac{n_b}{4\pi}\frac{L^{ab}}{v^2}\frac{1}{v_{Tb}}\left[f'(x_b) + \frac{f''(x_b)}{2x_b} + \frac{f(x_b)}{x_b^3} - \frac{f'(x_b)}{x_b^2}\right].$$

Since $f''(x_b) = -2x_b f'(x_b)$, see Eq. (3.36), we find that the first two terms in the square brackets cancel. We obtain then

$$\nu_{\parallel}^{ab}(v) = \frac{n_b}{4\pi}\frac{L^{ab}}{v^2}\frac{1}{v_{Tb}}\frac{2}{x_b}\left[\frac{f(x_b) - x_b f'(x_b)}{2x_b^2}\right] = 2\frac{n_b}{4\pi}\frac{L^{ab}}{v^3}G(x_b).$$

With $v = x_a v_{Ta}$ and the definition of L^{ab} (3.31) we find

$$\nu_{\parallel}^{ab}(v) = 2\frac{n_b}{4\pi}\left(\frac{q_a q_b}{\varepsilon m_a}\right)^2 \ln\Lambda \frac{1}{x^3 v_{Ta}^3} G(x_b) = 2\bar{\nu}_{ab}\frac{G(x_b)}{x_a^3}. \tag{3.40}$$

B. Let us now consider now $\nu_D^{ab}(v)$ with

$$\nu_D^{ab}(v) = -\frac{2L^{ab}}{v^3}\psi_b'(v) = \frac{n_b}{4\pi}\frac{L^{ab}}{v^3}\left[f(x_b) - G(x_b)\right],$$

where we used again the definition $\psi_b'(v) = -n_b/8\pi\,[f(x_b) - G(x_b)]$. With $v = x_a v_{Ta}$ and the definition of L^{ab} (3.31) we find

$$\nu_D^{ab}(v) = \frac{n_b}{4\pi}\left(\frac{q_a q_b}{\varepsilon m_a}\right)^2 \ln\Lambda \frac{1}{v_{Ta}^3}\frac{f(x_b) - G(x_b)}{x_a^3} = \bar{\nu}_{ab}\frac{f(x_b) - G(x_b)}{x_a^3}.$$

3.6 Collision Operator for Fast Ions

Problem 3.10 Suppose that energetic particles are introduced as an isotropic distribution with speed U at a rate η per unit volume. Since the energetic particles are isotropically distributed in velocity space, the kinetic equation may be expressed as

$$\frac{\partial f_\alpha}{\partial t} = \frac{1}{v^2\tau_s}\frac{\partial}{\partial v}\left[(v^3 + v_c^3)f_\alpha(v)\right] + \eta\frac{\delta(v - U)}{4\pi U^2}. \tag{3.41}$$

Subject to the boundary condition $f_\alpha(v > U) = 0$, show that the steady-state energetic particle distribution function is given by

$$f_\alpha(v) = \frac{\eta\tau_s}{4\pi\left(v^3 + v_c^3\right)} \qquad \text{for } v < U.$$

(Hint: Choose appropriate integration limits a and b, so that the velocity of the energetic particles U lies in the interval $[a, b]$.)

Solution In the steady state case we have $\partial f_\alpha/\partial t = 0$. In this case Eq. (3.41) becomes an ordinary differential equation,

$$\frac{d}{dv}\left[\left(v^3 + v_c^3\right) f_\alpha(v)\right] = -\frac{\eta\tau_s}{4\pi U^2} v^2 \delta(v - U).$$

Integrating this expression gives

$$\int_v^{v_\infty} \frac{d}{dv'}\left(v'^3 + v_c^3\right) f_\alpha(v') dv' = -\frac{\eta\tau_s}{4\pi U^2} \int_v^{v_\infty} v'^2 \delta(v' - U) dv'.$$

We changed the integration variable from $v \to v'$, since we want to denote the lower integration limit with v. The integration limits are then chosen in such a way that $v < U < v_\infty$. We can now easily integrate the right-hand side and obtain U^2. For the left-hand side we integrate by parts and obtain

$$\left(v_\infty^3 + v_c^3\right) f_\alpha(v_\infty) - \left(v^3 + v_c^3\right) f_\alpha(v) = -\frac{\eta\tau_s}{4\pi}.$$

Since $v_\infty > U$, it follows $f_\alpha(v_\infty) = 0$ due to the boundary condition. Thus, we obtain

$$f_\alpha(v) = \frac{\eta\tau_s}{4\pi\left(v^3 + v_c^3\right)} \qquad \text{for } v < U.$$

3.7 Transport Equations for a Collisional Electron-Proton Plasma

By considering a plasma comprising electron and protons only we may develop a transport theory in the presence of proton-proton, electron-proton, and electron-electron collisions. Since the electrons do not collide with a stationary background, we need to transform the kinetic equation for each species a to a coordinate frame moving with the mean or bulk flow velocity $\boldsymbol{u}_a(\boldsymbol{r}, t)$ of each species. This requires the coordinate transformation

(continued)

$(\boldsymbol{x}, \boldsymbol{v}, t) \to (\boldsymbol{x}, \boldsymbol{c}_a, t)$, where $\boldsymbol{c}_a = \boldsymbol{v} - \boldsymbol{u}_a$. The dominant time scales are those associated with the collision frequency and the gyrofrequency. For electrons, we can order the kinetic equation as (see [5])

$$C_{ee}(f_e) + C^0_{ep}(f_e) + \left(\frac{e}{m_e}\boldsymbol{c}_e \times \boldsymbol{B}\right) \cdot \nabla_{c_e} f_e = \frac{\partial f_e}{\partial t} + \boldsymbol{u}_e \cdot \nabla f_e + \boldsymbol{c}_e \cdot \nabla f_e$$
$$- \left[\frac{e}{m_e}\boldsymbol{E}' + \left(\frac{\partial}{\partial t} + \boldsymbol{u}_e \cdot \nabla\right)\boldsymbol{u}_e\right] \nabla_{c_e} f_{e0} - c'_{aj}\frac{\partial u_{ek}}{\partial x_j}\frac{\partial f_a}{\partial c_{ek}} - C^1_{ep}(f_e),$$

where the higher order correction to the collision operator has been included on the right-hand side because it acts more slowly than the leading order term.

Following the Chapman-Enskog expansion procedure (see also Sect. 2.4), we solve this equation by expanding the distribution function as $f_e = f_{e0} + f_{e1} + \ldots$. To the lowest order, the left-hand side must vanish, which requires the distribution function to be a Maxwell-Boltzmann distribution at rest in the moving frame (see also Eq. (2.36) in Sect. 2.3)

$$f_{e0} = n_e \left(\frac{m_e}{2\pi k T_e}\right)^{3/2} \exp\left(-\frac{m_e}{2kT_e}v^2\right) = n_e \left(\frac{\beta}{\pi}\right)^{3/2} e^{-\beta v^2},$$

with $\beta = m_e/2kT_e$. On using the zeroth order solution on the right-hand side we obtain an equation for the next order solution f_{e1},

$$C_{ee}(f_{e1}) + C^0_{ep}(f_{e1}) + \left(\frac{e}{m_e}\boldsymbol{v} \times \boldsymbol{B}\right) \cdot \nabla_v f_{e1} = \left(\frac{\partial}{\partial t} + \boldsymbol{u}_e \cdot \nabla\right) \ln n_e f_{e0}$$
$$+ \left(\beta v^2 - \frac{3}{2}\right)\left(\frac{\partial}{\partial t} + \boldsymbol{u}_e \cdot \nabla\right) \ln kT_e f_{e0} + \boldsymbol{v} \cdot \nabla \ln n_e f_{e0}$$
$$+ \left(\beta v^2 - \frac{3}{2}\right) \boldsymbol{v} \cdot \nabla \ln kT_e f_{e0} + \frac{m_e \boldsymbol{v}}{kT_e} \cdot \left[\frac{e}{m_e}\boldsymbol{E}' + \left(\frac{\partial}{\partial t} + \boldsymbol{u}_e \cdot \nabla\right)\boldsymbol{u}_e\right] f_{e0}$$
$$+ \frac{m_e v_j v_k}{kT_e}\frac{\partial u_{ek}}{\partial r_j} f_{e0} + \nu_{ep}\frac{m_e \boldsymbol{v} \cdot (\boldsymbol{u}_e - \boldsymbol{u}_p)}{kT_e} f_{e0}, \tag{3.42}$$

where, for notational convenience, we have written $\boldsymbol{v}$ for $\boldsymbol{c}_e$. It is convenient to make use of the convective derivative (see also Eq. (2.48) in Problem 2.10),

$$\frac{D}{Dt} = \frac{\partial}{\partial t} + \boldsymbol{u}_e \cdot \nabla. \tag{3.43}$$

Note that the last term on the right hand side of Eq. (3.42), the term proportional to ν_{ep}, corresponds to $-C^1_{ep}$ (see [5]). By pulling this term to

(continued)

the left hand side and by using $C_{ep}(f_{e1}) = C_{ep}^0 + C_{ep}^1$, we obtain

$$C_{ee}(f_{e1}) + C_{ep}(f_{e1}) + \left(\frac{e}{m_e}\boldsymbol{v} \times \boldsymbol{B}\right) \cdot \nabla_v f_{e1} = \frac{D \ln n_e}{Dt} f_{e0}$$
$$+ \left(\beta v^2 - \frac{3}{2}\right) \frac{D \ln kT_e}{Dt} f_{e0} + \boldsymbol{v} \cdot \nabla \ln n_e f_{e0} + \left(\beta v^2 - \frac{3}{2}\right) \boldsymbol{v} \cdot \nabla \ln kT_e f_{e0}$$
$$+ \frac{m_e \boldsymbol{v}}{kT_e} \cdot \left[\frac{e}{m_e}\boldsymbol{E}' + \frac{D\boldsymbol{u}_e}{Dt}\right] f_{e0} + \frac{m_e v_j v_k}{kT_e} \frac{\partial u_{ek}}{\partial r_j} f_{e0}. \tag{3.44}$$

Also note that we used a Chapman-Enskog expansion to derive Eq. (3.44), which gives the following constraints for the first order correction f_{e1} to the Maxwell-Boltzmann distribution f_{e0}:

$$\int f_{e1}\, d^3v = 0 \qquad \int \boldsymbol{v} f_{e1}\, d^3v = 0 \qquad \int v^2 f_{e1}\, d^3v = 0. \tag{3.45}$$

The velocity is given by $v^2 = v_x^2 + v_y^2 + v_z^2 = \sum v_i^2$. We recall from Problem 2.12 that

$$\int_{-\infty}^{\infty} x^n e^{-\beta x^2}\, dx = 0 \qquad \text{for } n = 1, 3, 5, \ldots \tag{3.46}$$

For even functions under the integral we have

$$\int_{-\infty}^{\infty} e^{-\beta x^2}\, dx = \sqrt{\frac{\pi}{\beta}} \qquad \int_{-\infty}^{\infty} x^2 e^{-\beta x^2}\, dx = \frac{\sqrt{\pi}}{2\beta^{3/2}}$$
$$\int_{-\infty}^{\infty} x^4 e^{-\beta x^2}\, dx = \frac{3\sqrt{\pi}}{4\beta^{5/2}} \qquad \int_{-\infty}^{\infty} x^6 e^{-\beta x^2}\, dx = \frac{15\sqrt{\pi}}{8\beta^{7/2}}. \tag{3.47}$$

In the following we will make frequent use of

$$\int f_{e0}\, d^3v = n_e \left(\frac{\beta}{\pi}\right)^{3/2} \int e^{-\beta v^2} d^3v = n_e \tag{3.48}$$

$$\int v_i^2 f_{e0}\, d^3v = n_e \left(\frac{\beta}{\pi}\right)^{3/2} \int v_i^2 e^{-\beta v^2} d^3v = \frac{n_e}{2\beta} \tag{3.49}$$

$$\int v_i^4 f_{e0}\, d^3v = n_e \left(\frac{\beta}{\pi}\right)^{3/2} \int v_i^4 e^{-\beta v^2} d^3v = \frac{3n_e}{4\beta^2}, \tag{3.50}$$

where $d^3v = dv_i dv_j dv_k$ and the integration is taken from $-\infty$ to ∞.

Problem 3.11 Show that integrating Eq. (3.44) over velocity space yields the continuity equation

$$\frac{\partial n_e}{\partial t} + \nabla \cdot (n_e \boldsymbol{u}_e) .$$

Solution Starting with Eq. (3.44) we take the zeroth moment, i.e., we integrate this equation with respect to velocity v. For convenience we consider each term separately.

A. The first term on the left-hand side is given by

$$\int C_{ee}(f_{e1})d^3v = 0,$$

since the collision operator for particles of the same species has to satisfy this condition. See also Eq. (3.5a) in Sect. 3.1 and compare with Sect. 4.1 in [5].

B. The second term on the left-hand side is

$$\int C_{ep}(f_{e1})d^3v = 0,$$

since the collision operator for electron-ion collisions has to satisfy this relation. Compare with Eq. (3.19a) in Problem 3.5. See also Sect. 4.1 in [5].

C. The third term on the left-hand side gives

$$\begin{aligned} \int (\boldsymbol{v} \times \boldsymbol{B}) \cdot \nabla_v f_{e1}\, d^3v &= \int dv_i dv_j dv_k \epsilon_{ijk} v_j B_k \frac{\partial f_{e1}}{\partial v_i} \\ &= \epsilon_{ijk} \int dv_j dv_k \left[v_j B_k f|_{v_i=-\infty}^{v_i=+\infty} - \int dv_i \frac{\partial v_j B_k}{\partial v_i} \right] \\ &= 0, \end{aligned}$$

where we neglected the constant e/m_e. Note that i, j, k are mutually distinct. In the first step we used Einstein's summation convention (summation over i), and expressed the crossproduct by $(\boldsymbol{v} \times \boldsymbol{B})_i = \epsilon_{ijk} v_j B_k$, where ϵ_{ijk} is the Levi-Civita tensor. In the second step we integrated by parts, where the first term vanishes because any physical distribution function f vanishes for $v_i = \pm\infty$. The second term on the right-hand side is zero, since B_k does not depend on the velocity and $\partial v_j / \partial v_i = 0$ because $i \neq j$, or, in other words, the ϵ_{ijk} tensor is zero for repeated indices. As a vector description one can also write $\nabla_v \cdot (\boldsymbol{v} \times \boldsymbol{B}) = 0$ (see also Problem 3.1 A, the derivation of the zeroth moment of the Fokker-Planck equation).

D. The first term on the right-hand side is given by (using the convective derivative)

$$\left(\frac{\partial \ln n_e}{\partial t} + \boldsymbol{u}_e \cdot \nabla \ln n_e\right) \int f_{e0}\, d^3 v = \frac{\partial n_e}{\partial t} + \boldsymbol{u}_e \cdot \nabla n_e,$$

where we used Eq. (3.48) and $\partial \ln n_e/\partial t = (1/n_e)\partial n_e/\partial t$.

E. The second term on the right-hand side is given by

$$\int \left(\beta v^2 - \frac{3}{2}\right) f_{e0}\, d^3 v = \beta \int v^2 f_{e0}\, d^3 v - \frac{3}{2} \int f_{e0}\, d^3 v$$
$$= \beta \left(\frac{n_e}{2\beta} + \frac{n_e}{2\beta} + \frac{n_e}{2\beta}\right) - \frac{3n_e}{2} = 0,$$

where we neglected the convective derivative $D \ln kT_e/Dt$, and used Eq. (3.49) with $v^2 = v_x^2 + v_y^2 + v_z^2$ for the evaluation of the integrals.

F. For the third, fourth, and fifth term on the right-hand side we find

$$\int \left[\boldsymbol{v} \cdot \nabla \ln n_e + \left(\beta v^2 - \frac{3}{2}\right) \boldsymbol{v} \cdot \nabla \ln kT_e \right.$$
$$\left. + \frac{m_e \boldsymbol{v}}{kT_e} \cdot \left(\frac{e}{m_e}\boldsymbol{E}' + \frac{D\boldsymbol{u}_e}{Dt}\right)\right] f_{e0}\, d^3 v = 0.$$

The integral is zero, since the function under the integral is odd in v, see Eq. (3.46).

G. For the sixth term on the right-hand side we find for $j = k$,

$$\frac{\partial u_{ej}}{\partial r_j} \frac{m_e}{kT_e} \int v_j^2 f_{e0}\, d^3 v = \frac{\partial u_{ej}}{\partial r_j} \frac{m_e}{kT_e} \frac{n_e}{2\beta} = n_e \nabla \cdot \boldsymbol{u}_e,$$

where we used $\beta = m_e/2kT_e$. Obviously, for $j \neq k$ the integral is zero.

By adding up all results we find the continuity equation

$$\frac{\partial n_e}{\partial t} + \boldsymbol{u}_e \cdot \nabla n_e + n_e \nabla \cdot \boldsymbol{u}_e = \frac{\partial n_e}{\partial t} + \nabla \cdot (n_e \boldsymbol{u}_e) = 0,$$

which shows that the convective derivative can be replaced by

$$\frac{D \ln n_e}{Dt} = -\nabla \cdot \boldsymbol{u}_e. \tag{3.51}$$

Problem 3.12 Show that the first moment of Eq. (3.44) yields the momentum equation (3.16) from Problem 3.2 without the viscous term (the term including the viscosity tensor π), and that

$$\left(\frac{\partial}{\partial t} + \boldsymbol{u}_e \cdot \nabla\right) \boldsymbol{u}_e + \frac{e\boldsymbol{E}'}{m_e} = \frac{\boldsymbol{R}_e - \nabla\left(n_e k T_e\right)}{m_e n_e}. \tag{3.52}$$

Solution In this case we multiply Eq. (3.44) with $\boldsymbol{v}$ and integrate with respect to v. For simplicity we consider here only the i-th component of the vector, i.e., v_i. We also use the results from the previous problem.

A. The first term on the left-hand side is given by

$$\int v_i C_{ee}(f_{e1}) d^3 v = 0,$$

since the collision operator for particles of the same species has to satisfy this condition. See also Eq. (3.5b) in Sect. 3.1 and compare with the previous Problem 3.11!

B. The second term on the left-hand side is given by

$$\int \boldsymbol{v} C_{ep}(f_{e1})\, d^3 v = \frac{\boldsymbol{R}_e}{m_e}.$$

See Eq. (3.7) in Sect. 3.1 and compare with Sect. 4.1 in [5].

C. The third term on the left-hand side is given by

$$\frac{e}{m_e} \int v_i \left[(\boldsymbol{v} \times \boldsymbol{B}) \cdot \nabla_v f_{e1}\right] d^3 v = \frac{e}{m_e} \int v_i \epsilon_{lmn} v_m B_n \frac{\partial f_{e1}}{\partial v_l} d^3 v,$$

where we used $(\boldsymbol{v} \times \boldsymbol{B})_l = \epsilon_{lmn} v_m B_n$ and Einstein's summation convention, i.e., summation over index l. In the following we consider the cases $i = m, i = n, i = l$.

- For $i = m$ we have

$$\begin{aligned}
&\epsilon_{lin} \frac{eB_n}{m_e} \int_{-\infty}^{\infty} dv_i v_i^2 \int_{-\infty}^{\infty} dv_n \int_{-\infty}^{\infty} dv_l \frac{\partial f_{e1}}{\partial v_l} \\
&= \epsilon_{lin} \frac{eB_n}{m_e} \int_{-\infty}^{\infty} dv_i v_i^2 \int_{-\infty}^{\infty} dv_n \left[f_{e1}\right]_{v_l=-\infty}^{v_l=+\infty} = 0,
\end{aligned}$$

  since any physical distribution function has to vanish as $v_l \to \pm\infty$.

- For $i = n$ we have similarly

$$\epsilon_{lmi}\frac{eB_n}{m_e}\int_{-\infty}^{\infty} dv_m v_m \int_{-\infty}^{\infty} dv_i v_i \int_{-\infty}^{\infty} dv_l \frac{\partial f_{e1}}{\partial v_l} d^3v$$
$$= \epsilon_{lmi}\frac{eB_n}{m_e}\int_{-\infty}^{\infty} dv_m v_m \int_{-\infty}^{\infty} dv_i v_i \left[f_{e1}\right]_{v_l=-\infty}^{v_l=+\infty} = 0.$$

- For $i = l$, however, we find

$$\epsilon_{imn}\frac{eB_n}{m_e}\int_{-\infty}^{\infty} dv_m v_m \int_{-\infty}^{\infty} dv_n \int_{-\infty}^{\infty} dv_i v_i \frac{\partial f_{e1}}{\partial v_i} d^3v$$
$$= \epsilon_{imn}\frac{eB_n}{m_e}\int_{-\infty}^{\infty} dv_m v_m \int_{-\infty}^{\infty} dv_n \left[(f_{e1}v_i)_{-\infty}^{\infty} - \int_{-\infty}^{\infty} dv_i f_{e1}\right]$$
$$= -\epsilon_{imn}\frac{eB_n}{m_e}\int d^3v\, v_m f_{e1} = 0,$$

where we used $(f_{e1}v_l)_{-\infty}^{\infty} = 0$ in the second line and relation (3.45) in the last line.

D. The first term on the right-hand side is given by

$$\left(\frac{\partial \ln n_e}{\partial t} + \boldsymbol{u}_e \cdot \nabla \ln n_e\right)\int v_i f_{e0}\, d^3v = 0,$$

since the function under the integral is odd with respect to v.

E. The second term on right-hand side is given by

$$\frac{D \ln kT_e}{Dt}\int \left(\beta v^2 - \frac{3}{2}\right) v_i f_{e0}\, d^3v = 0,$$

since the function under the integral is odd with respect to v, and where we again used the convective derivative.

F. The third term on right-hand side is given by

$$\int v_i\,[\boldsymbol{v}\cdot\nabla \ln n_e] f_{e0}\, d^3v = \frac{\partial \ln n_e}{\partial r_k}\int v_i v_k f_{e0}\, d^3v,$$

where we used $[\boldsymbol{v}\cdot\nabla \ln n_e] = v_k \partial \ln n_e/\partial r_k$ and where we used Einstein's summation convention, i.e., summation over index k. For $i \neq k$ this integral is zero. For $i = k$ we have

$$\int v_i\,[\boldsymbol{v}\cdot\nabla \ln n_e] f_{e0}\, d^3v = \frac{1}{n_e}\frac{\partial n_e}{\partial r_i}\int v_i^2 f_{e0}\, d^3v = \frac{1}{2\beta}\frac{\partial n_e}{\partial r_i} = \frac{kT_e}{m_e}\left[\nabla n_e\right]_i,$$

where we used Eq. (3.49). Note that we do *not* sum over i, since we consider the i-th component of the vector $\boldsymbol{v}$. It is rather that the summation over k contributes only if $k = i$, i.e., all terms of the summation over k are zero, except the term that has the same index as the vector component.

G. The fourth term is given by

$$\begin{aligned}
&\int v_i \left(\beta v^2 - \frac{3}{2}\right) [\boldsymbol{v} \cdot \nabla \ln kT_e] f_{e0}\, d^3 v \\
&= \int v_i \left(\beta v^2 - \frac{3}{2}\right) v_k \frac{\partial \ln kT_e}{\partial r_k} f_{e0}\, d^3 v \\
&= \frac{1}{T_e} \frac{\partial T_e}{\partial r_k} \int v_i v_k \left(\beta v^2 - \frac{3}{2}\right) f_{e0}\, d^3 v,
\end{aligned}$$

where $[\boldsymbol{v} \cdot \nabla \ln kT_e] = v_k \partial \ln kT_e / \partial r_k$ and where we used Einstein's summation convention, i.e., summation over k. Note that $\partial \ln kT_e / \partial r_k = 1/T_e \partial T_e / \partial r_k$ is independent of v and can be pulled in front of the integral. With $v^2 = v_x^2 + v_y^2 + v_z^2$ this integral is obviously zero for $i \neq k$, since the function under the integral is odd in v. The integral only contributes for $i = k$. For convenience we consider both terms in brackets separately. The first term gives

$$\begin{aligned}
\frac{\beta}{T_e} \frac{\partial T_e}{\partial r_i} \int v_i^2 v^2 f_{e0}\, d^3 v &= \frac{\beta}{T_e} \frac{\partial T_e}{\partial r_i} \int\int\int v_i^2 \left(v_i^2 + v_j^2 + v_z^2\right) f_{e0}\, dv_i dv_j dv_k \\
&= \frac{\beta}{T_e} \frac{\partial T_e}{\partial r_i} \left[\frac{3}{4} \frac{n_e}{\beta^2} + \frac{n_e}{4\beta^2} + \frac{n_e}{4\beta^2}\right] = \frac{5}{2} \frac{n_e}{m_e} [\nabla kT_e]_i ,
\end{aligned}$$

where we used Eqs. (3.47)–(3.50). The Boltzmann constant k was pulled under the derivative. The second term in the brackets gives

$$-\frac{3}{2} \frac{1}{T_e} \frac{\partial T_e}{\partial r_i} \int v_i^2 f_{e0}\, d^3 v = -\frac{3}{2} \frac{1}{T_e} \frac{\partial T_e}{\partial r_i} \frac{n_e}{2\beta} = -\frac{3}{2} \frac{n_e}{m_e} [\nabla kT_e]_i ,$$

so that we eventually get

$$\int v_i \left(\beta v^2 - \frac{3}{2}\right) [\boldsymbol{v} \cdot \nabla \ln T_e] f_{e0}\, d^3 v = \frac{n_e}{m_e} [\nabla kT_e]_i .$$

H. The fifth term is given by

$$\frac{m_e}{kT_e} \left[\frac{e}{m_e} \boldsymbol{E}' + \frac{D\boldsymbol{u}_e}{Dt}\right]_k \int v_i v_k f_{e0}\, d^3 v = \frac{m_e}{kT_e} \boldsymbol{A}_k \int v_i v_k f_{e0}\, d^3 v,$$

where the vector $\boldsymbol{A}$ corresponds to the expression in the brackets,

$$\boldsymbol{A} = \left[\frac{e}{m_e}\boldsymbol{E}' + \left(\frac{\partial}{\partial t} + \boldsymbol{u}_e \cdot \nabla\right)\boldsymbol{u}_e\right], \tag{3.53}$$

so that the k-th component is given by $\boldsymbol{A}_k$. For $i \neq k$ the integral is zero, since the integral is odd in v. For $i = k$ we find

$$\frac{m_e}{kT_e}\boldsymbol{A}_i \int v_i^2 f_{e0}\, d^3v = \frac{m_e}{kT_e}\boldsymbol{A}_i \frac{n_e}{2\beta} = \boldsymbol{A}_i n_e.$$

I. The sixth term gives

$$\frac{\partial u_{ek}}{\partial r_j}\frac{m_e}{kT_e}\int v_i v_j v_k f_{e0}\, d^3v = 0$$

for all combinations of i, j, k, since the integral is odd in v.

By adding up all results we find

$$\frac{\boldsymbol{R}_e}{m_e} = \frac{kT_e}{m_e}\nabla n_e + \frac{n_e}{m_e}\nabla kT_e + \left[\frac{e}{m_e}\boldsymbol{E}' + \left(\frac{\partial}{\partial t} + \boldsymbol{u}_e \cdot \nabla\right)\boldsymbol{u}_e\right] n_e.$$

The first and second term on the right-hand side can be simplified to $\nabla(n_e kT_e)$ and we find Eq. (3.52).

Problem 3.13 Show that the second moment of Eq. (3.44) yields the energy equation (3.17) without the heat conduction (the term including $\nabla \cdot \boldsymbol{q}$), viscous heating ($\pi : \nabla \boldsymbol{u}$), and energy exchange terms (Q), and hence that

$$\frac{3}{2}\left(\frac{\partial}{\partial t} + \boldsymbol{u}_e \cdot \nabla\right) \ln kT_e + \nabla \cdot \boldsymbol{u}_e = 0.$$

Solution Here we multiply Eq. (3.44) with v^2 and integrate with respect to v. We consider each term separately.

A. The first term on the left-hand side is given by

$$\int v^2 C_{ee}(f_{e1}) d^3v = 0,$$

since the collision operator for particles of the same species has to satisfy this condition. See also Eq. (3.5c) in Sect. 3.1 and compare with the previous Problem 3.11!

B. The second term of the left-hand side is given by

$$\int v^2 C_{ep}(f_{e1})\, d^3v = 0.$$

The scattering between electrons and protons is perfectly elastic, therefore, no transfer of momentum or energy is possible. Thus, the terms Q and $\boldsymbol{R}$ in Eq. (3.8) are zero.

C. The third term on the left-hand side is given by

$$\frac{e}{m_e}\int v^2\,(\boldsymbol{v}\times\boldsymbol{B})\cdot\frac{\partial f_{e1}}{\partial \boldsymbol{v}} d^3v$$
$$= \left[f_{e1}v^2\,(\boldsymbol{v}\times\boldsymbol{B})\right]_{-\infty}^{\infty} - \int f_{e1}\left[2\boldsymbol{v}\cdot(\boldsymbol{v}\times\boldsymbol{B}) + v^2\frac{\partial}{\partial \boldsymbol{v}}\,(\boldsymbol{v}\times\boldsymbol{B})\right] = 0,$$

where we used integration by parts. The term $\left[f_{e1}v^2\,(\boldsymbol{v}\times\boldsymbol{B})\right]$ vanishes since the distribution function goes to zero for $v \to \pm\infty$. The term $\boldsymbol{v}\cdot(\boldsymbol{v}\times\boldsymbol{B})$ is zero since $\boldsymbol{v}\perp(\boldsymbol{v}\times\boldsymbol{B})$. The last term vanishes for the same reasons as in Problem 3.11.

D. The first term on the right-hand side is given by

$$\left(\frac{\partial \ln n_e}{\partial t} + \boldsymbol{u}_e\cdot\nabla\ln n_e\right)\int v^2 f_{e0}\, d^3v = \frac{D\ln n_e}{Dt}\int (v_x^2+v_y^2+v_z^2)\, f_{e0}\, d^3v$$
$$= \frac{D\ln n_e}{Dt} 3\int v_i^2 f_{e0}\, d^3v.$$

With Eq. (3.49) we find

$$\frac{D\ln n_e}{Dt}\int v^2 f_{e0}\, d^3v = \frac{3}{2}\frac{n_e}{\beta}\frac{D\ln n_e}{Dt} = -\frac{3}{2}\frac{n_e}{\beta}\nabla\cdot\boldsymbol{u}_e,$$

where the convective derivative $D\ln n_e/Dt$ was be replaced by $-\nabla\cdot\boldsymbol{u}_e$.

E. The second term on the right-hand side is given by

$$\frac{D\ln T_e}{Dt}\int\left(\beta v^2 - \frac{3}{2}\right) v^2 f_{e0}\, d^3v = \frac{D\ln T_e}{Dt}\int\left(\beta v^4 - \frac{3}{2}v^2\right) f_{e0}\, d^3v.$$

With $v^2 = (v_x^2+v_y^2+v_z^2)$ we find

$$v^4 = v_x^4 + v_y^4 + v_z^4 + 2v_x^2v_y^2 + 2v_x^2v_z^2 + 2v_y^2v_z^2.$$

Substituting this expression into the equation we find

$$\frac{D\ln T_e}{Dt}\int\left(\beta v^2 - \frac{3}{2}\right) v^2 f_{e0}\, d^3v$$

$$= \frac{D\ln T_e}{Dt} \int \left(3\beta v_i^4 + 6\beta v_i^2 v_j^2 - \frac{3}{2}v^2\right) f_{e0}\, d^3v$$

$$= \frac{D\ln T_e}{Dt} \left(\frac{9}{4}\frac{n_e}{\beta} + \frac{6}{4}\frac{n_e}{\beta} - \frac{9}{4}\frac{n_e}{\beta}\right) = \frac{3}{2}\frac{n_e}{\beta}\frac{D\ln T_e}{Dt},$$

where $i \neq j$.

F. For the third, fourth, and fifth term we find

$$\int f_{e0} v^2 \left[[\boldsymbol{v} \cdot \nabla \ln n_e] + \left(\beta v^2 - \frac{3}{2}\right) [\boldsymbol{v} \cdot \nabla \ln T_e] + \frac{m_e}{T_e} \boldsymbol{A} \cdot \boldsymbol{v} \right] d^3v = 0,$$

where we again used the vector $\boldsymbol{A}$ to simplify the expression. All terms vanish since the integrand is an odd function with respect to v.

G. The sixth term on the right-hand side gives

$$\frac{\partial u_{ek}}{\partial r_j} \frac{m_e}{T_e} \int v^2 v_j v_k f_{e0}\, d^3v = 0$$

for $j \neq k$, since the integrand is an odd function in v. For $j = k$ we find

$$\frac{\partial u_{ej}}{\partial r_j} \frac{m_e}{T_e} \int v^2 v_j^2 f_{e0}\, d^3v = \frac{\partial u_{ej}}{\partial r_j} \frac{m_e}{T_e} \int \left(v_j^4 + v_j^2 v_k^2 + v_j^2 v_l^2\right) f_{e0}\, dv_j dv_k dv_l$$

$$= \frac{\partial u_{ej}}{\partial r_j} \frac{m_e}{T_e} \left(\frac{3}{4}\frac{n_e}{\beta^2} + \frac{1}{2}\frac{n_e}{\beta^2}\right) = \frac{\partial u_{ej}}{\partial r_j} \frac{m_e}{T_e} \frac{5}{4}\frac{n_e}{\beta^2},$$

where the integration over $v_j^2 v_k^2$ gives the same result as the integration over $v_j^2 v_l^2$. With $\beta = m_e/2T_e$ we find

$$\frac{\partial u_{ej}}{\partial r_j} \frac{m_e}{T_e} \int v^2 v_j^2 f_{e0}\, d^3v = \frac{5}{2}\frac{n_e}{\beta} \nabla \cdot \boldsymbol{u}_e.$$

By summarizing all results we find

$$-\frac{3}{2}\frac{n_e}{\beta} \nabla \cdot \boldsymbol{u}_e + \frac{3}{2}\frac{n_e}{\beta}\frac{D\ln T_e}{Dt} + \frac{5}{2}\frac{n_e}{\beta} \nabla \cdot \boldsymbol{u}_e = 0,$$

and after simplification we obtain

$$\frac{3}{2}\frac{D\ln T_e}{Dt} + \nabla \cdot \boldsymbol{u}_e = \frac{3}{2}\left(\frac{\partial}{\partial t} + \boldsymbol{u}_e \cdot \nabla\right) \ln T_e + \nabla \cdot \boldsymbol{u}_e = 0.$$

Problem 3.14 Eliminate the time derivatives in (3.42) using the results from the Exercises above to derive Eq. (3.57).

Solution For convenience, let us summarize the results from Problems 3.11–3.13. From the zeroth moment we find

$$\frac{D\ln n_e}{Dt} = \left(\frac{\partial}{\partial t} + \boldsymbol{u}_e \cdot \nabla\right)\ln n_e = -\nabla \cdot \boldsymbol{u}_e,$$

which shows that the convective derivative $D\ln n_e/Dt$ can be replaced by $-\nabla \cdot \boldsymbol{u}_e$. The first moment gives

$$\frac{e}{m_e}\boldsymbol{E}' + \left(\frac{\partial}{\partial t} + \boldsymbol{u}_e \cdot \nabla\right)\boldsymbol{u}_e = \frac{\boldsymbol{R}_e - \nabla(n_e kT_e)}{n_e m_e}$$

and the second moment

$$\frac{D\ln kT_e}{Dt} = \left(\frac{\partial}{\partial t} + \boldsymbol{u}_e \cdot \nabla\right)\ln(kT_e) = -\frac{2}{3}\nabla \cdot \boldsymbol{u}_e.$$

Substituting these results into Eq. (3.44) we can eliminate all time derivatives and obtain

$$C_{ee}(f_{e1}) + C_{ei}(f_{e1}) + \left(\frac{e}{m_e}\boldsymbol{v} \times \boldsymbol{B}\right) \cdot \nabla_v f_{e1} = -\nabla \cdot \boldsymbol{u}_e f_{e0}$$
$$-\left(\beta v^2 - \frac{3}{2}\right)\frac{2}{3}\nabla \cdot \boldsymbol{u}_e f_{e0} + \boldsymbol{v} \cdot \nabla \ln n_e f_{e0} + \left(\beta v^2 - \frac{3}{2}\right)\boldsymbol{v} \cdot \nabla \ln(kT_e) f_{e0}$$
$$+ \frac{m_e \boldsymbol{v}}{kT_e} \cdot \frac{\boldsymbol{R}_e - \nabla(n_e kT_e)}{n_e m_e} f_{e0} + \frac{m_e v_j v_k}{kT_e}\frac{\partial u_{ek}}{\partial r_j} f_{e0} + \nu_{ei}\frac{m_e \boldsymbol{v} \cdot (\boldsymbol{u}_e - \boldsymbol{u}_i)}{kT_e} f_{e0}.$$

The first term on the right hand side cancels with the second part of the second term. To simplify we also expand the third term on the right-hand side ($\boldsymbol{v} \cdot \nabla \ln n_e f_{e0}$) and the expression $\nabla(n_e T_e)$ in the fifth term. We obtain

$$C_{ee}(f_{e1}) + C_{ei}(f_{e1}) + \left(\frac{e}{m_e}\boldsymbol{v} \times \boldsymbol{B}\right) \cdot \nabla_v f_{e1} = -\frac{2}{3}\beta v^2 \nabla \cdot \boldsymbol{u}_e f_{e0}$$
$$+ \frac{\boldsymbol{v}}{n_e} \cdot \nabla n_e f_{e0} + \left(\beta v^2 - \frac{3}{2}\right)\boldsymbol{v} \cdot \nabla \ln(kT_e) f_{e0} - \frac{\boldsymbol{v}}{T_e} \cdot \nabla T_e f_{e0} - \frac{\boldsymbol{v}}{n_e} \cdot \nabla n_e f_{e0}$$
$$+ \frac{\boldsymbol{v}}{kT_e} \cdot \frac{\boldsymbol{R}_e}{n_e} f_{e0} + \frac{m_e v_j v_k}{kT_e}\frac{\partial u_{ek}}{\partial r_j} f_{e0} + \nu_{ei}\frac{m_e \boldsymbol{v} \cdot (\boldsymbol{u}_e - \boldsymbol{u}_i)}{kT_e} f_{e0}. \tag{3.54}$$

The second and the fifth term on the right-hand side cancel each other. By writing $\nabla T_e/T_e = \nabla \ln(kT_e)$ in the fourth term we find that the third and the fourth term can be combined. By rearranging the terms on the right-hand side we find

$$C_{ee}(f_{e1}) + C_{ei}(f_{e1}) + \left(\frac{e}{m_e}\boldsymbol{v}\times\boldsymbol{B}\right)\cdot\nabla_v f_{e1} = \left(\beta v^2 - \frac{5}{2}\right)\boldsymbol{v}\cdot\nabla\ln(kT_e)f_{e0}$$
$$+\boldsymbol{v}\cdot\left[\frac{\boldsymbol{R}_e}{kT_e n_e} + \nu_{ei}\frac{m_e(\boldsymbol{u}_e - \boldsymbol{u}_i)}{kT_e}\right]f_{e0} + \frac{m_e v_j v_k}{kT_e}\frac{\partial u_{ek}}{\partial r_j}f_{e0} - \frac{2}{3}\beta v^2\nabla\cdot\boldsymbol{u}_e f_{e0}.$$

The last two terms can be expressed through the rate-of-strain tensor W^a_{jk} by

$$\frac{m_e v_j v_k}{kT_e}\frac{\partial u_{ek}}{\partial r_j} - \frac{m}{3kT_e}v^2\nabla\cdot\boldsymbol{u}_e = \frac{m}{2kT_e}\left[v_j v_k - \frac{v^2}{3}\delta_{jk}\right]W_{jk}, \tag{3.55}$$

where we neglected the distribution function f_{e0} and substituted $\beta = m_e/2kT_e$. The rate-of-strain tensor W^a_{jk} is given by

$$W_{jk} = \frac{\partial u_j}{\partial r_k} + \frac{\partial u_k}{\partial r_j} - \frac{2}{3}\nabla\cdot\boldsymbol{u}_e\delta_{jk}. \tag{3.56}$$

To show this identity we substitute the rate-of-strain tensor in the expression on the right-hand side and obtain

$$\frac{m}{2kT_e}\left(v_j v_k - \frac{v^2}{3}\delta_{jk}\right)W_{jk} = \frac{m}{2kT_e}\left[v_j v_k\frac{\partial u_j}{\partial r_k} + v_j v_k\frac{\partial u_k}{\partial r_j} - v_j v_k\frac{2}{3}(\nabla\cdot\boldsymbol{u}_e)\,\delta_{jk}\right.$$
$$\left.-\frac{v^2}{3}\delta_{jk}\frac{\partial u_j}{\partial r_k} - \frac{v^2}{3}\delta_{jk}\frac{\partial u_k}{\partial r_j} + \frac{2v^2}{9}\delta_{jk}\nabla\cdot\boldsymbol{u}_e\right].$$

Note that we use Einstein's summation convention and implicitly sum over the indices j and k. Therefore, we can change the indices in the first term (the summation does not change) and combine the first and second term. The third term contributes only for $j = k$, so that $v_j v_k = v_j^2$. Since we sum over that index we obtain $v^2 = \sum v_j^2 = v_x^2 + v_y^2 + v_z^2$. The fourth and the fifth term also contribute only for $j = k$, and with $\sum \partial u_j/\partial r_j = \nabla\cdot\boldsymbol{u}$ we can rewrite both terms through the dot product. The last term also contributes only for $j = k$. Since we have to sum this term also over x, y, z we obtain a factor 3.

$$\frac{m}{2kT_e}\left(v_j v_k - \frac{v^2}{3}\delta_{jk}\right)W_{jk}$$
$$= \frac{m}{2kT_e}\left[2v_j v_k\frac{\partial u_k}{\partial r_j} - v^2\frac{2}{3}\nabla\cdot\boldsymbol{u}_e - \frac{v^2}{3}\nabla\cdot\boldsymbol{u}_e - \frac{v^2}{3}\nabla\cdot\boldsymbol{u}_e + 3\frac{2v^2}{9}\nabla\cdot\boldsymbol{u}_e\right].$$

The last four terms can be combined to

$$\frac{m}{2kT_e}\left(v_jv_k - \frac{v^2}{3}\delta_{jk}\right)W_{jk} = \frac{m}{2kT_e}\left[2v_jv_k\frac{\partial v_k}{\partial r_j} - \frac{2}{3}v^2\nabla\cdot\boldsymbol{u}_e\right].$$

After multiplying the squared brackets with the factor $1/2$ we obtain the exact same expression as in Eq. (3.55), and, therefore,

$$C_{ee}(f_{e1}) + C_{ei}(f_{e1}) + \left(\frac{e}{m_e}\boldsymbol{v}\times\boldsymbol{B}\right)\cdot\nabla_v f_{e1} = \left(\beta v^2 - \frac{5}{2}\right)\boldsymbol{v}\cdot\nabla\ln(kT_e)f_{e0}$$
$$+\boldsymbol{v}\cdot\left[\frac{\boldsymbol{R}_e}{kT_en_e} + \nu_{ei}\frac{m_e(\boldsymbol{u}_e-\boldsymbol{u}_i)}{kT_e}\right]f_{e0} + \frac{m}{2kT_e}\left(v_jv_k - \frac{v^2}{3}\delta_{jk}\right)W_{jk}. \tag{3.57}$$

3.8 Application 1: Transport Perpendicular to a Mean Magnetic Field

One can also consider an alternative approach to the transport of particles, momentum and energy across a mean magnetic field in a plasma. By approximating the collision operator by the BGK operator $C(f) = \nu(f_0 - f)$, the kinetic equation can be expressed in components as

$$\frac{\partial f}{\partial t} + \frac{\partial}{\partial x_i}[(u_i + c_i)f] + \frac{\partial}{\partial v_i}\left[\left(\frac{eE_i}{m} + \epsilon_{ijk}(u_j + c_j)\Omega_k\right)f\right]$$
$$= \nu\,(f_0 - f), \tag{3.58}$$

where $\boldsymbol{c} = \boldsymbol{v} - \boldsymbol{u}$ is the particle velocity relative to the bulk velocity, and $\Omega_k = eB_k/m$ the particle gyrofrequency. Note that $\boldsymbol{c} = \boldsymbol{c}(\boldsymbol{x}, t)$ and $\boldsymbol{u}(\boldsymbol{x}, t)$ are functions of time and space.

By taking the c_lc_k moment of the kinetic equation one can derive an evolution equation (3.63) for the pressure tensor P_{lm}, where the pressure tensor and the conductive heat flux are defined as

$$P_{lm} = m\int c_lc_mf\,d^3v \qquad q_{ilm} = m\int c_ic_lc_mf\,d^3v.$$

For the following calculations we will make frequent use of the relation

$$\int c_if\,d^3v = \int (v_i - u_i)f\,d^3v = \int v_if\,d^3v - u_i\int f\,d^3v$$
$$= nu_i - u_in = 0. \tag{3.59}$$

Problem 3.15 By taking the $c_l c_k$-moment of the kinetic equation, derive the evolution equation (3.63) for the pressure tensor P_{lm}.

Solution We start by multiplying Eq. (3.58) with mass m and $c_l c_m$ and integrate then with respect to v. Considering each term each term separately we find:

A. The first term on the left hand side is given by

$$\begin{aligned}
\int m c_l c_m \frac{\partial f}{\partial t} d^3 v
&= m \int \frac{\partial}{\partial t}(f c_l c_m)\, d^3 v - m \int f \frac{\partial}{\partial t}(c_l c_m)\, d^3 v \\
&= \frac{\partial}{\partial t} m \int (f c_l c_m)\, d^3 v - m \int f c_m \frac{\partial c_l}{\partial t}\, d^3 v - m \int f c_l \frac{\partial c_m}{\partial t}\, d^3 v \\
&= \frac{\partial P_{lm}}{\partial t}.
\end{aligned}$$

In the first step we used the chain rule to transform the function under the integral. Note that (as stated above) the random velocities c_i and the bulk velocity u_i depend on time. In the second step we used $\partial c_s/\partial t = -\partial u_s/\partial t$, since v_i is an independent variable, and does, therefore, not depend on time. $\partial u_s/\partial t$ can then be pulled in front of the integral (because u_s is independent of v_s). We are left with integrals of the form given in Eq. (3.59), which are zero. Only the first term remains.

B. For the second term on the left-hand side let us consider first the following derivative

$$\begin{aligned}
\frac{\partial c_l c_m (u_i + c_i) f}{\partial x_i} &= c_l c_m \frac{\partial (u_i + c_i) f}{\partial x_i} + (u_i + c_i) f \left[c_l \frac{\partial c_m}{\partial x_i} + c_m \frac{\partial c_l}{\partial x_i} \right] \\
&= c_l c_m \frac{\partial (u_i + c_i) f}{\partial x_i} - (u_i + c_i) f \left[c_l \frac{\partial u_m}{\partial x_i} + c_m \frac{\partial u_l}{\partial x_i} \right].
\end{aligned}$$

In the first step we used the chain rule twice. In the second step we used $\boldsymbol{c} = \boldsymbol{v} - \boldsymbol{u}$, where $\boldsymbol{v}$ is an independent variable (does not depend on the spatial coordinate). The second term on the left-hand side of Eq. (3.58) can then be written as

$$\begin{aligned}
\int m c_l c_m \frac{\partial}{\partial x_i} [(u_i + c_i) f]\, d^3 v =& m \int \frac{\partial c_l c_m (u_i + c_i) f}{\partial x_i} d^3 v \\
&+ m \int (u_i + c_i) f \left[c_l \frac{\partial u_m}{\partial x_i} + c_m \frac{\partial u_l}{\partial x_i} \right] d^3 v.
\end{aligned}$$

In the first integral on the right-hand side we can pull out the derivative with respect to x_i, so that

$$m\int \frac{\partial c_l c_m (u_i + c_i) f}{\partial x_i} d^3 v = \frac{\partial}{\partial x_i} m \int c_l c_m (u_i + c_i) f d^3 v$$
$$= \frac{\partial}{\partial x_i} \left[u_i P_{lm} + q_{ilm} \right],$$

where we pulled out u_i in the first term and where we used the definitions for the pressure tensor and the conductive heat flux. Therefore, the second term in the integral above can be rewritten as

$$m \int (u_i + c_i) f \left[c_l \frac{\partial u_m}{\partial x_i} + c_m \frac{\partial u_l}{\partial x_i} \right] d^3 v$$
$$= m u_i \int f \left[c_l \frac{\partial u_m}{\partial x_i} + c_m \frac{\partial u_l}{\partial x_i} \right] d^3 v + m \int c_i f \left[c_l \frac{\partial u_m}{\partial x_i} + c_m \frac{\partial u_l}{\partial x_i} \right] d^3 v$$
$$= \frac{\partial u_m}{\partial x_i} P_{il} + \frac{\partial u_l}{\partial x_i} P_{im}.$$

Note that the first term in the second line is zero, since the derivative $\partial u_s / \partial x_s$ can be pulled in front of the integral and we are left with integrals of the form given in Eq. (3.59). The second term in the second line yields the results for the pressure tensor. By combining both results we obtain

$$\int m c_l c_m \frac{\partial}{\partial x_i} \left[(u_i + c_i) f \right] d^3 c = \frac{\partial}{\partial x_i} \left[u_i P_{lm} + q_{ilm} \right] + \frac{\partial u_m}{\partial x_i} P_{il} + \frac{\partial u_l}{\partial x_i} P_{im}.$$

C. The first part of the third term in Eq. (3.58) can be written as

$$\frac{eE_i}{m} m \int c_l c_m \frac{\partial f}{\partial v_i} d^3 v = \frac{eE_i}{m} m \left[\int \frac{\partial f c_l c_m}{\partial v_i} d^3 v - \int f \frac{\partial c_l c_m}{\partial v_i} d^3 v \right],$$

where we integrated by parts. Note that the first term on the right-hand side vanishes for all possible combinations of l, m, i, since the distribution function vanishes for $v_i = \pm\infty$. In the second integral we rewrite the derivative as

$$\frac{\partial c_l c_m}{\partial v_i} = c_l \delta_{im} + c_m \delta_{il}, \tag{3.60}$$

where we again used the chain rule with $c_i = v_i - u_i$. We obtain then

$$\frac{eE_i}{m} m \int c_l c_m \frac{\partial f}{\partial v_i} d^3 v = -\frac{eE_i}{m} m \int f \left(c_l \delta_{im} + c_m \delta_{il} \right) d^3 v = 0. \tag{3.61}$$

This expression is zero for both cases $l \neq m$ (because the delta function is zero), and $l = m$ (because of Eq. (3.59)).

D. For the second part of the third term in Eq. (3.58) we consider both terms separately. We begin with

$$\epsilon_{ijk} u_j \Omega_k m \int c_l c_m \frac{\partial f}{\partial v_i} d^3 v = 0,$$

which vanishes for the same reasons as we have shown under part C of this Problem. Note that u_j and Ω_k do not depend on the velocity and can therefore be pulled in front of the derivative and the integral.

E. The next part of the third term in Eq. (3.58) is then given by

$$\epsilon_{ijk} \Omega_k m \int c_l c_m \frac{\partial c_j f}{\partial v_i} d^3 v = \epsilon_{ijk} \Omega_k m \int c_l c_m c_j \frac{\partial f}{\partial v_i} d^3 v.$$

Note that the ϵ tensor is zero for $i = j$. Let us consider now the following derivative

$$\frac{\partial f c_j c_l c_m}{\partial v_i} = c_l c_m c_j \frac{\partial f}{\partial v_i} + f \frac{\partial c_j c_l c_m}{\partial v_i} = c_l c_m c_j \frac{\partial f}{\partial v_i} + f c_j \left[c_l \delta_{im} + c_m \delta_{il} \right],$$

where we used the relation (3.60) and again the fact that the ϵ tensor is zero for $i = j$. With that we can rewrite the expression above as

$$\begin{aligned} &\epsilon_{ijk} \Omega_k m \int c_l c_m \frac{\partial c_j f}{\partial v_i} d^3 c \\ &= \epsilon_{ijk} \Omega_k m \left[\int \frac{\partial f c_j c_l c_m}{\partial v_i} d^3 v - \int f c_j \left[c_l \delta_{im} + c_m \delta_{il} \right] d^3 v \right]. \end{aligned}$$

The first term on the right hand side is zero for any combination of l, m, i and j, so that we are left with

$$\epsilon_{ijk} \Omega_k m \int c_l c_m \frac{\partial c_j f}{\partial v_i} d^3 v = -\epsilon_{ijk} \Omega_k m \int f c_j \left[c_l \delta_{im} + c_m \delta_{il} \right] d^3 v.$$

We have to distinguish the following cases:

- $l \neq m$: Here we have to distinguish between $m = i$ and $l = i$.

  A. For $l = i$ we have

$$\begin{aligned} \epsilon_{ijk} \Omega_k m \int c_l c_m \frac{\partial c_j f}{\partial v_i} d^3 v &= -\epsilon_{ljk} \Omega_k m \int f c_j c_m d^3 v \\ &= -\epsilon_{ljk} \Omega_k P_{jm} \end{aligned}$$

and likewise

B. $m = i$

$$\epsilon_{ijk}\Omega_k m \int c_l c_m \frac{\partial c_j f}{\partial v_i} d^3v = -\epsilon_{mjk}\Omega_k m \int f c_j c_l d^3v$$
$$= -\epsilon_{mjk}\Omega_k P_{jl}.$$

Both results have to be added up to obtain

$$\epsilon_{ijk}\Omega_k m \int c_l c_m \frac{\partial c_j f}{\partial v_i} d^3v = -\Omega_k \left[\epsilon_{ljk} P_{jm} + \epsilon_{mjk} P_{jl}\right]. \tag{3.62}$$

- $l = m$: In this case we have

$$\epsilon_{ijk}\Omega_k m \int c_l c_m \frac{\partial c_j f}{\partial v_i} d^3v = -2\epsilon_{ijk}\Omega_k m \int f c_j c_l \delta_{il} d^3v$$
$$= -2\epsilon_{ijk}\Omega_k m \int f c_j c_i d^3v = -2\epsilon_{ijk}\Omega_k P_{ji}.$$

This result can also be derived from Eq. (3.62) by setting $l = m = i$.

We summarize that

$$\epsilon_{ijk}\Omega_k m \int c_l c_m \frac{\partial c_j f}{\partial v_i} d^3v = -\Omega_k \left[\epsilon_{ljk} P_{jm} + \epsilon_{mjk} P_{jl}\right],$$

which is valid for $l = m$ **and** $l \neq m$.

F. The collision term on the right hand side becomes

$$\nu \int c_l c_m (f_0 - f) d^3v = \nu \int c_l c_m f_0 d^3v - \nu \int c_l c_m f d^3v = \nu \left(p\delta_{lm} - P_{lm}\right).$$

Here f_0 denotes the Maxwellian distribution. Therefore, the first integral on the right-hand side vanishes for $l \neq m$, since we have an odd function in c under the integral. The first term contributes then only for $l = m$, where p is the pressure (see [5]). Compare also with Problem 2.8 part E. For the second term we use the definition for the pressure tensor P_{lm} given above.

By combining all results we obtain the evolution equation for the pressure tensor

$$\frac{\partial P_{lm}}{\partial t} + \frac{\partial}{\partial x_i}\left[u_i P_{lm} + q_{ilm}\right] + \frac{\partial u_m}{\partial x_i} P_{il} + \frac{\partial u_l}{\partial x_i} P_{im}$$
$$- \Omega_k \left[\epsilon_{ljk} P_{jm} + \epsilon_{mjk} P_{jl}\right] = \nu \left(p\delta_{lm} - P_{lm}\right). \tag{3.63}$$

Problem 3.16 Complete the steps in deriving the Kaufmann representation for S_{lm} and the K-operator.

Solution Here we begin with the result of the preceding problem, Eq. (3.63). By invoking Braginskii's short-mean-free-path orderings to the pressure tensor evolution equation we find that terms proportional to ν and Ω are larger than all other terms, i.e., they are dominating. That means that, to lowest order, the distribution function becomes Maxwellian, $f = f_0$, so that

$$P_{lm} = p\delta_{lm} \qquad\qquad q_{ilm} = 0.$$

For the next order we substitute these expressions in the small terms of Eq. (3.63), i.e., terms that are *not* proportional to ν and Ω. In this case we obtain

$$\begin{aligned}&\left[\frac{\partial p}{\partial t} + \frac{\partial u_i p}{\partial x_i}\right]\delta_{lm} + \left(\frac{\partial u_m}{\partial x_l} + \frac{\partial u_l}{\partial x_m}\right)p + \nu\left(P_{lm} - p\delta_{lm}\right)\\ &= \Omega_k\left[\epsilon_{ljk}P_{jm} + \epsilon_{mjk}P_{jl}\right],\end{aligned} \tag{3.64}$$

where we pulled the term proportional to Ω to the right-hand side and the term proportional to ν to the left-hand side (note that we changed the order to compensate the sign). We denote the left hand side as S_{lm} and the right hand side as $K(P)\Omega$, so that Eq. (3.64) can be written as

$$K(P) = \frac{S}{\Omega}.$$

Problem 3.17 Show that for a Maxwell-Boltzmann distribution,

$$H_{ij} \equiv \int \frac{m}{2} v^2 v_i v_j f \, d^3v = \frac{5}{2m} p k_B T \delta_{ij}.$$

Solution The Maxwell-Boltzmann distribution is given by (see also Eq. (3.38) in Problem 3.8)

$$f(\boldsymbol{v}) = n\left(\frac{\beta}{\pi}\right)^{3/2} e^{-\beta v^2} \qquad \text{where } \beta = \frac{m}{2k_B T} \tag{3.65}$$

and $v^2 = v_i v_i = \sum v_i^2$. We evaluate the integral for the two cases: $i \neq j$ and $i = j$.

A. The case $i \neq j$: Here the integral is zero since we have odd functions of v under the integral, see also Eq. (3.46),

$$\int \frac{m}{2} v^2 v_i v_j f \, d^3v = 0.$$

B. The case $i = j$: Here the integral takes the form

$$
\begin{aligned}
H_{ii} &= \frac{m}{2}\int v^2 v_i^2 f\, d^3v = \frac{m}{2}\int \left(v_i^2 + v_j^2 + v_k^2\right) v_i^2 f\, d^3v \\
&= \frac{m}{2}\left[\int v_i^4 f\, d^3v + \int v_j^2 v_i^2 f\, d^3v + \int v_k^2 v_i^2 f\, d^3v\right] \\
&= n\frac{m}{2}\left(\frac{\beta}{\pi}\right)^{3/2}\left[\int v_i^4 e^{-\beta v^2}\, d^3v + \int v_j^2 v_i^2 e^{-\beta v^2}\, d^3v + \int v_k^2 v_i^2 e^{-\beta v^2}\, d^3v\right] \\
&= n\frac{m}{2}\left(\frac{\beta}{\pi}\right)^{3/2}\left[\frac{3}{4}\frac{\pi^{3/2}}{\beta^{7/2}} + \frac{1}{4}\frac{\pi^{3/2}}{\beta^{7/2}} + \frac{1}{4}\frac{\pi^{3/2}}{\beta^{7/2}}\right] \\
&= n\frac{m}{2}\frac{5}{4}\frac{1}{\beta^2},
\end{aligned}
$$

where we used the integrals from Eq. (3.47). Substituting β we obtain

$$
H_{ii} = n\frac{m}{2}\frac{5}{4}\frac{4k_B^2T^2}{m^2} = \frac{5}{2m}nk_B^2T^2 = \frac{5}{2m}pk_BT,
$$

where we used $p = nk_BT$. It follows for the general expression that

$$
H_{ij} = \frac{5}{2m}pk_BT\delta_{ij}.
$$

3.9 Application 2: The Equations of Magnetohydrodynamics

The *ideal MHD equations* are given by

$$
\frac{\partial \rho}{\partial t} + \nabla\cdot(\rho \boldsymbol{u}) = 0 \qquad \text{(MHD-1)}
$$

$$
\rho\left(\frac{\partial \boldsymbol{u}}{\partial t} + \boldsymbol{u}\cdot\nabla\boldsymbol{u}\right) = -\nabla P + \boldsymbol{J}\times\boldsymbol{B} \qquad \text{(MHD-2)}
$$

$$
\frac{\partial \boldsymbol{B}}{\partial t} = \nabla\times(\boldsymbol{u}\times\boldsymbol{B}) \qquad \text{(MHD-3)}
$$

$$
\nabla\times\boldsymbol{B} = \mu_0\boldsymbol{J} \qquad \text{(MHD-4)}
$$

(continued)

$$\nabla \cdot \boldsymbol{B} = 0 \tag{MHD-5}$$

$$\frac{\partial P}{\partial t} + \boldsymbol{u} \cdot \nabla P + \gamma P \nabla \cdot \boldsymbol{u} = 0, \tag{MHD-6}$$

where ρ is the mass density, $\boldsymbol{u}$ the center-of-flow velocity, $\boldsymbol{J}$ the current density, P the total pressure, $\boldsymbol{B}$ the magnetic field, and γ the adiabatic index.

Problem 3.18 Linearize the ideal MHD equations with $\rho = \rho_0 = const.$, $\boldsymbol{B} = B_0\hat{\mathbf{z}} + \delta B\hat{\mathbf{y}}$, $\boldsymbol{u} = \delta u\hat{\mathbf{y}}$, $\boldsymbol{J} = \delta J\hat{\mathbf{x}}$. Seek solutions of the linearized 1D MHD equations in the form $\exp[i(\omega t - kz)]$, where ω is the wave frequency and k the corresponding wave number and derive the Alfvén wave dispersion relation.

Solution We have

$$\rho = \rho_0 \qquad \boldsymbol{B} = \begin{pmatrix} 0 \\ \delta B \\ B_0 \end{pmatrix} \qquad \boldsymbol{u} = \begin{pmatrix} 0 \\ \delta u \\ 0 \end{pmatrix} \qquad \boldsymbol{J} = \begin{pmatrix} \delta J \\ 0 \\ 0 \end{pmatrix}.$$

A. Since the density is constant we find for the first equation (MHD-1),

$$\rho_0 \frac{\partial \delta u}{\partial y} = 0, \tag{3.66}$$

where we already evaluated the dot product. The time derivative vanishes because the density $\rho = \rho_0$ is constant.

B. The second equation (MHD-2) becomes

$$\rho_0 \frac{\partial}{\partial t} \begin{pmatrix} 0 \\ \delta u \\ 0 \end{pmatrix} = \begin{pmatrix} \delta J \\ 0 \\ 0 \end{pmatrix} \times \begin{pmatrix} 0 \\ \delta B \\ B_0 \end{pmatrix} = \begin{pmatrix} 0 \\ -B_0 \delta J \\ \delta J \delta B \end{pmatrix},$$

where we neglected the term $\boldsymbol{u} \cdot \nabla \boldsymbol{u} = \delta u \partial_y \delta u = \mathcal{O}(\delta u^2)$, since it is of second order in the velocity turbulence. Note also that due to the constant density there is no pressure gradient, i.e., $\nabla P = 0$. We also find $\delta J \delta B = 0$ (last component of the vector). From the second component we obtain

$$\rho_0 \frac{\partial \delta u}{\partial t} = -B_0 \delta J. \tag{3.67}$$

C. The third equation (MHD-3) becomes

$$\frac{\partial}{\partial t}\begin{pmatrix}0\\ \delta B\\ B_0\end{pmatrix} = \nabla\times\left[\begin{pmatrix}0\\ \delta u\\ 0\end{pmatrix}\times\begin{pmatrix}0\\ \delta B\\ B_0\end{pmatrix}\right] = \begin{pmatrix}0\\ B_0\frac{\partial\delta u}{\partial z}\\ -B_0\frac{\partial\delta u}{\partial y}\end{pmatrix}.$$

The time derivation of B_0 is zero and from Eq. (3.66) we know that $\partial\delta u/\partial y = 0$. From the second component we obtain the single equation

$$\frac{\partial\delta B}{\partial t} = B_0\frac{\partial\delta u}{\partial z}. \tag{3.68}$$

D. The fourth equation (MHD-4) becomes

$$\nabla\times\begin{pmatrix}0\\ \delta B\\ B_0\end{pmatrix} = \begin{pmatrix}\frac{\partial B_0}{\partial y}-\frac{\partial\delta B}{\partial z}\\ -\frac{\partial B_0}{\partial x}\\ \frac{\partial\delta B}{\partial x}\end{pmatrix} = \mu_0\begin{pmatrix}\delta J\\ 0\\ 0\end{pmatrix}.$$

From the first component we find

$$-\frac{\partial\delta B}{\partial z} = \mu_0\delta J, \tag{3.69}$$

where $\partial B_0/\partial y = 0$, since $B_0 = const.$

E. The fifth equation (MHD-5) becomes

$$\nabla\cdot\boldsymbol{B} = \nabla\cdot\begin{pmatrix}0\\ \delta B\\ B_0\end{pmatrix} = 0 \qquad\Rightarrow\qquad \frac{\partial\delta B}{\partial y} = 0.$$

F. The sixth equation (MHD-6) becomes then

$$\frac{\partial P}{\partial t} + \gamma P\nabla\cdot\begin{pmatrix}0\\ \delta u\\ 0\end{pmatrix} = 0 \qquad\Rightarrow\qquad \frac{\partial P}{\partial t} + \gamma P\frac{\partial\delta u}{\partial y} = 0.$$

From Eq. (3.66) we know that $\partial\delta u/\partial y = 0$, so that $\partial P/\partial t = 0$.

The set of linearized 1D ideal MHD equations can then be summarized as

$$\rho_0\frac{\partial\delta u}{\partial t} = -B_0\delta J \qquad \frac{\partial\delta B}{\partial t} = B_0\frac{\partial\delta u}{\partial z} \qquad -\frac{\partial\delta B}{\partial z} = \mu_0\delta J.$$

By combining Eqs. (3.67) and (3.69) we obtain with Eq. (3.68) the following set of equations

$$\rho_0 \frac{\partial \delta u}{\partial t} = \frac{B_0}{\mu_0} \frac{\partial \delta B}{\partial z} \qquad \text{and} \qquad \frac{\partial \delta B}{\partial t} = B_0 \frac{\partial \delta u}{\partial z}. \tag{3.70}$$

We are now seeking solutions of the form

$$\delta \Psi = \delta \bar{\Psi} e^{i(\omega t - kz)},$$

where Ψ is either δu or δB. In general we obtain for the derivations

$$\frac{\partial \delta \Psi}{\partial t} = i\omega \delta \Psi \qquad \text{and} \qquad \frac{\partial \delta \Psi}{\partial z} = -ik\delta \Psi,$$

so that the set of Eqs. (3.70) becomes

$$\rho_0 i\omega \delta u = -ik\frac{B_0}{\mu_0}\delta B \qquad \text{and} \qquad i\omega \delta B = -ikB_0 \delta u. \tag{3.71}$$

Combining both equations leads to

$$\rho_0 \omega \delta u = k^2 \frac{B_0^2}{\mu_0 \omega} \delta u \quad \Rightarrow \quad \frac{\omega^2}{k^2} = \frac{B_0^2}{\mu_0 \rho_0} \quad \Rightarrow \quad v_A = \frac{\omega}{k} = \pm \frac{B_0}{\sqrt{\mu_0 \rho_0}},$$

which is the dispersion relation for the Alfvén velocity.

Problem 3.19 For $\boldsymbol{B} = B\hat{z}$, linearize the ideal 1D (say for the x coordinate) MHD equations about a stationary constant state. Seek solutions of the linearized 1D MHD equations in the form $\exp[i(\omega t - kx)]$, where ω is the wave frequency and k the corresponding wave number and derive the magnetosonic wave dispersion relation.

Solution To linearize about a constant state we define

$$\rho = \rho_0 + \delta\rho \qquad \boldsymbol{B} = \boldsymbol{B}_0 + \delta\boldsymbol{B} \qquad \boldsymbol{u} = \delta\boldsymbol{u} \qquad \boldsymbol{J} = \delta\boldsymbol{J} \qquad P = P_0 + \delta P,$$

where the index 0 describes the time invariant and homogeneous background plasma at rest, and $\delta\rho$, $\delta\boldsymbol{B}$, and $\delta\boldsymbol{u}$ are small perturbations to the background plasma. Note that $\boldsymbol{u} = \delta\boldsymbol{u}$ since we linearize about a stationary constant state, hence we have no background flow. Also, since $\boldsymbol{J} = e(n_i\boldsymbol{u}_i + n_e\boldsymbol{u}_e)/\rho_q$ we find $\boldsymbol{J} = \delta\boldsymbol{J}$. The first MHD equation (MHD-1) becomes

$$\frac{\partial \delta\rho}{\partial t} + \rho_0 \nabla \cdot \delta\boldsymbol{u} = 0,$$

where we neglected the term $\nabla \cdot \delta\rho\delta\boldsymbol{u}$. The second ideal MHD equation (MHD-2) becomes then

$$\rho_0 \frac{\partial \delta \boldsymbol{u}}{\partial t} + \nabla \delta P = \delta \boldsymbol{J} \times \boldsymbol{B}_0, \tag{3.72}$$

where we again neglected the term $\boldsymbol{u} \cdot \nabla \boldsymbol{u} = \delta u \partial_y \delta u = \mathcal{O}(\delta u^2)$. The third and fourth MHD equations (MHD-3) and (MHD-4) cane written as

$$\begin{aligned} \frac{\partial \delta \boldsymbol{B}}{\partial t} &= \nabla \times (\delta \boldsymbol{u} \times \boldsymbol{B}_0) \\ \nabla \times \delta \boldsymbol{B} &= \mu_0 \delta \boldsymbol{J}. \end{aligned} \tag{3.73}$$

For the divergence free magnetic field and the pressure we find

$$\begin{aligned} \nabla \cdot \delta \boldsymbol{B} &= 0 \\ \frac{\partial \delta P}{\partial t} + \gamma P_0 \nabla \cdot \delta \boldsymbol{u} &= 0. \end{aligned}$$

By combining Eqs. (3.72) and (3.73) we obtain the following set of equations

$$\begin{aligned} \frac{\partial \delta \rho}{\partial t} + \rho_0 \nabla \cdot \delta \boldsymbol{u} &= 0 \\ \rho_0 \frac{\partial \delta \boldsymbol{u}}{\partial t} + \nabla \delta P + \frac{1}{\mu_0} \boldsymbol{B}_0 \times (\nabla \times \delta \boldsymbol{B}) &= 0 \\ \frac{\partial \delta \boldsymbol{B}}{\partial t} - \nabla \times (\delta \boldsymbol{u} \times \boldsymbol{B}_0) &= 0 \\ \frac{\partial \delta P}{\partial t} + \gamma P_0 \nabla \cdot \delta \boldsymbol{u} &= 0. \end{aligned}$$

For the magnetic field and the variations in velocity and pressure, we have

$$\boldsymbol{B} = \begin{pmatrix} 0 \\ 0 \\ B_0 + \delta B_z(x,t) \end{pmatrix} \qquad \delta \boldsymbol{u} = \begin{pmatrix} \delta u_x(x,t) \\ 0 \\ 0 \end{pmatrix} \qquad \delta P = \delta P(x,t).$$

Note that all variations depend on the x-coordinate solely! We obtain the 1D set of equations

$$\begin{aligned} \frac{\partial \delta \rho}{\partial t} + \rho_0 \frac{\partial \delta u_x}{\partial x} &= 0 \\ \rho_0 \frac{\partial \delta u_x}{\partial t} + \frac{\partial \delta P}{\partial x} + \frac{B_0}{\mu_0} \frac{\partial \delta B_z}{\partial x} &= 0 \end{aligned}$$

$$\frac{\partial \delta B_z}{\partial t} + B_0 \frac{\partial \delta u_x}{\partial x} = 0$$

$$\frac{\partial \delta P}{\partial t} + \gamma P_0 \frac{\partial \delta u_x}{\partial x} = 0.$$

We are seeking solutions of the form

$$\delta\Psi = \delta\bar{\Psi} e^{i(\omega t - kx)},$$

where Ψ is either δu, δB or δP. In general we obtain for the derivations

$$\frac{\partial \delta\Psi}{\partial t} = i\omega\delta\Psi \qquad \text{and} \qquad \frac{\partial \delta\Psi}{\partial x} = -ik\delta\Psi,$$

so that the set of equations can be expressed by

$$\omega\delta\rho - k\rho_0\delta u_x = 0 \tag{3.74a}$$

$$\omega\rho_0\delta u_x - k\delta P - k\frac{B_0}{\mu_0}\delta B_z = 0 \tag{3.74b}$$

$$\omega\delta B_z - kB_0\delta u_x = 0, \tag{3.74c}$$

$$\omega\delta P - k\gamma P_0\delta u_x = 0, \tag{3.74d}$$

where we divided by the complex number i.

- **Alternative 1:** By substituting Eqs. (3.74d) and (3.74c) in Eq. (3.74b), we find

$$\omega^2\delta u_x - k^2\frac{\gamma P_0}{\rho_0}\delta u_x - k^2\frac{B_0^2}{\rho_0\mu_0}\delta u_x = 0,$$

  where we also divided by ρ_0. With $v_A = \sqrt{B_0^2/\rho_0\mu_0}$ and $v_s = \sqrt{\gamma P_0/\rho_0}$, it follows that

$$\frac{\omega^2}{k^2} - v_s^2 - v_A^2 = 0 \qquad \Rightarrow \qquad v_{ms}^2 = \frac{\omega^2}{k^2} = v_s^2 + v_A^2, \tag{3.75}$$

  where the velocity of the magnetosonic wave is the geometric mean of the sound speed of the particles and the Alfvén speed, $v_{ms} = \sqrt{v_s^2 + v_A^2}$.
- **Alternative 2** We may rewrite the set of equations in matrix form as:

$$\begin{pmatrix} \omega & -k\rho_0 & 0 & 0 \\ 0 & \omega\rho_0 & -k & -k\frac{B_0}{\mu_0} \\ 0 & -kB_0 & 0 & \omega \\ 0 & -k\gamma P_0 & \omega & 0 \end{pmatrix} \cdot \begin{pmatrix} \delta\rho \\ \delta u_x \\ \delta P \\ \delta B_z \end{pmatrix} = 0$$

Now, the determinant of the matrix has to be zero. Developing the determinant for the first column leads to

$$\det M = \omega \left[-\omega\rho_0(\omega^2) + k(\omega k\gamma P_0) + k\frac{B_0}{\mu_0}(\omega k B_0) \right] = 0.$$

Simplifying and dividing by ρ_0 yields

$$\det M = \omega^4 - \omega^2 k^2 \frac{\gamma P_0}{\rho_0} - \omega^2 k^2 \frac{B_0^2}{\rho_0 \mu_0} = 0.$$

With $v_A = \sqrt{B_0^2/\rho_0\mu_0}$ and $v_s = \sqrt{\gamma P_0/\rho_0}$ we obtain

$$\omega^4 = \omega^2 k^2 (v_s^2 + v_A^2) \qquad \Rightarrow \qquad v_{ms}^2 = \frac{\omega^2}{k^2} = v_s^2 + v_A^2,$$

where the velocity of the magnetosonic wave is the geometric mean of the sound speed of the particles and the Alfvén speed, $v_{ms} = \sqrt{v_s^2 + v_A^2}$.

Problem 3.20 Derive the frozen-in field equation,

$$\frac{d}{dt}\left(\frac{\boldsymbol{B}}{\rho}\right) = \frac{1}{\rho}(\boldsymbol{B}\cdot\nabla\boldsymbol{u} - \boldsymbol{B}\nabla\cdot\boldsymbol{u}) + \frac{\boldsymbol{B}}{\rho}\nabla\cdot\boldsymbol{u} = \frac{\boldsymbol{B}}{\rho}\cdot\nabla\boldsymbol{u}$$

by using the Lagrangian form for the magnetic flux, or induction equation,

$$\frac{d\boldsymbol{B}}{dt} = \frac{\partial \boldsymbol{B}}{\partial t} + \boldsymbol{u}\cdot\nabla\boldsymbol{B} = \boldsymbol{B}\cdot\nabla\boldsymbol{u} - \boldsymbol{B}\nabla\cdot\boldsymbol{u} \tag{3.76}$$

and the conservation of mass

$$\frac{\partial\rho}{\partial t} + \nabla\cdot(\rho\boldsymbol{u}) = \frac{\partial\rho}{\partial t} + \rho\nabla\cdot\boldsymbol{u} + \boldsymbol{u}\cdot\nabla\rho = 0 \quad \leftrightarrow \quad \frac{d\rho}{dt} = -\rho\nabla\cdot\boldsymbol{u}, \tag{3.77}$$

where we used the convective derivative (see Eq. (3.43) in Chap. 3.7)

$$\frac{d}{dt} = \frac{\partial}{\partial t} + \boldsymbol{u}\cdot\nabla, \tag{3.78}$$

which represents the time derivative seen in the local rest frame of the fluid (see also Eq. (2.48) of Problem 2.10).

Solution We calculate the derivative

$$\frac{d}{dt}\left(\frac{\boldsymbol{B}}{\rho}\right) = \frac{1}{\rho}\frac{d\boldsymbol{B}}{dt} - \frac{\boldsymbol{B}}{\rho^2}\frac{d\rho}{dt} = \frac{1}{\rho}(\boldsymbol{B}\cdot\nabla\boldsymbol{u} - \boldsymbol{B}\nabla\cdot\boldsymbol{u}) + \frac{\boldsymbol{B}}{\rho}\nabla\cdot\boldsymbol{u} = \frac{\boldsymbol{B}}{\rho}\cdot\nabla\boldsymbol{u},$$

where we substituted (3.76) and (3.77) for the convective derivatives with respect to $\boldsymbol{B}$ and ρ, respectively.

Problem 3.21 Show, using the kinematic equation of motion for a volume element, that mass in a fluid element $dM \equiv \rho\, dV$ is conserved.

Solution If the mass in a fluid element is constant, then the total time derivative of dM has to be zero. The time derivative of the mass in a fluid element is given by

$$\frac{d}{dt}(dM) = \frac{d}{dt}(\rho\, dV) = \frac{d\rho}{dt}dV + \rho\frac{d}{dt}(dV).$$

The kinematic motion of a fluid element dV is given by (see [5])

$$\frac{d}{dt}(dV) = \nabla\cdot\boldsymbol{u}\, dV, \tag{3.79}$$

which results from the conservation of mass. Together with expression (3.77), $d\rho/dt = -\rho\nabla\cdot\boldsymbol{u}$, we find

$$\frac{d}{dt}(dM) = -\rho\nabla\cdot\boldsymbol{u}\, dV + \rho\nabla\cdot\boldsymbol{u}\, dV = 0.$$

Therefore, the mass in a fluid element is conserved.

Problem 3.22 Show that the momentum of a fluid element $(\rho\boldsymbol{u})dV$ is *not* constant.

Solution The time derivative of the momentum of a fluid element is given by

$$\begin{aligned}\frac{d}{dt}(dp) = \frac{d}{dt}\left[(\rho\boldsymbol{u})dV\right] &= \frac{d(\rho\boldsymbol{u})}{dt}dV + \rho\boldsymbol{u}\frac{d}{dt}(dV)\\ &= \boldsymbol{u}\frac{d\rho}{dt}dV + \rho\frac{d\boldsymbol{u}}{dt}dV + \rho\boldsymbol{u}\nabla\cdot\boldsymbol{u}\, dV,\\ &= -\boldsymbol{u}\rho\nabla\cdot\boldsymbol{u}dV + \rho\frac{d\boldsymbol{u}}{dt}dV + \rho\boldsymbol{u}\nabla\cdot\boldsymbol{u}\, dV,\end{aligned}$$

where we substituted Eq. (3.79) for the derivative of the volume element in the second step and the time derivative by the convective derivative, $d\rho/dt = -\rho\nabla\cdot\boldsymbol{u}$, in the third step (see also expression (3.77)). Obviously, the first and last term cancel each other and we obtain

$$\frac{d}{dt}(dp) = \rho\frac{d\boldsymbol{u}}{dt}dV.$$

Since $d\boldsymbol{u}/dt \neq 0$, see for example Eq. (MHD-2), we find that the momentum of a fluid element is *not* conserved.

Problem 3.23 Show that the total energy density of a fluid element $\varepsilon = \rho u^2/2 + P/(\gamma - 1) + B^2/2\mu_0$ is *not* constant.

Solution Consider the time derivative of the total energy density

$$\frac{d}{dt}(\varepsilon\, dV) = \frac{d}{dt}\left[\frac{1}{2}\rho u^2 dV + \frac{P}{\gamma - 1}dV + \frac{B^2}{2\mu_0}dV\right]. \tag{3.80}$$

A. The first term is then

$$\begin{aligned}\frac{d}{dt}\left[\frac{1}{2}\rho u^2 dV\right] &= \boldsymbol{u}\cdot\frac{d\boldsymbol{u}}{dt}\rho\, dV + \frac{u^2}{2}dV\frac{d\rho}{dt} + \frac{u^2}{2}\rho\frac{d}{dt}(dV)\\ &= \boldsymbol{u}\cdot\frac{d\boldsymbol{u}}{dt}\rho\, dV + \frac{u^2}{2}[-\rho dV\nabla\cdot\boldsymbol{u} + \rho dV\nabla\cdot\boldsymbol{u}]\\ &= \boldsymbol{u}\cdot\frac{d\boldsymbol{u}}{dt}\rho\, dV \neq 0,\end{aligned} \tag{3.81}$$

where we used the conservation of mass, Eq. (3.77), to substitute $d\rho/dt$ and Eq. (3.79) to substitute the derivative of the volume element. Obviously, the second and third term cancel; the first term remains, since $d\boldsymbol{u}/dt \neq 0$.

B. The second term is

$$\begin{aligned}\frac{d}{dt}\left[\frac{P}{\gamma - 1}dV\right] &= \frac{1}{\gamma - 1}\frac{dP}{dt}dV + \frac{1}{\gamma - 1}P\frac{d}{dt}(dV)\\ &= -\frac{\gamma}{\gamma - 1}P\nabla\cdot\boldsymbol{u}dV + \frac{1}{\gamma - 1}PdV\nabla\cdot\boldsymbol{u}\\ &= -P\nabla\cdot\boldsymbol{u}dV\end{aligned} \tag{3.82}$$

where we used $dP/dt = -\gamma P\nabla\cdot\boldsymbol{u}$, compare with Eq. (MHD-6), and Eq. (3.79) for the derivative of the fluid element.

C. The last term is given by

$$\begin{aligned}\frac{d}{dt}\left[\frac{B^2}{2\mu_0}dV\right] &= \frac{\boldsymbol{B}}{\mu_0}\cdot\frac{d\boldsymbol{B}}{dt}dV + \frac{B^2}{2\mu_0}\frac{d}{dt}(dV)\\ &= \frac{1}{\mu_0}\boldsymbol{B}\cdot[\boldsymbol{B}\cdot\nabla\boldsymbol{u} - \boldsymbol{B}\nabla\cdot\boldsymbol{u}]\, dV + \frac{B^2}{2\mu_0}\nabla\cdot\boldsymbol{u}\, dV\\ &= \frac{1}{\mu_0}\boldsymbol{B}\cdot(\boldsymbol{B}\cdot\nabla\boldsymbol{u}) - \frac{B^2}{2\mu_0}\nabla\cdot\boldsymbol{u}\, dV, \neq 0,\end{aligned} \tag{3.83}$$

where we used the induction equation (3.76) for $d\boldsymbol{B}/dt$ and the derivative of the fluid element (3.79).

Combining all results we find that the total energy density is not constant,

$$\frac{d}{dt}\left(\varepsilon\, dV\right) \neq 0. \tag{3.84}$$

3.10 Application 3: MHD Shock Waves

Problem 3.24 Starting from the equation for the shock adiabatic,

$$0 = \left(M_{A1}^2 - r\right)^2 \left[M_{A1}^2 - 2r\frac{a_{s1}^2}{V_{An1}^2}\frac{1}{r+1-\gamma\,(r-1)}\right] - rM_{A1}^2\left[\frac{2r-\gamma\,(r-1)}{r+1-\gamma\,(r-1)}M_{A1}^2 - r\right]\tan^2\theta_1 \tag{3.85}$$

derive the shock polar relation or shock cubic (3.87), by using $r \equiv M_{A1}^2/M_{A2}^2$, and $\sec^2\theta_1 = 1 + \tan^2\theta_1$.

Solution As a starting point we substitute $r = M_{A1}^2/M_{A2}^2$ in Eq. (3.85). In doing so, we consider first the denominator of the fractions occurring in the equation,

$$\begin{aligned} r+1-\gamma\,(r-1) &= \frac{M_{A1}^2}{M_{A2}^2} + 1 - \gamma\left(\frac{M_{A1}^2}{M_{A2}^2} - 1\right) \\ &= \frac{1}{M_{A2}^2}\left[M_{A1}^2(1-\gamma) + M_{A2}^2(1+\gamma)\right], \end{aligned}$$

Secondly, we consider the numerator of the fractions,

$$2r-\gamma\,(r-1) = 2\frac{M_{A1}^2}{M_{A2}^2} - \gamma\left(\frac{M_{A1}^2}{M_{A2}^2} - 1\right) = \frac{1}{M_{A2}^2}\left[M_{A1}^2(2-\gamma) + M_{A2}^2\gamma\right],$$

Also,

$$\left(M_{A1}^2 - r\right)^2 = \frac{M_{A1}^4}{M_{A2}^4}\left(M_{A2}^2 - 1\right)^2.$$

With these relations we obtain for Eq. (3.85) the following expression,

$$\begin{aligned} 0 = {} & \frac{M_{A1}^4}{M_{A2}^4}\left(M_{A2}^2 - 1\right)^2\left[M_{A1}^2 - 2\frac{a_{s1}^2}{V_{An1}^2}\frac{M_{A1}^2}{M_{A2}^2}\frac{M_{A2}^2}{M_{A1}^2(1-\gamma) + M_{A2}^2(1+\gamma)}\right] \\ & - \frac{M_{A1}^2}{M_{A2}^2}M_{A1}^2\left[\frac{M_{A1}^2(2-\gamma) + M_{A2}^2\gamma}{M_{A1}^2(1-\gamma) + M_{A2}^2(1+\gamma)}M_{A1}^2 - \frac{M_{A1}^2}{M_{A2}^2}\right]\tan^2\theta_1. \end{aligned}$$

In the first squared brackets we cancel M_{A2}^2 in the second term. Assuming that $M_{A1}^2 \neq 0$ and $M_{A2}^2 \neq 0$, we multiply the equation by M_{A2}^4 and divide the entire equation by M_{A1}^4, obtaining

$$0 = \left(M_{A2}^2 - 1\right)^2 \left[M_{A1}^2 - 2\frac{a_{s1}^2}{V_{An1}^2}\frac{M_{A1}^2}{M_{A1}^2(1-\gamma) + M_{A2}^2(1+\gamma)}\right]$$
$$- M_{A2}^2\left[\frac{M_{A1}^2(2-\gamma) + M_{A2}^2\gamma}{M_{A1}^2(1-\gamma) + M_{A2}^2(1+\gamma)}M_{A1}^2 - \frac{M_{A1}^2}{M_{A2}^2}\right]\tan^2\theta_1.$$

Note that each term (in squared brackets) contains M_{A1}^2. Under the assumption that $M_{A1}^2 \neq 0$, we divide this expression by M_{A1}^2. We also pull M_{A2}^2 into the second term with squared brackets,

$$0 = \left(M_{A2}^2 - 1\right)^2 \left[1 - 2\frac{a_{s1}^2}{V_{An1}^2}\frac{1}{M_{A1}^2(1-\gamma) + M_{A2}^2(1+\gamma)}\right]$$
$$- \left[\frac{M_{A1}^2(2-\gamma) + M_{A2}^2\gamma}{M_{A1}^2(1-\gamma) + M_{A2}^2(1+\gamma)}M_{A2}^2 - 1\right]\tan^2\theta_1.$$

Now, multiply the entire equation with the denominator $M_{A1}^2(1-\gamma) + M_{A2}^2(1+\gamma)$ to obtain

$$0 = \left(M_{A2}^2 - 1\right)^2 \left\{\left[M_{A1}^2(1-\gamma) + M_{A2}^2(1+\gamma)\right] - 2\frac{a_{s1}^2}{V_{An1}^2}\right\}$$
$$- \left\{\left[M_{A1}^2(2-\gamma) + M_{A2}^2\gamma\right]M_{A2}^2 - \left[M_{A1}^2(1-\gamma) + M_{A2}^2(1+\gamma)\right]\right\}\tan^2\theta_1.$$

By pulling $\left(M_{A2}^2 - 1\right)^2$ into the first curly brackets and $\tan^2\theta_1$ into the second curly brackets, we obtain

$$0 = \left(M_{A2}^2 - 1\right)^2\left[M_{A1}^2(1-\gamma) + M_{A2}^2(1+\gamma)\right] - 2\frac{a_{s1}^2}{V_{An1}^2}\left(M_{A2}^2 - 1\right)^2$$
$$- \left[M_{A1}^2(2-\gamma) + M_{A2}^2\gamma\right]M_{A2}^2\tan^2\theta_1$$
$$+ \left[M_{A1}^2(1-\gamma) + M_{A2}^2(1+\gamma)\right]\tan^2\theta_1.$$

Now, we need to pull all terms that contain M_{A1}^2 to the left-hand side. For simplicity, we first expand all squared brackets and obtain

$$0 = M_{A1}^2(1-\gamma)\left(M_{A2}^2 - 1\right)^2 + M_{A2}^2(1+\gamma)\left(M_{A2}^2 - 1\right)^2 - \frac{2a_{s1}^2}{V_{An1}^2}\left(M_{A2}^2 - 1\right)^2$$
$$- M_{A1}^2(2-\gamma)M_{A2}^2\tan^2\theta_1 - M_{A2}^4\gamma\tan^2\theta_1$$
$$+ M_{A1}^2(1-\gamma)\tan^2\theta_1 + M_{A2}^2(1+\gamma)\tan^2\theta_1,$$

and then pull the first term of each line to the left-hand side (since they all include M_{A1}^2); we obtain

$$\begin{aligned}
&-M_{A1}^2(1-\gamma)\left(M_{A2}^2-1\right)^2+M_{A1}^2(2-\gamma)M_{A2}^2\tan^2\theta_1-M_{A1}^2(1-\gamma)\tan^2\theta_1\\
&=M_{A2}^2(1+\gamma)\left(M_{A2}^2-1\right)^2-2\frac{a_{s1}^2}{V_{An1}^2}\left(M_{A2}^2-1\right)^2\\
&-M_{A2}^4\gamma\tan^2\theta_1+M_{A2}^2(1+\gamma)\tan^2\theta_1 .
\end{aligned}$$

By simplifying this expression we obtain the shock cubic or shock polar relation,

$$\begin{aligned}
&M_{A1}^2\left\{(\gamma-1)\left(M_{A2}^2-1\right)^2+\left[(2-\gamma)M_{A2}^2-(1-\gamma)\right]\tan^2\theta_1\right\} \qquad (3.86)\\
&=\left[M_{A2}^2(1+\gamma)-2\frac{a_{s1}^2}{V_{An1}^2}\right]\left(M_{A2}^2-1\right)^2-M_{A2}^2\left[M_{A2}^2\gamma-(1+\gamma)\right]\tan^2\theta_1 .
\end{aligned}$$

Note that on the left-hand side we changed $(1-\gamma)\rightarrow-(\gamma-1)$. Since V_{An1} is the Alfvén velocity normal to the shock front we find

$$\frac{1}{V_{An1}^2}=\frac{1}{V_{A1}^2\cos^2\theta}=\frac{1}{V_{A1}^2}\sec^2\theta,$$

where we used $\sec\theta=1/\cos\theta$. We find the shock polar relation

$$\begin{aligned}
&M_{A1}^2\left\{(\gamma-1)\left(M_{A2}^2-1\right)^2+\left[(2-\gamma)M_{A2}^2-(1-\gamma)\right]\tan^2\theta_1\right\} \qquad (3.87)\\
&=\left[M_{A2}^2(1+\gamma)-2\frac{a_{s1}^2}{V_{A1}^2}\sec^2\theta\right]\left(M_{A2}^2-1\right)^2-M_{A2}^2\left[M_{A2}^2\gamma-(1+\gamma)\right]\tan^2\theta_1 .
\end{aligned}$$

Problem 3.25 Starting from the shock polar relation (3.86), derive the alternative form

$$\begin{aligned}
&M_{A1}^2-M_{A2}^2\\
&=\frac{2(M_{A2}^2-1)(M_{A2}^2-M_{1+}^2)(M_{A2}^2-M_{1-}^2)}{(\gamma-1)(M_{A2}^2-1)^2+(M_{1+}^2-1)(1-M_{1-}^2)\left[(2-\gamma)M_{A2}^2-(1-\gamma)\right]}
\end{aligned}$$

by using the relations

$$\frac{a_s^2}{V_{An}^2}=M_+^2M_-^2 \qquad \tan^2\theta_1=(M_{1+}^2-1)(1-M_{1-}^2). \qquad (3.88)$$

Solution First, we divide Eq. (3.86) by the term in curly brackets, so that

$$M_{A1}^2 = \frac{\left[M_{A2}^2(1+\gamma) - 2\frac{a_{s1}^2}{V_{An1}^2}\right]\left(M_{A2}^2 - 1\right)^2 - M_{A2}^2\left[M_{A2}^2\gamma - (1+\gamma)\right]\tan^2\theta_1}{(\gamma-1)\left(M_{A2}^2-1\right)^2 + \left[(2-\gamma)M_{A2}^2 - (1-\gamma)\right]\tan^2\theta_1}.$$

Obviously, on substituting $\tan^2\theta_1$ by relation (3.88) we find that the denominator of this expression already corresponds to the desired result,

$$\begin{aligned}\beta &= (\gamma-1)\left(M_{A2}^2-1\right)^2 + \left[(2-\gamma)M_{A2}^2 - (1-\gamma)\right]\tan^2\theta_1\\ &= (\gamma-1)\left(M_{A2}^2-1\right)^2 + (M_{1+}^2-1)(1-M_{1-}^2)\left[(2-\gamma)M_{A2}^2 - (1-\gamma)\right],\end{aligned}$$

so that

$$M_{A1}^2 = \frac{\left[M_{A2}^2(1+\gamma) - 2\frac{a_{s1}^2}{V_{An1}^2}\right]\left(M_{A2}^2 - 1\right)^2 - M_{A2}^2\left[M_{A2}^2\gamma - (1+\gamma)\right]\tan^2\theta_1}{\beta}.$$

Consider now the numerator. On substituting the relations (3.88), we obtain

$$\begin{aligned}M_{A1}^2 = &\frac{\left[M_{A2}^2(1+\gamma) - 2M_{1+}^2M_{1-}^2\right]\left(M_{A2}^2-1\right)^2}{\beta}\\ &- \frac{M_{A2}^2\left[M_{A2}^2\gamma - (1+\gamma)\right](M_{1+}^2-1)(1-M_{1-}^2)}{\beta}.\end{aligned}$$

Now we subtract M_{A2}^2 on both sides and obtain

$$\begin{aligned}M_{A1}^2 - M_{A2}^2 = &\frac{\left[M_{A2}^2(1+\gamma) - 2M_{1+}^2M_{1-}^2\right]\left(M_{A2}^2-1\right)^2}{\beta}\\ &- \frac{M_{A2}^2\left[M_{A2}^2\gamma - (1+\gamma)\right](M_{1+}^2-1)(1-M_{1-}^2)}{\beta} - \frac{M_{A2}^2\beta}{\beta},\end{aligned}$$

where M_{A2}^2 on the right-hand side has been multiplied with $1 = \beta/\beta$ in order to calculate the numerator. The numerator of the entire right-hand side can then be written as

$$\begin{aligned}\alpha = &\left[M_{A2}^2(1+\gamma) - 2M_{1+}^2M_{1-}^2\right]\left(M_{A2}^2-1\right)^2\\ &- M_{A2}^2\left[M_{A2}^2\gamma - (1+\gamma)\right](M_{1+}^2-1)(1-M_{1-}^2) - M_{A2}^2\beta.\end{aligned}$$

On substituting β we obtain

$$\begin{aligned}
\alpha = &\left[M_{A2}^2(1+\gamma) - 2M_{1+}^2 M_{1-}^2\right]\left(M_{A2}^2 - 1\right)^2 && \text{(SP-1)}\\
&- M_{A2}^2\left[M_{A2}^2\gamma - (1+\gamma)\right](M_{1+}^2 - 1)(1 - M_{1-}^2) && \text{(SP-2)}\\
&- M_{A2}^2(\gamma - 1)\left(M_{A2}^2 - 1\right)^2 && \text{(SP-3)}\\
&- M_{A2}^2(M_{1+}^2 - 1)(1 - M_{1-}^2)\left[(2-\gamma)M_{A2}^2 - (1-\gamma)\right] && \text{(SP-4)}
\end{aligned}$$

Combining now (SP-1) and (SP-3) yields

$$\text{(SP-1)} + \text{(SP-3)} = 2\left(M_{A2}^2 - 1\right)^2\left(M_{A2}^2 - M_{1+}^2 M_{1-}^2\right).$$

Combining (SP-2) and (SP-4) yields

$$\text{(SP-2)} + \text{(SP-4)} = -2M_{A2}^2(M_{1+}^2 - 1)(1 - M_{1-}^2)\left(M_{A2}^2 - 1\right).$$

We can then find

$$\begin{aligned}
\alpha&\\
&= 2\left(M_{A2}^2 - 1\right)^2\left(M_{A2}^2 - M_{1+}^2 M_{1-}^2\right) - 2M_{A2}^2(M_{1+}^2 - 1)(1 - M_{1-}^2)\left(M_{A2}^2 - 1\right)\\
&= 2\left(M_{A2}^2 - 1\right)\left[\left(M_{A2}^2 - M_{1+}^2 M_{1-}^2\right)\left(M_{A2}^2 - 1\right) - M_{A2}^2(M_{1+}^2 - 1)(1 - M_{1-}^2)\right]\\
&= 2\left(M_{A2}^2 - 1\right)\left[M_{A2}^4 - M_{A2}^2 M_{1+}^2 M_{1-}^2 - M_{A2}^2 + M_{1+}^2 M_{1-}^2\right.\\
&\qquad \left. - M_{A2}^2 M_{1+}^2 + M_{A2}^2 M_{1+}^2 M_{1-}^2 + M_{A2}^2 - M_{A2}^2 M_{1-}^2\right].
\end{aligned}$$

Obviously,

$$\begin{aligned}
\alpha &= 2\left(M_{A2}^2 - 1\right)\left[M_{A2}^4 + M_{1+}^2 M_{1-}^2 - M_{A2}^2 M_{1+}^2 - M_{A2}^2 M_{1-}^2\right]\\
&= 2\left(M_{A2}^2 - 1\right)\left[M_{A2}^4 - M_{A2}^2 M_{1+}^2 + M_{1+}^2 M_{1-}^2 - M_{A2}^2 M_{1-}^2\right]\\
&= 2\left(M_{A2}^2 - 1\right)\left[M_{A2}^2\left(M_{A2}^2 - M_{1+}^2\right) - M_{1-}^2\left(M_{A2}^2 - M_{1+}^2\right)\right]\\
&= 2\left(M_{A2}^2 - 1\right)\left(M_{A2}^2 - M_{1+}^2\right)\left(M_{A2}^2 - M_{1-}^2\right).
\end{aligned}$$

Putting all results together we obtain

$$\begin{aligned}
M_{A1}^2 - M_{A2}^2 &= \frac{\alpha}{\beta}\\
&= \frac{2\left(M_{A2}^2 - 1\right)\left(M_{A2}^2 - M_{1+}^2\right)\left(M_{A2}^2 - M_{1-}^2\right)}{(\gamma - 1)\left(M_{A2}^2 - 1\right)^2 + (M_{1+}^2 - 1)(1 - M_{1-}^2)\left[(2-\gamma)M_{A2}^2 - (1-\gamma)\right]}.
\end{aligned}$$

Problem 3.26 Solve the shock polar relation numerically and plot curves for $\beta_{p1} = 0.1$ and $\beta_{p1} = 4$ and with $\gamma = 5/3$.

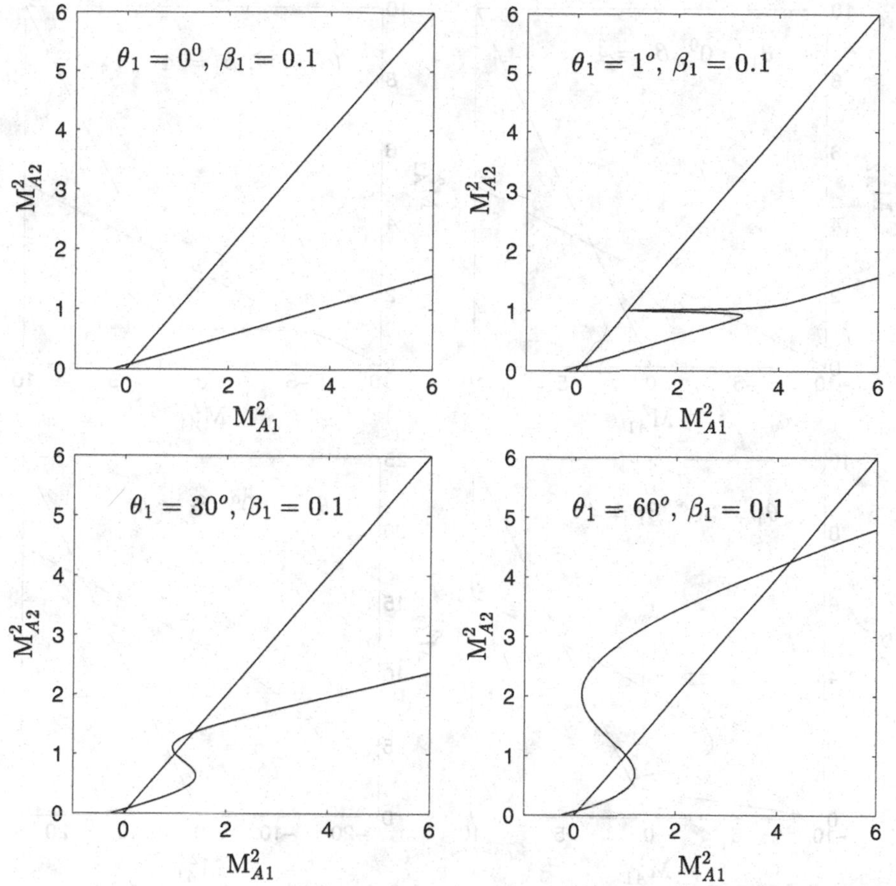

Fig. 3.2 Shown is how the Alfvén Mach number shock polar (3.89) changes as the upstream magnetic field angle θ_1 varies ($\theta_1 = 0°, 1°, 30°$, and $60°$) for $\beta_{p1} = 0.1$

Solution We rewrite Eq. (3.87) as

$$\begin{aligned} &M_{A1}^2\left\{(\gamma-1)\left(M_{A2}^2-1\right)^2+\left[(2-\gamma)M_{A2}^2-(1-\gamma)\right]\tan^2\theta_1\right\} \\ &\quad=\left[M_{A2}^2(1+\gamma)-\gamma\beta_p\sec^2\theta\right]\left(M_{A2}^2-1\right)^2-M_{A2}^2\left[M_{A2}^2\gamma-(1+\gamma)\right]\tan^2\theta_1. \end{aligned} \tag{3.89}$$

where we used

$$\frac{a_{s1}^2}{V_{A1}^2}=\frac{\gamma}{2}\beta_p. \tag{3.90}$$

Equation (3.89) is solved numerically and the results are plotted in Figs. 3.2 and 3.3. Shown is the change of the upstream Alfvén Mach number M_{A2}^2 depending on the

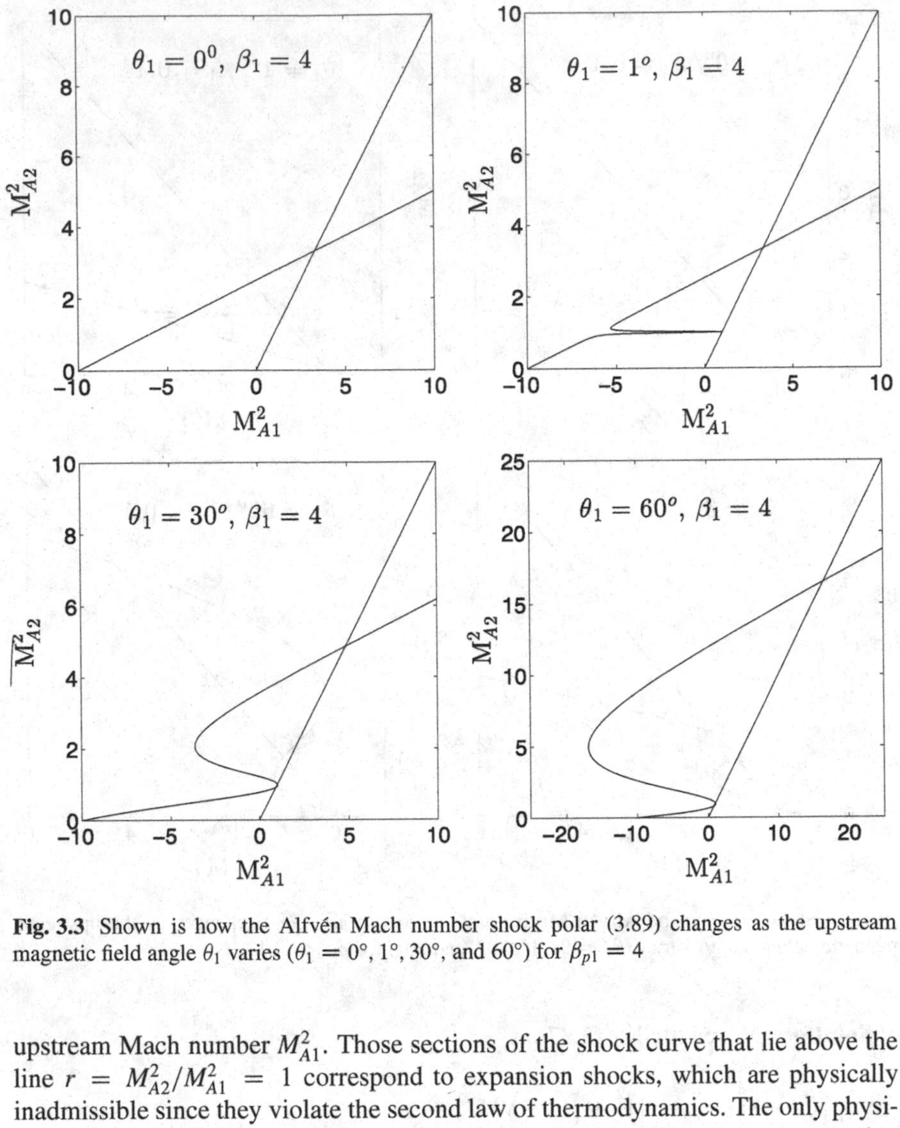

Fig. 3.3 Shown is how the Alfvén Mach number shock polar (3.89) changes as the upstream magnetic field angle θ_1 varies ($\theta_1 = 0°, 1°, 30°$, and $60°$) for $\beta_{p1} = 4$

upstream Mach number M^2_{A1}. Those sections of the shock curve that lie above the line $r = M^2_{A2}/M^2_{A1} = 1$ correspond to expansion shocks, which are physically inadmissible since they violate the second law of thermodynamics. The only physically acceptable solutions are those below the line $r = 1$; curves that correspond to compressive solutions. For further information about the interpretation of the curves we refer to Zank [5].

Chapter 4
Charged Particle Transport in a Collisionless Magnetized Plasma

4.1 The Focussed Transport Equation

Here we start with the non-relativistic Boltzmann equation (2.1),

$$\frac{\partial f}{\partial t} + \boldsymbol{v} \cdot \nabla_r f + \frac{\boldsymbol{F}}{m} \cdot \nabla_v f = \left(\frac{\delta f}{\delta t}\right)_s + S. \tag{4.1}$$

Note that, here, this equation has been implicitly separated into mean and fluctuating parts, with the fluctuating components being treated as *scattering* terms and have been relegated to the right-hand side. The particle source term is denoted by S. The force term in the Boltzmann equation (2.1) can be quite general. Here we restrict ourselves to $\boldsymbol{F} = q(\boldsymbol{E} + \boldsymbol{v} \times \boldsymbol{B}/c_s)$; the electromagnetic force on a particle with mass m and charge q, where c_s is the speed of light.

Consider now a frame of reference that propagates in the inertial *rest* frame at a velocity $\boldsymbol{u}$ (with the transformation (2.3), $\boldsymbol{c} = \boldsymbol{v} - \boldsymbol{u}$), in which the motional electric field is described by $\boldsymbol{E} = -\boldsymbol{u} \times \boldsymbol{B}/c_s$. By transforming into the moving frame the motional electric field cancels exactly the electric field and leaves $\boldsymbol{F} = q\boldsymbol{c} \times \boldsymbol{B}/c_s$. Following the transformations from Problem 2.2 we find for the Boltzmann equation in a mixed phase space

$$\frac{\partial f}{\partial t} + (u_i + c_i)\frac{\partial f}{\partial r_i} - \left(\frac{\partial u_i}{\partial t} + (u_j + c_j)\frac{\partial u_i}{\partial r_j} - \frac{q}{mc_s}(\boldsymbol{c} \times \boldsymbol{B})_i\right)\frac{\partial f}{\partial c_i} = \left(\frac{\delta f}{\delta t}\right)_s, \tag{4.2}$$

(continued)

A. Dosch, G.P. Zank, *Transport Processes in Space Physics and Astrophysics*, Lecture Notes in Physics 918, DOI 10.1007/978-3-319-24880-6_4

where we neglected particle sources (see also Eq. (2.2)). Assuming that the particle gyroradius is much smaller than any other spatial length scales in the system and, similarly, that their gyroperiod is much smaller than any other time scales in the system, then, the particle distribution may be regarded as nearly *gyrotropic*, i.e., $f(\boldsymbol{x}, \boldsymbol{v}, t) = f(\boldsymbol{x}, c, \mu, \phi, t)$ is essentially independent of gyrophase ϕ, so that $f(\boldsymbol{x}, c, \mu, \phi, t) \simeq f(\boldsymbol{x}, c, \mu, t)$, where $\mu \equiv \cos\theta = \boldsymbol{c} \cdot \boldsymbol{b}/c$ is the particle pitch angle and $\boldsymbol{b} = \boldsymbol{B}/B$ is the unit vector along the large-scale magnetic field. Since we are assuming gyrotropy we may average Eq. (4.2) over gyrophase, see the following Problem 4.1.

Problem 4.1 By collecting all the terms associated with the gyro-phase averaging, derive the general form of the gyro-phase averaged transport equation.

Solution We start with Eq. (4.2) and transform first into spherical coordinates. The distribution function in Eq. (4.2) is a function of position $\boldsymbol{r}$, random velocity $\boldsymbol{c}$, and time t, so that $f = f(\boldsymbol{r}, \boldsymbol{c}, t)$. By transforming into spherical coordinates we have $\boldsymbol{c} \to (c, \mu, \phi)$, where c is the absolute value of $\boldsymbol{c}$, $\mu = \boldsymbol{c} \cdot \boldsymbol{b}/c$ is the cosine of the pitch angle, and ϕ is the gyrophase. Here, $\boldsymbol{b} = \boldsymbol{B}/|\boldsymbol{B}|$ is a unit vector along the large scale magnetic field. With a spatially varying magnetic field $\boldsymbol{B}$ it follows that μ and ϕ depend on the location of the particle (since the magnetic field changes with different particle positions). Thus, $\mu = \mu(\boldsymbol{r})$ and $\phi = \phi(\boldsymbol{r})$. With that we have to substitute the gradient of f by

$$\frac{\partial f}{\partial r_i} \Longrightarrow \frac{\partial f}{\partial r_i} + \frac{\partial \mu}{\partial r_i}\frac{\partial f}{\partial \mu} + \frac{\partial \phi}{\partial r_i}\frac{\partial f}{\partial \phi}. \tag{4.3}$$

For the substitution of the velocity gradient we transform the random velocity $\boldsymbol{c}$ into spherical coordinates. For that we may write

$$\boldsymbol{c} = c_1\boldsymbol{e}_1 + c_2\boldsymbol{e}_2 + c_3\boldsymbol{e}_3.$$

Here $\boldsymbol{e}_1, \boldsymbol{e}_2$, and $\boldsymbol{e}_3$ are arbitrary but Cartesian unit vectors with $|\boldsymbol{e}_i| = 1$ and c_1, c_2, and c_3 are the coordinates. Note that $\boldsymbol{e}_1, \boldsymbol{e}_2$, and $\boldsymbol{e}_3$ are orthogonal but *not* necessarily $(1, 0, 0)$; $(0, 1, 0)$; $(0, 0, 1)$. The coordinate system, in which the random velocity $\boldsymbol{c}$ is defined, is continuously changing in such a way that $\boldsymbol{e}_3$ always coincides with the direction of the magnetic field $\boldsymbol{b}$, i.e., $\boldsymbol{e}_3 = \boldsymbol{b}$. For the transformation into spherical coordinates we use

$$c_1 = c\sin\theta\cos\phi \qquad c_2 = c\sin\theta\sin\phi \qquad c_3 = c\cos\theta = c\mu.$$

The curvilinear basis vectors are calculated by $\boldsymbol{e}_\alpha = (\partial \boldsymbol{c}/\partial\alpha)/|\partial \boldsymbol{c}/\partial\alpha|$, where $\alpha = c, \theta, \phi$. We find

$$\begin{aligned}
\boldsymbol{e}_c &= \sin\theta\cos\phi\boldsymbol{e}_1 + \sin\theta\sin\phi\boldsymbol{e}_2 + \cos\theta\boldsymbol{e}_3 \\
\boldsymbol{e}_\theta &= \cos\theta\cos\phi\boldsymbol{e}_1 + \cos\theta\sin\phi\boldsymbol{e}_2 - \sin\theta\boldsymbol{e}_3 \\
\boldsymbol{e}_\phi &= -\sin\phi\boldsymbol{e}_1 + \cos\phi\boldsymbol{e}_2 .
\end{aligned}$$

The random velocity can therefore also be written as

$$\boldsymbol{c} = c\boldsymbol{e}_c = c\left(\sin\theta\cos\phi\boldsymbol{e}_1 + \sin\theta\sin\phi\boldsymbol{e}_2 + \cos\theta\boldsymbol{b}\right). \tag{4.4}$$

Note that we substituted $\boldsymbol{e}_3 = \boldsymbol{b}$ in the last term. Substituting the spatial derivative by Eq. (4.3) we can rewrite Eq. (4.2) as

$$\begin{aligned}
&\frac{\partial f}{\partial t} + (u_i + c_i)\left[\frac{\partial f}{\partial r_i} + \frac{\partial\mu}{\partial r_i}\frac{\partial f}{\partial\mu} + \frac{\partial\phi}{\partial r_i}\frac{\partial f}{\partial\phi}\right] \\
&- \left(\frac{\partial u_i}{\partial t} + (u_j + c_j)\frac{\partial u_i}{\partial r_j} - \Omega\epsilon_{ijk}c_j b_k\right)\frac{\partial f}{\partial c_i} = \left(\frac{\delta f}{\delta t}\right)_s ,
\end{aligned} \tag{4.5}$$

where we used the gyrofrequency $\Omega = q|\boldsymbol{B}|/mc_s$ and the Levi-Civita tensor ϵ_{ijk}. For the transformation of the velocity gradient we consider

$$\frac{\partial f}{\partial c_i} = \frac{\partial f}{\partial c}\frac{\partial c}{\partial c_i} + \frac{\partial f}{\partial\mu}\frac{\partial\mu}{\partial c_i} + \frac{\partial f}{\partial\phi}\frac{\partial\phi}{\partial c_i}.$$

- In the first term on the right-hand side we consider the derivative of c with respect to c_i and find

$$\frac{\partial c}{\partial c_i} = \frac{\partial\sqrt{c_j c_j}}{\partial c_i} = \frac{1}{2c}2c_j\frac{\partial c_j}{\partial c_i} = \frac{c_j}{c}\delta_{ij} = e_{ci},$$

  where $c_i/c = e_{ci}$. In other words, the derivative of $\partial c/\partial c_i$ is the i-th component of the unit vector of the random velocity.
- In the second term we have, with $\mu = \boldsymbol{e}_c \cdot \boldsymbol{b} = e_{cj}b_j$,

$$\frac{\partial\mu}{\partial c_i} = \frac{\partial e_{cj}b_j}{\partial c_i} = b_j\frac{\partial}{\partial c_i}\left(\frac{c_j}{c}\right) = b_j\left[\frac{1}{c}\delta_{ij} - \frac{e_{ci}e_{cj}}{c}\right] = \left[\frac{b_i}{c} - \frac{\mu e_{ci}}{c}\right],$$

  where $\boldsymbol{b}$ is independent of $\boldsymbol{c}$.

At this point we do not consider the gyrophase term and obtain for the velocity gradient of the distribution function

$$\frac{\partial f}{\partial c_i} = \frac{\partial f}{\partial c} e_{ci} + \left[\frac{b_i}{c} - \frac{\mu e_{ci}}{c} \right] \frac{\partial f}{\partial \mu} + \frac{\partial f}{\partial \phi} \frac{\partial \phi}{\partial c_i}. \tag{4.6}$$

Substituting the velocity gradient in Eq. (4.5) we find the general form of the transport equation in spherical coordinates,

$$\frac{\partial f}{\partial t} + (u_i + c_i) \left[\frac{\partial f}{\partial r_i} + \frac{\partial \mu}{\partial r_i} \frac{\partial f}{\partial \mu} + \frac{\partial \phi}{\partial r_i} \frac{\partial f}{\partial \phi} \right] - \left(\frac{\partial u_i}{\partial t} + (u_j + c_j) \frac{\partial u_i}{\partial r_j} - \Omega \epsilon_{ijk} c_j b_k \right)$$
$$\times \left[\frac{\partial f}{\partial c} e_{ci} + \left[\frac{b_i}{c} - \frac{\mu e_{ci}}{c} \right] \frac{\partial f}{\partial \mu} + \frac{\partial f}{\partial \phi} \frac{\partial \phi}{\partial c_i} \right] = \left(\frac{\delta f}{\delta t} \right)_s .$$

Under the assumption of a nearly gyrotropic distribution function we may neglect terms proportional to $\partial f / \partial \phi$ and obtain

$$\frac{\partial f}{\partial t} + (u_i + c_i) \left[\frac{\partial f}{\partial r_i} + \frac{\partial \mu}{\partial r_i} \frac{\partial f}{\partial \mu} \right] - \left[\frac{\partial u_i}{\partial t} + (u_j + c_j) \frac{\partial u_i}{\partial r_j} - \Omega \epsilon_{ijk} c_j b_k \right] \tag{4.7}$$
$$\times \left[\frac{\partial f}{\partial c} e_{ci} + \left[\frac{b_i}{c} - \frac{\mu e_{ci}}{c} \right] \frac{\partial f}{\partial \mu} \right] = \left(\frac{\delta f}{\delta t} \right)_s .$$

Now, we introduce a gyrophase average

$$\langle \ldots \rangle_\phi = \frac{1}{2\pi} \int_0^{2\pi} d\phi,$$

where

$$\langle \sin \phi \rangle_\phi = \frac{1}{2\pi} \int_0^{2\pi} d\phi \sin \phi = 0, \quad \langle \cos \phi \rangle_\phi = \frac{1}{2\pi} \int_0^{2\pi} d\phi \cos \phi = 0, \tag{4.8}$$

and average Eq. (4.7) over gyrophase. For that we consider each term separately:

A. For the time derivative we obtain immediately

$$\left\langle \frac{\partial f}{\partial t} \right\rangle = \frac{\partial f}{\partial t},$$

since the distribution function is independent of gyrophase ϕ.

B. The second term is described by

$$\left\langle (u_i + c_i)\frac{\partial f}{\partial r_i}\right\rangle_\phi = (u_i + \langle c_i \rangle_\phi)\frac{\partial f}{\partial r_i},$$

where we average over c_i, since $\boldsymbol{u}$ and the spatial gradient of the distribution function f are independent of ϕ. Consider now expression (4.4), where the average is written as $\langle \boldsymbol{c} \rangle_\phi = c \langle \boldsymbol{e}_c \rangle_\phi$, so that

$$c \langle \boldsymbol{e}_c \rangle_\phi = c \langle \sin\theta \cos\phi \boldsymbol{e}_1 + \sin\theta \sin\phi \boldsymbol{e}_2 + \cos\theta \boldsymbol{b} \rangle_\phi = c\mu \boldsymbol{b}. \tag{4.9}$$

According to the relations (4.8) the averages $\langle \sin\phi \rangle_\phi$ and $\langle \cos\phi \rangle_\phi$ vanish and it remains only the $\boldsymbol{e}_3 = \boldsymbol{b}$ vector. It follows immediately for the i-th component of that vector that $\langle c_i \rangle_\phi = c\mu b_i$ and, therefore,

$$\left\langle (u_i + c_i)\frac{\partial f}{\partial r_i}\right\rangle_\phi = (u_i + c\mu b_i)\frac{\partial f}{\partial r_i}.$$

C. For the third term we have to calculate

$$\left\langle (u_i + c_i)\frac{\partial \mu}{\partial r_i}\frac{\partial f}{\partial \mu}\right\rangle_\phi = \left\langle u_i \frac{\partial \mu}{\partial r_i}\frac{\partial f}{\partial \mu}\right\rangle_\phi + \left\langle c_i \frac{\partial \mu}{\partial r_i}\frac{\partial f}{\partial \mu}\right\rangle_\phi. \tag{4.10}$$

Let us consider each term separately:

a. The first term on the right-hand side of Eq. (4.10) is given by

$$\left\langle u_i \frac{\partial \mu}{\partial r_i}\frac{\partial f}{\partial \mu}\right\rangle_\phi = u_i \left\langle \frac{\partial \mu}{\partial r_i}\right\rangle_\phi \frac{\partial f}{\partial \mu}.$$

Remember that the pitch angle $\mu = \boldsymbol{e}_c \cdot \boldsymbol{b} = e_{cj} b_j$, so that

$$\left\langle \frac{\partial \mu}{\partial r_i}\right\rangle_\phi = \left\langle \frac{\partial e_{cj} b_j}{\partial r_i}\right\rangle_\phi = \langle e_{cj} \rangle_\phi \frac{\partial b_j}{\partial r_i} = \mu b_j \frac{\partial b_j}{\partial r_i}.$$

Keep in mind that the average over ϕ cannot be pulled into the derivative, since ϕ is a function of $\boldsymbol{r}$. In the third step we used the results from Eq. (4.9) and find $\langle e_{cj} \rangle_\phi = \mu b_j$. In order to calculate $\partial b_j/\partial r_i$ consider first

$$\frac{1}{2}\frac{\partial(\boldsymbol{b}\cdot\boldsymbol{b})}{\partial r_i} = \frac{1}{2}\frac{\partial\, b_j b_j}{\partial r_i} = b_j \frac{\partial\, b_j}{\partial r_i} = 0.$$

The result must be zero, since $\boldsymbol{b}$ is a unit vector with $\boldsymbol{b}\cdot\boldsymbol{b} = 1$ per definition. Because of this result we have

$$\left\langle u_i \frac{\partial \mu}{\partial r_i}\frac{\partial f}{\partial \mu}\right\rangle_\phi = 0.$$

b. The second term on the right-hand side of Eq. (4.10) can be written as

$$\left\langle c_i \frac{\partial \mu}{\partial r_i}\frac{\partial f}{\partial \mu}\right\rangle_\phi = c\left\langle e_{ci}\frac{\partial e_{cj}b_j}{\partial r_i}\right\rangle_\phi \frac{\partial f}{\partial \mu} = c\left\langle e_{ci}e_{cj}\right\rangle_\phi \frac{\partial b_j}{\partial r_i}\frac{\partial f}{\partial \mu},$$

where we used $\mu = \boldsymbol{e}_c \cdot \boldsymbol{b} = e_{cj}b_j$ and $c_i = ce_{ci}$. In a vector description we find for the averaging

$$\langle \boldsymbol{e}_c \cdot \nabla (\boldsymbol{e}_c \cdot \boldsymbol{b})\rangle_\phi = \left[\langle \boldsymbol{e}_c \otimes \boldsymbol{e}_c\rangle_\phi \nabla\right]\cdot \boldsymbol{b}. \tag{4.11}$$

where the symbol $\otimes$ denotes a dyadic product. In the following we will omit the symbol and refer to $\boldsymbol{e}_c \otimes \boldsymbol{e}_c = \boldsymbol{e}_c\boldsymbol{e}_c$ as a dyadic product, see also Sect. 5.1. For the average we find

$$\begin{aligned}
\langle \boldsymbol{e}_c\boldsymbol{e}_c\rangle_\phi &= \frac{1}{2\pi}\int_0^{2\pi} \boldsymbol{e}_c\boldsymbol{e}_c \, d\phi \\
&= \frac{1}{2\pi}\int_0^{2\pi} (\sin\theta\cos\phi\boldsymbol{e}_1 + \sin\theta\sin\phi\boldsymbol{e}_2 + \cos\theta\boldsymbol{e}_3) \\
&\quad \times (\sin\theta\cos\phi\boldsymbol{e}_1 + \sin\theta\sin\phi\boldsymbol{e}_2 + \cos\theta\boldsymbol{e}_3)\; d\phi \\
&= \frac{1}{2\pi}\int_0^{2\pi} (\sin^2\theta\cos^2\phi\; \boldsymbol{e}_1\boldsymbol{e}_1 + \sin^2\theta\sin^2\phi\; \boldsymbol{e}_2\boldsymbol{e}_2 \\
&\quad + \cos^2\theta\; \boldsymbol{e}_3\boldsymbol{e}_3 + 2\sin^2\theta\sin\phi\cos\phi\boldsymbol{e}_1\boldsymbol{e}_2 \\
&\quad + 2\sin\theta\cos\theta\cos\phi\boldsymbol{e}_1\boldsymbol{e}_3 + 2\sin\theta\cos\theta\sin\phi\boldsymbol{e}_2\boldsymbol{e}_3)\; d\phi,
\end{aligned}$$

where we substituted Eq. (4.9) for $\boldsymbol{e}_c$. With the results given by Eq. (4.8) and with

$$\frac{1}{2\pi}\int_0^{2\pi} \sin^2\phi\; d\phi = \frac{1}{2}; \qquad \frac{1}{2\pi}\int_0^{2\pi} \cos^2\phi\; d\phi = \frac{1}{2}$$

we find for the average

$$\langle \boldsymbol{e}_c\boldsymbol{e}_c\rangle_\phi = \frac{1-\mu^2}{2}\boldsymbol{e}_1\boldsymbol{e}_1 + \frac{1-\mu^2}{2}\boldsymbol{e}_2\boldsymbol{e}_2 + \mu^2\boldsymbol{e}_3\boldsymbol{e}_3. \tag{4.12}$$

As a side note: Since the base vectors are orthogonal, we have $\boldsymbol{e}_1\boldsymbol{e}_1 + \boldsymbol{e}_2\boldsymbol{e}_2 + \boldsymbol{e}_3\boldsymbol{e}_3 = \mathbb{1}$, where $\mathbb{1}$ is the identity (unity) matrix. And with $\boldsymbol{e}_3 = \boldsymbol{b}$ we can also write

$$\langle \boldsymbol{e}_c\boldsymbol{e}_c \rangle_\phi = \frac{1-\mu^2}{2}[\mathbb{1} - \boldsymbol{bb}] + \mu^2\boldsymbol{bb}.$$

By substituting this result back into Eq. (4.11) we obtain

$$\begin{aligned}\langle \boldsymbol{e}_c \cdot \nabla(\boldsymbol{e}_c \cdot \boldsymbol{b}) \rangle_\phi &= \left[\left(\frac{1-\mu^2}{2}[\mathbb{1} - \boldsymbol{bb}] + \mu^2\boldsymbol{bb}\right)\nabla\right] \cdot \boldsymbol{b} \\ &= \left[\frac{1-\mu^2}{2}(\delta_{ij} - b_ib_j) + \mu^2 b_ib_j\right]\frac{\partial}{\partial r_i}b_j.\end{aligned}$$

Consider the last two terms on the right-hand side of this expression, which are essentially given by

$$b_ib_j\frac{\partial b_j}{\partial r_i} = \frac{b_i}{2}\frac{\partial b_jb_j}{\partial r_i} = \frac{b_i}{2}\frac{\partial \boldsymbol{b}\cdot\boldsymbol{b}}{\partial r_i} = 0,$$

so that

$$\langle \boldsymbol{e}_c \cdot \nabla(\boldsymbol{e}_c \cdot \boldsymbol{b}) \rangle_\phi = \left[\frac{1-\mu^2}{2}\delta_{ij}\right]\frac{\partial b_j}{\partial r_i} = \frac{1-\mu^2}{2}\frac{\partial b_i}{\partial r_i} = \frac{1-\mu^2}{2}\nabla\cdot\boldsymbol{b},$$

and therefore

$$\left\langle \boldsymbol{c}\cdot\nabla\mu\frac{\partial f}{\partial \mu}\right\rangle_\phi = c\frac{1-\mu^2}{2}\frac{\partial b_i}{\partial r_i}\frac{\partial f}{\partial \mu}.$$

By combining both results we find for Eq. (4.10)

$$\left\langle (u_i + c_i)\frac{\partial \mu}{\partial r_i}\frac{\partial f}{\partial \mu}\right\rangle_\phi = c\frac{1-\mu^2}{2}\frac{\partial b_i}{\partial r_i}\frac{\partial f}{\partial \mu}.$$

D. For the next term we use

$$\left\langle \frac{\partial u_i}{\partial t}\left[\frac{\partial f}{\partial c}e_{ci} + \left(\frac{b_i}{c} - \frac{\mu e_{ci}}{c}\right)\frac{\partial f}{\partial \mu}\right]\right\rangle_\phi = \frac{\partial u_i}{\partial t}\left[\frac{\partial f}{\partial c}\mu b_i + \frac{1-\mu^2}{c}b_i\frac{\partial f}{\partial \mu}\right],$$

where the average operator acts only on $\langle e_{ci} \rangle_\phi = \mu b_i$.

E. The second last term is given by

$$f_1 + f_2 = -\left\langle (u_j + c_j)\frac{\partial u_i}{\partial r_j}\left[\frac{\partial f}{\partial c}e_{ci} + \left(\frac{b_i}{c} - \frac{\mu e_{ci}}{c}\right)\frac{\partial f}{\partial \mu}\right]\right\rangle_\phi,$$

where

$$f_1 = -\left\langle u_j \frac{\partial u_i}{\partial r_j}\left[\frac{\partial f}{\partial c} e_{ci} + \left(\frac{b_i}{c} - \frac{\mu e_{ci}}{c}\right)\frac{\partial f}{\partial \mu}\right]\right\rangle_\phi$$

$$f_2 = -\left\langle c_j \frac{\partial u_i}{\partial r_j}\left[\frac{\partial f}{\partial c} e_{ci} + \left(\frac{b_i}{c} - \frac{\mu e_{ci}}{c}\right)\frac{\partial f}{\partial \mu}\right]\right\rangle_\phi .$$

The first average, f_1, can easily be solved with the results from the preceding term, and we obtain

$$f_1 = -u_j \frac{\partial u_i}{\partial r_j}\left[\frac{\partial f}{\partial c}\mu b_i + \frac{1-\mu^2}{c} b_i \frac{\partial f}{\partial \mu}\right].$$

The second average, f_2, can be simplified by using $c_j = c e_{cj}$ to obtain

$$\begin{aligned} f_2 &= -\left\langle c_j \frac{\partial u_i}{\partial r_j}\left[\frac{\partial f}{\partial c} e_{ci} + \left(\frac{b_i}{c} - \frac{\mu e_{ci}}{c}\right)\frac{\partial f}{\partial \mu}\right]\right\rangle_\phi \\ &= -\left\langle \frac{\partial u_i}{\partial r_j}\left[\frac{\partial f}{\partial c} c e_{cj} e_{ci} + \left(e_{cj} b_i - \mu e_{cj} e_{ci}\right)\frac{\partial f}{\partial \mu}\right]\right\rangle_\phi . \end{aligned}$$

With the previous result

$$\left\langle e_{ci} e_{cj} \right\rangle_\phi = \left(\delta_{ij} - b_i b_j\right)\frac{1-\mu^2}{2} + \mu^2 b_i b_j \tag{4.13}$$

we find

$$\begin{aligned} &-\left\langle c_j \frac{\partial u_i}{\partial r_j}\left[\frac{\partial f}{\partial c} e_{ci} + \left(\frac{b_i}{c} - \frac{\mu e_{ci}}{c}\right)\frac{\partial f}{\partial \mu}\right]\right\rangle_\phi \\ &= -\frac{\partial u_i}{\partial r_j}\left\{\frac{\partial f}{\partial c} c\left[\left(\delta_{ij} - b_i b_j\right)\frac{1-\mu^2}{2} + \mu^2 b_i b_j\right]\right. \\ &\quad \left. + \left[\mu b_j b_i - \mu\left(\left(\delta_{ij} - b_i b_j\right)\frac{1-\mu^2}{2} + \mu^2 b_i b_j\right)\right]\frac{\partial f}{\partial \mu}\right\} . \end{aligned} \tag{4.14}$$

F. The last term can be written as

$$\left\langle \Omega \epsilon_{ijk} c_j b_k \left[\frac{\partial f}{\partial c} e_{ci} + \left[\frac{b_i}{c} - \frac{\mu e_{ci}}{c}\right]\frac{\partial f}{\partial \mu}\right]\right\rangle_\phi = 0.$$

The last term vanishes for the following reasons:

a. For the first term we obtain

$$\left\langle \Omega \epsilon_{ijk} c_j b_k \frac{\partial f}{\partial c} e_{ci} \right\rangle_\phi = \Omega \epsilon_{ijk} c \left\langle e_{ci} e_{cj} \right\rangle_\phi b_k \frac{\partial f}{\partial c} = 0,$$

where we used $c_j = c e_{cj}$. This term vanishes for $i = j$, since the Levi-Civita tensor is zero for two identical indices, $\epsilon_{iik} = 0$. From the considerations made under point (C0b) we find, that $\left\langle e_{ci} e_{cj} \right\rangle_\phi$ also vanishes for any $i \neq j \neq k$. As an example, consider

$$\begin{aligned} \epsilon_{ijk} \left\langle e_{ci} e_{cj} \right\rangle_\phi b_k &= \left\langle e_{c1} e_{c2} \right\rangle_\phi b_3 - \left\langle e_{c1} e_{c3} \right\rangle_\phi b_2 + \left\langle e_{c3} e_{c1} \right\rangle_\phi b_2 \\ &\quad - \left\langle e_{c3} e_{c2} \right\rangle_\phi b_1 + \left\langle e_{c2} e_{c3} \right\rangle_\phi b_1 - \left\langle e_{c2} e_{c1} \right\rangle_\phi b_3 = 0, \end{aligned}$$

where we used Einstein's summation convention.

b. The second term is described by

$$\left\langle \Omega \epsilon_{ijk} c_j b_k \frac{b_i}{c} \frac{\partial f}{\partial \mu} \right\rangle_\phi = \Omega \epsilon_{ijk} \left\langle e_{cj} \right\rangle_\phi b_k b_i \frac{\partial f}{\partial \mu} = \Omega \epsilon_{ijk} \mu b_j b_k b_i \frac{\partial f}{\partial \mu} = 0,$$

where we used $\left\langle e_{cj} \right\rangle_\phi = \mu b_j$. Obviously this term is zero, since $\epsilon_{ijk} b_i b_j b_k = 0$ for any combination of i, j, and k.

c. The last term can be written as

$$\left\langle \Omega \epsilon_{ijk} c_j b_k \frac{\mu e_{ci}}{c} \frac{\partial f}{\partial \mu} \right\rangle_\phi = \Omega \epsilon_{ijk} \left\langle e_{ci} e_{cj} \right\rangle_\phi b_k \mu \frac{\partial f}{\partial \mu} = 0,$$

where we used $c_j = c e_{cj}$. This term is zero for the same reason as under point a.

By combining all results we find

$$\begin{aligned} &\frac{\partial f}{\partial t} + (u_i + c\mu b_i) \frac{\partial f}{\partial r_i} + c \frac{1-\mu^2}{2} \frac{\partial b_i}{\partial r_i} \frac{\partial f}{\partial \mu} - \frac{\partial u_i}{\partial t} \left[\frac{\partial f}{\partial c} \mu b_i + \frac{1-\mu^2}{c} b_i \frac{\partial f}{\partial \mu} \right] \\ &- u_j \frac{\partial u_i}{\partial r_j} \left[\frac{\partial f}{\partial c} \mu b_i + \frac{1-\mu^2}{c} b_i \frac{\partial f}{\partial \mu} \right] - \frac{\partial u_i}{\partial r_j} \left\{ \frac{\partial f}{\partial c} c \left[(\delta_{ij} - b_i b_j) \frac{1-\mu^2}{2} + \mu^2 b_i b_j \right] \right. \\ &\left. + \left(\mu b_j b_i - \mu \left[(\delta_{ij} - b_i b_j) \frac{1-\mu^2}{2} + \mu^2 b_i b_j \right] \right) \frac{\partial f}{\partial \mu} \right\} = \left\langle \left. \frac{\delta f}{\delta t} \right|_s \right\rangle_\phi . \end{aligned}$$

In the second line the second term in brackets can be simplified. By sorting all terms with respect to μ and c derivatives we obtain

$$\frac{\partial f}{\partial t} + (u_i + c\mu b_i) \frac{\partial f}{\partial r_i} \tag{4.15}$$

$$+\frac{1-\mu^2}{2}\left[c\frac{\partial b_i}{\partial r_i}+\mu\frac{\partial u_i}{\partial r_i}-3\mu b_i b_j\frac{\partial u_i}{\partial r_j}-\frac{2b_i}{c}\left(\frac{\partial u_i}{\partial t}+u_j\frac{\partial u_i}{\partial r_j}\right)\right]\frac{\partial f}{\partial \mu}$$
$$+\left[\frac{1-3\mu^2}{2}b_i b_j\frac{\partial u_i}{\partial r_j}-\frac{1-\mu^2}{2}\frac{\partial u_i}{\partial r_i}-\frac{\mu b_i}{c}\left(\frac{\partial u_i}{\partial t}+u_j\frac{\partial u_i}{\partial r_j}\right)\right]c\frac{\partial f}{\partial c}=\left\langle\left.\frac{\delta f}{\delta t}\right|_s\right\rangle_\phi.$$

Problem 4.2 By assuming a constant radial flow velocity for the solar wind with $\boldsymbol{U} = U_r\boldsymbol{e}_r$ and a radial interplanetary magnetic field pointing away from the sun with $\boldsymbol{b} = \boldsymbol{e}_r$, derive the 1D focussed transport equation (4.17).

Solution The vector form of the gyrophase averaged transport equation (4.15) is

$$\frac{\partial f}{\partial t}+(\boldsymbol{U}+c\mu\boldsymbol{b})\cdot\nabla f$$
$$+\frac{1-\mu^2}{2}\left[c\nabla\cdot\boldsymbol{b}+\mu\nabla\cdot\boldsymbol{U}-3\mu\boldsymbol{b}\cdot[(\boldsymbol{b}\cdot\nabla)\,\boldsymbol{U}]-\frac{2\boldsymbol{b}}{c}\cdot\left[\frac{\partial \boldsymbol{U}}{\partial t}+(\boldsymbol{U}\cdot\nabla)\,\boldsymbol{U}\right]\right]\frac{\partial f}{\partial \mu}$$
$$-\left[\frac{\mu\boldsymbol{b}}{c}\cdot\left[\frac{\partial \boldsymbol{U}}{\partial t}+(\boldsymbol{U}\cdot\nabla)\,\boldsymbol{U}\right]+\frac{1-\mu^2}{2}\nabla\cdot\boldsymbol{U}-\frac{1-3\mu^2}{2}\boldsymbol{b}\cdot[(\boldsymbol{b}\cdot\nabla)\,\boldsymbol{U}]\right]c\frac{\partial f}{\partial c}$$
$$=\left\langle\left.\frac{\delta f}{\delta t}\right|_s\right\rangle_\phi,$$

where we used $\boldsymbol{U}$ for the background flow instead of $\boldsymbol{u}$. Note that the spherical coordinates of the background flow and magnetic field are given by

$$\boldsymbol{U}=\begin{pmatrix}U_r\\0\\0\end{pmatrix}\qquad \boldsymbol{b}=\boldsymbol{e}_r=\begin{pmatrix}b_r\\0\\0\end{pmatrix}=\begin{pmatrix}1\\0\\0\end{pmatrix},$$

with $U_r = const.$ and $U_\theta = U_\phi = 0$, since the flow is purely radial and has no ϕ- or θ- component. Of course, $\boldsymbol{b}$ is a *unit* vector in radial direction with length one, so that $b_r = 1$. Recall that the nabla operator in spherical coordinates is given by

$$\nabla=\boldsymbol{e}_r\frac{\partial}{\partial r}+\boldsymbol{e}_\theta\frac{1}{r}\frac{\partial}{\partial \theta}+\boldsymbol{e}_\phi\frac{1}{r\sin\theta}\frac{\partial}{\partial \phi}. \tag{4.16}$$

- The divergence of the vector $\boldsymbol{U}$ in spherical coordinates is then given by

$$\nabla\cdot\boldsymbol{U}=\frac{1}{r^2}\frac{\partial}{\partial r}\left[r^2U_r\right]+\frac{1}{r\sin\theta}\frac{\partial}{\partial\theta}\left[\sin\theta\;U_\theta\right]+\frac{1}{r\sin\theta}\frac{\partial U_\phi}{\partial\phi}=\frac{2U_r}{r}.$$

- It follows immediately that

$$\nabla\cdot\boldsymbol{b}=\frac{2}{r}.$$

- By considering the expression $(\boldsymbol{b}\cdot\nabla)\,\boldsymbol{U}$ we need to transform the nabla operator into spherical coordinates, bearing in mind that $\boldsymbol{b} = \boldsymbol{e}_r$, so that

$$\boldsymbol{b}\cdot\nabla = \frac{\partial}{\partial r},$$

since $\boldsymbol{e}_r\cdot\boldsymbol{e}_r = 1$, and $\boldsymbol{e}_r\cdot\boldsymbol{e}_\theta = 0$ and $\boldsymbol{e}_r\cdot\boldsymbol{e}_\phi = 0$. Consider now

$$(\boldsymbol{b}\cdot\nabla)\,\boldsymbol{U} = \frac{\partial}{\partial r}\boldsymbol{U} = U_r\frac{\partial\boldsymbol{e}_r}{\partial r} = 0,$$

since the unit vector $\boldsymbol{e}_r = (\sin\theta\cos\phi, \sin\theta\sin\phi, \cos\theta)$ is independent of r. Obviously, the same holds for

$$(\boldsymbol{U}\cdot\nabla)\,\boldsymbol{U} = U_r\frac{\partial}{\partial r}\boldsymbol{U} = U_r^2\frac{\partial\boldsymbol{e}_r}{\partial r} = 0,$$

- Lastly, consider the term

$$(\boldsymbol{U} + c\mu\boldsymbol{b})\cdot\nabla f = (U_r + c\mu)\,\frac{\partial f}{\partial r}.$$

With that we obtain

$$\frac{\partial f}{\partial t} + (U_r + c\mu)\,\frac{\partial f}{\partial r} + \frac{1-\mu^2}{r}\,(c + \mu U_r)\,\frac{\partial f}{\partial \mu} - \frac{1-\mu^2}{r}U_r c\frac{\partial f}{\partial c} = \left\langle \left.\frac{\delta f}{\delta t}\right|_s \right\rangle. \tag{4.17}$$

Problem 4.3 Assume that the one spatial dimensional gyrotropic distribution function can be expressed as

$$f(r, c, \mu, t) = f_-(r, c, t)H(-\mu) + f_+(r, c, t)H(\mu),$$

where $H(x)$ denotes the Heaviside step function and $f_\pm$ refers to anti-sunward (f_+) / sunward (f_-) hemispherical distributions. By substituting $f = f_+H(\mu) + f_-H(-\mu)$ in the 1D focused transport equation (4.17), where the scattering term on the right-hand side is given by

$$\left\langle \left.\frac{\delta f}{\delta t}\right|_s \right\rangle = \frac{\partial}{\partial\mu}\left(\nu\left(1-\mu^2\right)\frac{\partial f}{\partial\mu}\right),$$

and by integrating over μ separately from -1 to 0 and then from 0 to 1, show that

$$\frac{\partial f_\pm}{\partial t} + \left(U \pm \frac{c}{2}\right)\frac{\partial f_\pm}{\partial r} - \frac{2U}{r}\frac{c}{3}\frac{\partial f_\pm}{\partial c} + \frac{c}{r}(f_+ - f_-) = \mp\Gamma\,(f_+ - f_-),$$

where $\Gamma \equiv \nu(\mu = 0)$ gives the rate of scattering across $\mu = 0$. Note that the form of the scattering term is of diffusion in pitch-angle. The term ν is the scattering frequency.

Solution Here we use the following definition of the Heaviside step function,

$$H(x) = \begin{cases} 0 & x < 0 \\ 1 & x \geq 0 \end{cases}.$$

We will also make use of the following definitions,

$$\int H(x)\,dx = x\,H(x) + C_1$$

$$\int x\,H(x)\,dx = \frac{x^2}{2}H(x) + C_2$$

$$\int x^2\,H(x)\,dx = \frac{x^3}{3}H(x) + C_3.$$

First, we substitute the definition of the distribution function f and obtain

$$\frac{\partial f_+ H(\mu)}{\partial t} + \frac{\partial f_- H(-\mu)}{\partial t} + (U_r + c\mu)\,\frac{\partial f_+ H(\mu)}{\partial r} + (U_r + c\mu)\,\frac{\partial f_- H(-\mu)}{\partial r}$$
$$+ \frac{1-\mu^2}{r}(c + \mu U_r)\,\frac{\partial f_+ H(\mu)}{\partial \mu} + \frac{1-\mu^2}{r}(c + \mu U_r)\,\frac{\partial f_- H(-\mu)}{\partial \mu}$$
$$- \frac{1-\mu^2}{r}U_r c\frac{\partial f_+ H(\mu)}{\partial c} - \frac{1-\mu^2}{r}U_r c\frac{\partial f_- H(-\mu)}{\partial c}$$
$$= \frac{\partial}{\partial \mu}\left(\nu\left(1-\mu^2\right)\frac{\partial f_+ H(\mu)}{\partial \mu}\right) + \frac{\partial}{\partial \mu}\left(\nu\left(1-\mu^2\right)\frac{\partial f_- H(-\mu)}{\partial \mu}\right).$$

Since $f_\pm = f_\pm(r, c, t)$ is a function of $r, c,$ and t alone and the Heaviside step function is a function of μ alone, we find

$$H(\mu)\frac{\partial f_+}{\partial t} + H(-\mu)\frac{\partial f_-}{\partial t} + (U_r + c\mu)\,H(\mu)\frac{\partial f_+}{\partial r} + (U_r + c\mu)\,H(-\mu)\frac{\partial f_-}{\partial r}$$
$$+ \frac{1-\mu^2}{r}(c + \mu U_r) f_+ \delta(\mu) - \frac{1-\mu^2}{r}(c + \mu U_r) f_- \delta(\mu)$$
$$- \frac{1-\mu^2}{r}U_r c H(\mu)\frac{\partial f_+}{\partial c} - \frac{1-\mu^2}{r}U_r c H(-\mu)\frac{\partial f_-}{\partial c}$$
$$= f_+\frac{\partial}{\partial \mu}\left(\nu\left(1-\mu^2\right)\delta(\mu)\right) - f_-\frac{\partial}{\partial \mu}\left(\nu\left(1-\mu^2\right)\delta(\mu)\right)$$

$$= [f_+ - f_-] \frac{\partial}{\partial \mu} \left[\nu \left(1 - \mu^2\right) \delta(\mu) \right],$$

where we used $\partial H(\pm\mu)/\partial\mu = \pm\delta(\mu)$, and where $\delta(x)$ is the delta distribution.

A. Now we integrate this equation with respect to μ from -1 to 0. Obviously, terms that include $H(\mu)$ vanish. By considering each term separately we find the following results.

- For the second term on the left-hand side we find

$$\int_{-1}^{0} d\mu \, H(-\mu) \frac{\partial f_-}{\partial t} = \frac{\partial f_-}{\partial t} \int_{-1}^{0} d\mu \, H(-\mu) = \frac{\partial f_-}{\partial t} \int_{0}^{1} d\mu \, H(\mu)$$
$$= \frac{\partial f_-}{\partial t} \left[\mu H(\mu)\right]_0^1 = \frac{\partial f_-}{\partial t},$$

where we used the transformation $-\mu \to \mu$.

- The fourth term on the left-hand side is given by

$$\int_{-1}^{0} d\mu \, \left(U_r + c\mu\right) H(-\mu) \frac{\partial f_-}{\partial r} = \frac{\partial f_-}{\partial r} \int_{0}^{1} d\mu \, \left(U_r - c\mu\right) H(\mu)$$
$$= \frac{\partial f_-}{\partial r} \left(U_r - \frac{c}{2}\right).$$

- The fifth and sixth term on the left-hand side can be written as

$$\int_{-1}^{0} d\mu \, \frac{1-\mu^2}{r} \left(c + \mu U_r\right) f_+ \delta(\mu) - \int_{-1}^{0} d\mu \, \frac{1-\mu^2}{r} \left(c + \mu U_r\right) f_- \delta(\mu)$$
$$= \frac{c}{r} \left(f_+ - f_-\right),$$

since the delta distribution contributes only for $\mu = 0$.

- The eighth term on the left-hand side is

$$-\int_{-1}^{0} d\mu \, \frac{1-\mu^2}{r} U_r c H(-\mu) \frac{\partial f_-}{\partial c} = -\frac{\partial f_-}{\partial c} \frac{U_r c}{r} \int_{-1}^{0} d\mu \, \left(1 - \mu^2\right) H(-\mu)$$
$$= -\frac{\partial f_-}{\partial c} \frac{U_r c}{r} \left(1 - \frac{1}{3}\right) = -\frac{2}{3} \frac{\partial f_-}{\partial c} \frac{U_r c}{r}.$$

- The term on the right-hand side can be written as

$$[f_+ - f_-] \int_{-1}^{0} d\mu \, \frac{\partial}{\partial \mu} \left[\nu \left(1 - \mu^2\right) \delta(\mu) \right] = [f_+ - f_-] \, \nu(\mu = 0).$$

By combining all results we find for the integration from $\mu = -1$ to 0

$$\frac{\partial f_-}{\partial t} + \frac{\partial f_-}{\partial r}\left(U_r - \frac{c}{2}\right) + \frac{c}{r}\left(f_+ - f_-\right) - \frac{2}{3}\frac{\partial f_-}{\partial c}\frac{U_r c}{r} = \left[f_+ - f_-\right]\nu(\mu = 0).$$

B. Now we integrate this equation with respect to μ from 0 to 1. Similarly, terms that include $H(-\mu)$ vanish and by considering each term separately we find the following results.

- The first term is

$$\int_0^1 d\mu\, H(\mu)\frac{\partial f_+}{\partial t} = \frac{\partial f_+}{\partial t}\int_0^1 d\mu\, H(\mu) = \frac{\partial f_+}{\partial t}\left[\mu H(\mu)\right]_0^1 = \frac{\partial f_+}{\partial t}.$$

- The third term is

$$\int_0^1 d\mu\, \left(U_r + c\mu\right)H(\mu)\frac{\partial f_+}{\partial r} = \frac{\partial f_+}{\partial r}\int_0^1 d\mu\, \left(U_r + c\mu\right)H(\mu)$$
$$= \frac{\partial f_+}{\partial r}\left(U_r + \frac{c}{2}\right).$$

- The fifth and sixth term are similar to the case above and can be written as

$$\int_0^1 d\mu\, \frac{1-\mu^2}{r}\left(c + \mu U_r\right)f_+\delta(\mu) - \int_0^1 d\mu\, \frac{1-\mu^2}{r}\left(c + \mu U_r\right)f_-\delta(\mu)$$
$$= \frac{c}{r}\left(f_+ - f_-\right),$$

  since the delta distribution contributes only for $\mu = 0$.
- The eighth term is

$$-\int_0^1 d\mu\, \frac{1-\mu^2}{r}U_r c H(\mu)\frac{\partial f_+}{\partial c} = -\frac{\partial f_+}{\partial c}\frac{U_r c}{r}\int_0^1 d\mu\, \left(1-\mu^2\right)H(\mu)$$
$$= -\frac{\partial f_+}{\partial c}\frac{U_r c}{r}\left(1 - \frac{1}{3}\right) = -\frac{2}{3}\frac{\partial f_+}{\partial c}\frac{U_r c}{r}.$$

- The term on the right-hand side is given by

$$\left[f_+ - f_-\right]\int_0^1 d\mu\, \frac{\partial}{\partial \mu}\left[\nu\left(1-\mu^2\right)\delta(\mu)\right] = -\left[f_+ - f_-\right]\nu(\mu = 0),$$

  where (mathematically somewhat simple) the delta function contributes only for $\mu = 0$ and was set to 1.

By combining all results we find

$$\frac{\partial f_+}{\partial t} + \frac{\partial f_+}{\partial r}\left(U_r + \frac{c}{2}\right) + \frac{c}{r}(f_+ - f_-) - \frac{2}{3}\frac{\partial f_+}{\partial c}\frac{U_r c}{r} = -[f_+ - f_-]\,\nu(\mu = 0).$$

By combining all results from the integrations with respect to μ from -1 to 0 and from 0 to1, we obtain the two equations

$$\frac{\partial f_\pm}{\partial t} + \frac{\partial f_\pm}{\partial r}\left(U_r \pm \frac{c}{2}\right) + \frac{c}{r}(f_+ - f_-) - \frac{2}{3}\frac{\partial f_\pm}{\partial c}\frac{U_r c}{r} = \mp[f_+ - f_-]\,\Gamma,$$

where $\Gamma \equiv \nu(\mu = 0)$ gives the rate of scattering across $\mu = 0$.

4.2 Quasi-Linear Transport Theory of Charged Particles: Derivation of the Scattering Tensor

The relativistic Vlasov equation is given by

$$\frac{\partial f}{\partial t} + \frac{\boldsymbol{p}}{m}\cdot\nabla f + q\left(\boldsymbol{E} + \frac{\boldsymbol{p}\times\boldsymbol{B}}{m}\right)\cdot\frac{\partial f}{\partial \boldsymbol{p}} = 0. \tag{4.18}$$

Problem 4.4 Rewrite the relativistic Vlasov equation (4.18) using a mean field expansion for the electromagnetic variables, assuming that the particle distribution function is co-moving with the plasma (thus ensuring that $\boldsymbol{E}_0 = 0$), and neglecting the fluctuating electric field term. Hence derive (4.20) and (4.22).

Solution The electric field is given by $\boldsymbol{E} = \boldsymbol{E}_0 + \delta\boldsymbol{E}$ and by setting $\boldsymbol{E}_0 = 0$ and neglecting electric turbulence ($\delta\boldsymbol{E} \approx 0$) we find for the Vlasov equation

$$\frac{\partial f}{\partial t} + \frac{\boldsymbol{p}}{m}\cdot\nabla f + \frac{q}{m}\boldsymbol{p}\times\boldsymbol{B}\cdot\frac{\partial f}{\partial \boldsymbol{p}} = 0.$$

By using

$$\begin{aligned} &\boldsymbol{B} = \boldsymbol{B}_0 + \delta\boldsymbol{B} && \langle \boldsymbol{B}\rangle = \boldsymbol{B}_0 && \langle \delta\boldsymbol{B}\rangle = 0 \\ &f = f_0 + \delta f && \langle f\rangle = f_0 && \langle \delta f\rangle = 0, \end{aligned}$$

we can write for the Vlasov equation

$$\frac{\partial f_0}{\partial t} + \frac{\partial \delta f}{\partial t} + \frac{\boldsymbol{p}}{m} \cdot \nabla f_0 + \frac{\boldsymbol{p}}{m} \cdot \nabla \delta f + \frac{q}{m} \boldsymbol{p} \times \boldsymbol{B}_0 \cdot \frac{\partial f_0}{\partial \boldsymbol{p}}$$
$$+ \frac{q}{m} \boldsymbol{p} \times \delta \boldsymbol{B} \cdot \frac{\partial f_0}{\partial \boldsymbol{p}} + \frac{q}{m} \boldsymbol{p} \times \boldsymbol{B}_0 \cdot \frac{\partial \delta f}{\partial \boldsymbol{p}} + \frac{q}{m} \boldsymbol{p} \times \delta \boldsymbol{B} \cdot \frac{\partial \delta f}{\partial \boldsymbol{p}} = 0. \quad (4.19)$$

Now we average Eq. (4.19) over an ensemble of particles. Note that all terms with fluctuations (δf and $\delta \boldsymbol{B}$) vanish except the last term on the left hand side. We are left with

$$\frac{\partial f_0}{\partial t} + \frac{\boldsymbol{p}}{m} \cdot \nabla f_0 + \frac{q}{m} \boldsymbol{p} \times \boldsymbol{B}_0 \cdot \frac{\partial f_0}{\partial \boldsymbol{p}} + \frac{q}{m} \left\langle \boldsymbol{p} \times \delta \boldsymbol{B} \cdot \frac{\partial \delta f}{\partial \boldsymbol{p}} \right\rangle = 0. \quad (4.20)$$

By subtracting Eq. (4.20) from Eq. (4.19) we find

$$\frac{\partial \delta f}{\partial t} + \frac{\boldsymbol{p}}{m} \cdot \nabla \delta f + \frac{q}{m} \boldsymbol{p} \times \boldsymbol{B}_0 \cdot \frac{\partial \delta f}{\partial \boldsymbol{p}}$$
$$= -\frac{q}{m} \boldsymbol{p} \times \delta \boldsymbol{B} \cdot \frac{\partial f_0}{\partial \boldsymbol{p}} - \frac{q}{m} \boldsymbol{p} \times \delta \boldsymbol{B} \cdot \frac{\partial \delta f}{\partial \boldsymbol{p}} + \frac{q}{m} \left\langle \boldsymbol{p} \times \delta \boldsymbol{B} \cdot \frac{\partial \delta f}{\partial \boldsymbol{p}} \right\rangle. \quad (4.21)$$

We consider now the three terms on the right-hand side of Eq. (4.21). The expression in front of the derivations $\partial f_0 / \partial \boldsymbol{p}$ and $\partial \delta f / \partial \boldsymbol{p}$ can be interpreted as a force $\boldsymbol{F}$ (compare the units),

$$\boldsymbol{F} = \frac{q}{m} \boldsymbol{p} \times \delta \boldsymbol{B}.$$

Since the turbulent magnetic fields are small perturbations to the constant magnetic background field $\boldsymbol{B}_0$, we may consider the force term to be small as well. In fact, this force term has to be significantly small so that we can introduce the time scales

$$\tau_c \ll t \ll \tau_{f_0},$$

where τ_{f_0} represents the time scale on which the force term affects the evolution of the particle distribution f_0 and Eq. (4.21) cannot be applied. We can also find a time scale τ_c, where the two-point two-time correlation function for the magnetic fluctuations becomes negligibly small (see Chap. 5.4 in [5]).

Again, the force term has to be significantly small so that within the time scale t (where f_0 is unaffected by the force term) the variation δf, which is generated by the force term, remains much smaller than f_0. (For a detailed description see [4]) In this case the last two terms on the right-hand side of Eq. (4.21) can be neglected and we find

$$\frac{\partial \delta f}{\partial t} + \frac{\boldsymbol{p}}{m} \cdot \nabla \delta f + \frac{q}{m} \boldsymbol{p} \times \boldsymbol{B}_0 \cdot \frac{\partial \delta f}{\partial \boldsymbol{p}} = -\frac{q}{m} \boldsymbol{p} \times \delta \boldsymbol{B} \cdot \frac{\partial f_0}{\partial \boldsymbol{p}}. \quad (4.22)$$

Problem 4.5 Derive the relations (4.23a)–(4.23c) and hence show that

$$-\frac{q}{m}\left(\boldsymbol{p}\times\delta\boldsymbol{B}\cdot\nabla_p f_0\right)=-\frac{\Omega}{B}\left(\delta B_x\sin\phi-\delta B_y\cos\phi\right)\frac{\partial f_0}{\partial\theta}.$$

Solution By transforming the momentum vector $\boldsymbol{p}$ from Cartesian coordinates into spherical coordinates $(p_x, p_y, p_z)\to(p,\phi,\theta)$ we have for each component

$$p_x = p\sin\theta\cos\phi \qquad p_y = p\sin\theta\sin\phi \qquad p_z = p\cos\theta.$$

For the transformation of the Cartesian partial derivatives into spherical coordinates we use the Jacobian matrix with

$$J=\begin{pmatrix}\frac{\partial p_x}{\partial p} & \frac{\partial p_x}{\partial\theta} & \frac{\partial p_x}{\partial\phi}\\ \frac{\partial p_y}{\partial p} & \frac{\partial p_y}{\partial\theta} & \frac{\partial p_y}{\partial\phi}\\ \frac{\partial p_z}{\partial p} & \frac{\partial p_z}{\partial\theta} & \frac{\partial p_z}{\partial\phi}\end{pmatrix}=\begin{pmatrix}\sin\theta\cos\phi & p\cos\theta\cos\phi & -p\sin\theta\sin\phi\\ \sin\theta\sin\phi & p\cos\theta\sin\phi & p\sin\theta\cos\phi\\ \cos\theta & -p\sin\theta & 0\end{pmatrix}.$$

The inverse Jacobian matrix is then given by

$$J^{-1}=\begin{pmatrix}\sin\theta\cos\phi & \sin\theta\sin\phi & \cos\theta\\ \frac{\cos\theta\cos\phi}{p} & \frac{\cos\theta\sin\phi}{p} & -\frac{\sin\theta}{p}\\ -\frac{\sin\phi}{p\sin\theta} & \frac{\cos\phi}{p\sin\theta} & 0\end{pmatrix},$$

so that

$$\begin{pmatrix}\frac{\partial}{\partial p_x}\\ \frac{\partial}{\partial p_y}\\ \frac{\partial}{\partial p_z}\end{pmatrix}=\begin{pmatrix}\sin\theta\cos\phi & \frac{\cos\theta\cos\phi}{p} & -\frac{\sin\phi}{p\sin\theta}\\ \sin\theta\sin\phi & \frac{\cos\theta\sin\phi}{p} & \frac{\cos\phi}{p\sin\theta}\\ \cos\theta & -\frac{\sin\theta}{p} & 0\end{pmatrix}\begin{pmatrix}\frac{\partial}{\partial p}\\ \frac{\partial}{\partial\theta}\\ \frac{\partial}{\partial\phi}\end{pmatrix},$$

where we used the transposed Jacobian matrix $(J^{-1})^T$. We find the relations (see Eq. (5.33) in [5])

$$\frac{\partial}{\partial p_x}=\sin\theta\cos\phi\frac{\partial}{\partial p}+\frac{\cos\theta\cos\phi}{p}\frac{\partial}{\partial\theta}-\frac{\sin\phi}{p\sin\theta}\frac{\partial}{\partial\phi} \tag{4.23a}$$

$$\frac{\partial}{\partial p_y}=\sin\theta\sin\phi\frac{\partial}{\partial p}+\frac{\cos\theta\sin\phi}{p}\frac{\partial}{\partial\theta}+\frac{\cos\phi}{p\sin\theta}\frac{\partial}{\partial\phi} \tag{4.23b}$$

$$\frac{\partial}{\partial p_z}=\cos\theta\frac{\partial}{\partial p}-\frac{\sin\theta}{p}\frac{\partial}{\partial\theta}. \tag{4.23c}$$

With the momentum-vector in spherical coordinates and $\delta\boldsymbol{B} = (\delta B_x, \delta B_y, \delta B_z)$ we find for the cross product

$$\boldsymbol{p} \times \delta\boldsymbol{B} = \begin{pmatrix} p\sin\theta\sin\phi\delta B_z - p\cos\theta\delta B_y \\ p\cos\theta\delta B_x - p\sin\theta\cos\phi\delta B_z \\ p\sin\theta\cos\phi\delta B_y - p\sin\theta\sin\phi\delta B_x \end{pmatrix},$$

so that

$$\begin{aligned}(\boldsymbol{p} \times \delta\boldsymbol{B}) \cdot \nabla_p f_0 &= \left(p\sin\theta\sin\phi\delta B_z - p\cos\theta\delta B_y\right)\frac{\partial f_0}{\partial p_x} \\ &+ (p\cos\theta\delta B_x - p\sin\theta\cos\phi\delta B_z)\frac{\partial f_0}{\partial p_y} \\ &+ \left(p\sin\theta\cos\phi\delta B_y - p\sin\theta\sin\phi\delta B_x\right)\frac{\partial f_0}{\partial p_z}.\end{aligned}$$

Now we substitute the Cartesian partial derivatives by (4.23a)–(4.23c). On assuming that the averaged distribution function is gyrotropic, i.e., independent of the particle phase angle ϕ, we neglect any derivatives with respect to ϕ and find

$$\begin{aligned}&(\boldsymbol{p} \times \delta\boldsymbol{B}) \cdot \nabla_p f_0 \\ &= \left(p\sin\theta\sin\phi\delta B_z - p\cos\theta\delta B_y\right)\left(\sin\theta\cos\phi\frac{\partial}{\partial p} + \frac{\cos\theta\cos\phi}{p}\frac{\partial}{\partial\theta}\right)f_0 \\ &+ (p\cos\theta\delta B_x - p\sin\theta\cos\phi\delta B_z)\left(\sin\theta\sin\phi\frac{\partial}{\partial p} + \frac{\cos\theta\sin\phi}{p}\frac{\partial}{\partial\theta}\right)f_0 \\ &+ \left(p\sin\theta\cos\phi\delta B_y - p\sin\theta\sin\phi\delta B_x\right)\left(\cos\theta\frac{\partial}{\partial p} - \frac{\sin\theta}{p}\frac{\partial}{\partial\theta}\right)f_0.\end{aligned}$$

Note that the terms proportional to $\partial f_0/\partial p$ vanish, so that

$$\begin{aligned}(\boldsymbol{p} \times \delta\boldsymbol{B}) \cdot \nabla_p f_0 &= \left(\sin\theta\sin\phi\delta B_z - \cos\theta\delta B_y\right)\cos\theta\cos\phi\frac{\partial f_0}{\partial\theta} \\ &+ (\cos\theta\delta B_x - \sin\theta\cos\phi\delta B_z)\cos\theta\sin\phi\frac{\partial f_0}{\partial\theta} \\ &- \left(\sin\theta\cos\phi\delta B_y - \sin\theta\sin\phi\delta B_x\right)\sin\theta\frac{\partial f_0}{\partial\theta}.\end{aligned}$$

Furthermore, terms proportional to δB_z vanish. By expanding and collecting all remaining terms proportional to δB_x and δB_y we obtain

$$(\boldsymbol{p} \times \delta\boldsymbol{B}) \cdot \nabla_p f_0 = \left[\sin\phi\delta B_x - \cos\phi\delta B_y\right]\frac{\partial}{\partial\theta}.$$

With the gyrofrequency $|\Omega| = qB/m$ we obtain finally

$$-\frac{q}{m}(\boldsymbol{p} \times \delta \boldsymbol{B}) \cdot \nabla_p f_0 = -\frac{|\Omega|}{B}\left[\sin\phi \delta B_x - \cos\phi \delta B_y\right]\frac{\partial f_0}{\partial \theta}.$$

4.3 Hydrodynamic Description of Energetic Particles

Suppose that wave propagation is one-dimensional and is represented by a wave vector $\boldsymbol{k} = k\boldsymbol{e}_x$ in Cartesian coordinates (x, y, z) and that $\partial/\partial y = \partial/\partial z = 0$. By writing $\boldsymbol{u} = (u_x, u_y, u_z))$ and $\boldsymbol{B} = (B_x, B_y, B_z)$ we find for the MHD equations (MHD-1)–(MHD-6) (see Sect. 3.9) the 1D transport equations

$$\frac{\partial \rho}{\partial t} + \frac{\partial}{\partial x}(\rho u_x) = 0 \tag{T-1}$$

$$\rho \frac{du_x}{dt} = -\frac{\partial}{\partial x}(P_g + P_c) - \frac{1}{2\mu}\frac{\partial}{\partial x}\left(B_x^2 + B_z^2\right) \tag{T-2}$$

$$\rho \frac{du_y}{dt} = \frac{B_x}{\mu}\frac{\partial B_y}{\partial x} \tag{T-3}$$

$$\rho \frac{du_z}{dt} = \frac{B_x}{\mu}\frac{\partial B_z}{\partial x} \tag{T-4}$$

$$B_x = const. \tag{T-5}$$

$$\frac{\partial B_y}{\partial t} = \frac{\partial}{\partial x}\left[u_y B_x - u_x B_y\right] \tag{T-6}$$

$$\frac{\partial B_z}{\partial t} = \frac{\partial}{\partial x}\left[u_z B_x - u_x B_z\right] \tag{T-7}$$

$$\frac{dP_g}{dt} + \gamma_g P_g \frac{\partial u_x}{\partial x} = 0 \tag{T-8}$$

$$\frac{dP_c}{dt} + \gamma_c P_c \frac{\partial u_x}{\partial x} - \kappa \frac{\partial^2 P_c}{\partial x^2} = 0, \tag{T-9}$$

where the convective derivative is given by

$$\frac{d}{dt} = \frac{\partial}{\partial t} + u_x \frac{\partial}{\partial x}. \tag{4.24}$$

Problem 4.6 Derive the dispersion relation (4.27) for linear wave modes in a cosmic ray mediated plasma.

Solution First we linearize the above Eqs. (T-1)–(T-9) about a uniform equilibrium state, where

$$\rho = \rho_0 + \delta\rho \qquad P_c = P_{c0} + \delta P_c \qquad P_g = P_{g0} + \delta P_g$$

$$\boldsymbol{B} = \begin{pmatrix} B_x \\ B_{y0} + \delta B_y \\ B_{z0} + \delta B_z \end{pmatrix} \qquad \boldsymbol{u} = \begin{pmatrix} \delta u_x \\ \delta u_y \\ \delta u_z \end{pmatrix}.$$

Note that $\boldsymbol{u}_0 = 0$ and $B_x = const.$ and that any derivative $\delta u_x \partial/\partial x$ will be of second order in the small quantity and can therefore be neglected. Thus, the convective derivative reduces to $d/dt = \partial/\partial t$. The linearized 1D transport equations can then be written as

$$\frac{\partial \delta\rho}{\partial t} + \rho_0 \frac{\partial \delta u_x}{\partial x} = 0$$

$$\rho_0 \frac{\partial \delta u_x}{\partial t} + \frac{\partial \delta P_g}{\partial x} + \frac{\partial \delta P_c}{\partial x} + \frac{B_{y0}}{\mu}\frac{\partial \delta B_y}{\partial x} + \frac{B_{z0}}{\mu}\frac{\partial \delta B_z}{\partial x} = 0$$

$$\rho_0 \frac{\partial \delta u_y}{\partial t} - \frac{B_x}{\mu}\frac{\partial \delta B_y}{\partial x} = 0$$

$$\rho_0 \frac{\partial \delta u_z}{\partial t} - \frac{B_x}{\mu}\frac{\partial \delta B_z}{\partial x} = 0$$

$$\frac{\partial \delta B_y}{\partial t} - B_x \frac{\partial \delta u_y}{\partial x} + B_{y0}\frac{\partial \delta u_x}{\partial x} = 0$$

$$\frac{\partial \delta B_z}{\partial t} - B_x \frac{\partial \delta u_z}{\partial x} + B_{z0}\frac{\partial \delta u_x}{\partial x} = 0$$

$$\frac{\partial \delta P_g}{\partial t} + \gamma_g P_g \frac{\partial \delta u_x}{\partial x} = 0$$

$$\frac{\partial \delta P_c}{\partial t} + \gamma_c P_c \frac{\partial \delta u_x}{\partial x} - \kappa \frac{\partial^2 \delta P_c}{\partial x^2} = 0.$$

We seek solutions of the form $\Psi = \Psi_0 \exp[i(\omega t - kx)]$ (compare with Problem 2.10), so that the derivatives can be written as

$$\frac{\partial \Psi}{\partial t} = i\omega\Psi \qquad \frac{\partial \Psi}{\partial x} = -ik\Psi \qquad \frac{\partial^2 \Psi}{\partial x^2} = -k^2\Psi,$$

where Ψ can be substituted by any of the variables $\delta\rho, \delta B, \delta u$, and δP. By substituting these results back into the transport equations, multiplying each equation by the complex number $-i$, dividing by k, and using $V_p = \omega/k$ we obtain

$$V_p\delta\rho - \rho_0\delta u_x = 0$$

$$V_p\rho_0\delta u_x - \delta P_g - \delta P_c - \frac{B_{y0}}{\mu}\delta B_y - \frac{B_{z0}}{\mu}\delta B_z = 0$$

$$V_p\rho_0\delta u_y + \frac{B_x}{\mu}\delta B_y = 0 \tag{4.25a}$$

$$V_p\rho_0\delta u_z + \frac{B_x}{\mu}\delta B_z = 0 \tag{4.25b}$$

$$V_p\delta B_y + B_x\delta u_y - B_{y0}\delta u_x = 0 \tag{4.25c}$$

$$V_p\delta B_z + B_x\delta u_z - B_{z0}\delta u_x = 0 \tag{4.25d}$$

$$V_p\delta P_g - \gamma_g P_{g0}\delta u_x = 0$$

$$V_p\delta P_c - \gamma_c P_{c0}\delta u_x - i\kappa k\delta P_c = \left(V_p - i\kappa k\right)\delta P_c - \gamma_c P_{c0}\delta u_x = 0.$$

We combine now Eqs. (4.25a) and (4.25c) to substitute δu_y and Eqs. (4.25b) and (4.25d) to substitute δu_z and obtain

$$V_p B_{y0}\delta u_x - \left(V_p^2 - V_x^2\right)\delta B_y = 0$$

$$V_p B_{z0}\delta u_x - \left(V_p^2 - V_x^2\right)\delta B_z = 0,$$

where we used $V_x^2 = B_x^2/\mu\rho_0$. The set of equations reduces to

$$V_p\delta\rho - \rho_0\delta u_x = 0$$

$$V_p\rho_0\delta u_x - \delta P_g - \delta P_c - \frac{B_{y0}}{\mu}\delta B_y - \frac{B_{z0}}{\mu}\delta B_z = 0$$

$$V_p B_{y0}\delta u_x - \left(V_p^2 - V_x^2\right)\delta B_y = 0$$

$$V_p B_{z0}\delta u_x - \left(V_p^2 - V_x^2\right)\delta B_z = 0$$

$$V_p\delta P_g - \gamma_g P_{g0}\delta u_x = 0$$

$$V_p\delta P_c - \gamma_c P_{c0}\delta u_x - i\kappa k\delta P_c = \left(V_p - i\kappa k\right)\delta P_c - \gamma_c P_{c0}\delta u_x = 0.$$

It is convenient to rewrite this set of equations in matrix form

$$\underline{\boldsymbol{M}} = \begin{pmatrix} V_p & -\rho_0 & 0 & 0 & 0 & 0 \\ 0 & V_p\rho_0 & -B_{y0}/\mu & -B_{z0}/\mu & 0 & 0 \\ 0 & V_p B_{y0} & -(V_p^2 - V_x^2) & 0 & 0 & 0 \\ 0 & V_p B_{z0} & 0 & -(V_p^2 - V_x^2) & 0 & 0 \\ 0 & -\gamma_g P_{g0} & 0 & 0 & V_p & 0 \\ 0 & -\gamma_g P_{c0} & 0 & 0 & 0 & (V_p - i\kappa k) \end{pmatrix},$$

so that

$$\underline{\boldsymbol{M}} \cdot \delta \boldsymbol{\Psi} = 0,$$

where the vector $\delta\boldsymbol{\Psi} = (\delta\rho, \delta u_x, \delta B_y, \delta B_z, \delta P_g, \delta P_c)^T$. As in Problem 2.10, the trivial solution is given by $\delta\boldsymbol{\Psi} = 0$, and the non-trivial solutions are given by the *eigenvalues* of the matrix $\underline{\boldsymbol{M}}$. Therefore, we calculate the characteristic polynomial $\det(\underline{\boldsymbol{M}}) = 0$. Developing for the first column we find

$$\det(\underline{\boldsymbol{M}}) = V_p \det \begin{pmatrix} V_p\rho_0 & -B_{y0}/\mu & -B_{z0}/\mu & 0 & 0 \\ V_pB_{y0} & -(V_p^2 - V_x^2) & 0 & 0 & 0 \\ V_pB_{z0} & 0 & -(V_p^2 - V_x^2) & 0 & 0 \\ -\gamma_g P_{g0} & 0 & 0 & V_p & 0 \\ -\gamma_g P_{c0} & 0 & 0 & 0 & (V_p - \imath\kappa k) \end{pmatrix}.$$

The remaining determinant can readily be calculated and we obtain

$$\det(\underline{\boldsymbol{M}}) = V_p \Big[V_p^2\rho_0 \left(V_p^2 - V_x^2\right)^2 (V_p - \imath\kappa k) - \frac{B_{y0}^2}{\mu} V_p^2 (V_p^2 - V_x^2)(V_p - \imath\kappa k)$$
$$- \frac{B_{z0}^2}{\mu} V_p^2 (V_p^2 - V_x^2)(V_p - \imath\kappa k) - \gamma_g P_{g0} \left(V_p^2 - V_x^2\right)^2 (V_p - \imath\kappa k)$$
$$-\gamma_c P_{c0} \left(V_p^2 - V_x^2\right)^2 V_p \Big] = 0,$$

where the remaining determinant was developed for the first row. The factor $(V_p^2 - V_x^2)$ can be pulled out of the square brackets. We also combine the second and third term in square brackets, leading to

$$0 = V_p \left(V_p^2 - V_x^2\right) \left[V_p^2 \rho_0 \left(V_p^2 - V_x^2\right) (V_p - \imath\kappa k) - \frac{B_{y0}^2 + B_{z0}^2}{\mu} V_p^2 (V_p - \imath\kappa k) \right.$$
$$\left. - \gamma_g P_{g0} \left(V_p^2 - V_x^2\right) (V_p - \imath\kappa k) - \gamma_c P_{c0} \left(V_p^2 - V_x^2\right) V_p \right].$$

Recall that $V_x^2 = B_x^2/\mu\rho_0$, we can combine the V_x^2 summand in the first term with the second term by using $B_0^2 = B_x^2 + B_{y0}^2 + B_{z0}^2$. We also divide by ρ_0 to obtain

$$0 = V_p \left(V_p^2 - V_x^2\right) \left[V_p^4 (V_p - \imath\kappa k) - \frac{B_0^2}{\mu\rho_0} V_p^2 (V_p - \imath\kappa k) \right.$$
$$\left. - \frac{\gamma_g P_{g0}}{\rho_0} \left(V_p^2 - V_x^2\right) (V_p - \imath\kappa k) - \frac{\gamma_c P_{c0}}{\rho_0} \left(V_p^2 - V_x^2\right) V_p \right].$$

By using the definitions

$$V_A^2 = \frac{B_0^2}{\mu\rho_0} \qquad a_{g0}^2 = \frac{\gamma_g P_{g0}}{\rho_0} \qquad a_{c0}^2 = \frac{\gamma_c P_{c0}}{\rho_0} \qquad a_*^2 = a_{g0}^2 + a_{c0}^2 \tag{4.26}$$

we obtain

$$\begin{aligned} 0 = V_p\left(V_p^2 - V_x^2\right)\big[&V_p^4(V_p - i\kappa k) - V_A^2 V_p^2(V_p - i\kappa k) \\ &- a_{g0}^2\left(V_p^2 - V_x^2\right)(V_p - i\kappa k) - a_{c0}^2\left(V_p^2 - V_x^2\right)V_p\big]. \end{aligned}$$

We reorder the expressions in square brackets from highest to lowest order in V_p and obtain the dispersion relation

$$\begin{aligned} 0 = V_p\left(V_p^2 - V_x^2\right)\big[&V_p^5 - i\kappa k V_p^4 - \left(V_A^2 + a_*^2\right)V_p^3 \\ &+ i\kappa k\left(V_A^2 + a_{g0}^2\right)V_p^2 + a_*^2 V_x^2 V_p - i\kappa k a_{g0}^2 V_x^2\big]. \end{aligned} \tag{4.27}$$

Problem 4.7 By considering the *long wavelength* limit of the dispersion relation (4.27), show that the fast and slow magnetosonic modes are damped by cosmic rays, since the waves propagate approximately according to

$$V_p = V_{f,s} + i\kappa k\beta + O\left((\kappa k)^2\right), \tag{4.28}$$

where

$$\beta = \frac{a_{c0}^2\left(V_{f,s}^2 - V_x^2\right)}{2\left[\left(V_A^2 + a_*^2\right)V_{f,s}^2 - 2a_*^2 V_x^2\right]}. \tag{4.29}$$

Solution We begin by considering only the square bracket in the dispersion relation (4.27), which is given by

$$\begin{aligned} &V_p^5 - i\kappa k V_p^4 - \left(V_A^2 + a_*^2\right)V_p^3 \\ &\quad + i\kappa k\left(V_A^2 + a_{g0}^2\right)V_p^2 + a_*^2 V_x^2 V_p - i\kappa k a_{g0}^2 V_x^2 = 0. \end{aligned} \tag{4.30}$$

By setting $\kappa = 0$, i.e., no particle diffusion, we obtain

$$V_p\left[V_p^4 - \left(V_A^2 + a_*^2\right)V_p^2 + a_*^2 V_x^2\right] = 0, \tag{4.31}$$

which has the solutions $V_p = 0$ and (by solving the biquadratic formula)

$$V_p^2 = V_{f,s}^2 = \frac{1}{2}\left[V_A^2 + a_*^2 \pm \sqrt{\left(V_A^2 + a_*^2\right)^2 - 4a_*^2 V_x^2}\right], \tag{4.32}$$

the speed for the fast (+) and slow (−) magnetosonic modes. The difference between these wave modes and those derived in the usual MHD theory (see Problem 3.19) is the presence of the *mixed sound speed* $a_* = \sqrt{(\gamma_g P_{g0} + \gamma_c P_{c0})/\rho_0}$, indicating that the cosmic rays couple to the background plasma and alter the phase speed for these wave modes.

We assume now that the phase speed V_p (now with $\kappa \neq 0$) can be approximated by the fast/slow wave modes plus a small perturbation, $V_p = V_{f,s} + \varepsilon\beta$, where $\varepsilon = i\kappa k$ is a small but non-zero quantity and β (independent of κ) remains to be determined. All appearing orders (up to the fifth) can be written as

$$
\begin{aligned}
V_p &= V_{f,s} + \varepsilon\beta & V_p^4 &= V_{f,s}^4 + 4\varepsilon V_{f,s}^3\beta + O\left(\varepsilon^2\right) \\
V_p^2 &= V_{f,s}^2 + 2\varepsilon V_{f,s}\beta + O\left(\varepsilon^2\right) & V_p^5 &= V_{f,s}^5 + 5\varepsilon V_{f,s}^4\beta + O\left(\varepsilon^2\right) \\
V_p^3 &= V_{f,s}^3 + 3\varepsilon V_{f,s}^2\beta + O\left(\varepsilon^2\right).
\end{aligned}
$$

Substituting these approximations into Eq. (4.30) and using $\varepsilon \equiv i\kappa k$ we obtain

$$
\begin{aligned}
&V_{f,s}^5 + 5\varepsilon V_{f,s}^4\beta - \varepsilon\left(V_{f,s}^4 + 4\varepsilon V_{f,s}^3\beta\right) - \left(V_A^2 + a_*^2\right)\left(V_{f,s}^3 + 3\varepsilon V_{f,s}^2\beta\right) \\
&\quad + \varepsilon\left(V_A^2 + a_{g0}^2\right)\left(V_{f,s}^2 + 2\varepsilon V_{f,s}\beta\right) + a_*^2 V_x^2\left(V_{f,s} + \varepsilon\beta\right) - \varepsilon a_{g0}^2 V_x^2 = 0.
\end{aligned}
$$

The two terms of order $O(\varepsilon^2)$ may be neglected for further calculations. We rewrite the equation in terms of orders of ε, which leads to

$$
\begin{aligned}
&V_{f,s}\left[V_{f,s}^4 - \left(V_A^2 + a_*^2\right)V_{f,s}^2 + a_*^2 V_x^2\right] \\
&\quad - \varepsilon V_{f,s}^4 - \varepsilon a_{g0}^2 V_x^2 + \varepsilon\left(V_A^2 + a_{g0}^2\right)V_{f,s}^2 \\
&\quad + 5\varepsilon V_{f,s}^4\beta - 3\varepsilon\left(V_A^2 + a_*^2\right)V_{f,s}^2\beta + a_*^2 V_x^2\varepsilon\beta = 0.
\end{aligned}
$$

Obviously, the first three terms is zero since $V_{f,s}$ is a solution to Eq. (4.31). All remaining terms include ε, thus, we may divide the equation by $\varepsilon \neq 0$ and obtain

$$
-V_{f,s}^4 - a_{g0}^2 V_x^2 + \left(V_A^2 + a_{g0}^2\right)V_{f,s}^2 + \beta\left[5V_{f,s}^4 - 3\left(V_A^2 + a_*^2\right)V_{f,s}^2 + a_*^2 V_x^2\right] = 0.
$$

By pulling the first three terms to the right-hand side and dividing the equation by the expression in square brackets we obtain

$$
\beta = \frac{N}{D} = \frac{V_{f,s}^4 - \left(V_A^2 + a_{g0}^2\right)V_{f,s}^2 + a_{g0}^2 V_x^2}{5V_{f,s}^4 - 3\left(V_A^2 + a_*^2\right)V_{f,s}^2 + a_*^2 V_x^2}. \tag{4.33}
$$

The numerator can be simplified by using the relation $a_{g0}^2 + a_{c0}^2 = a_*^2$, leading to

$$
N = V_{f,s}^4 - \left(V_A^2 + a_{g0}^2\right)V_{f,s}^2 + a_{g0}^2 V_x^2
$$

$$= \left[V_{f,s}^4 - \left(V_A^2 + a_*^2\right) V_{f,s}^2 + a_*^2 V_x^2\right] + a_{c0}^2 \left(V_{f,s}^2 - V_x^2\right)$$
$$= a_{c0}^2 \left(V_{f,s}^2 - V_x^2\right), \tag{4.34}$$

where the term in square brackets is zero, since $V_{f,s}$ is a solution to Eq. (4.31) (see above). The dominator can be simplified in a similar manner to obtain

$$D = 5V_{f,s}^4 - 3\left(V_A^2 + a_*^2\right) V_{f,s}^2 + a_*^2 V_x^2$$
$$= 4V_{f,s}^4 - 2\left(V_A^2 + a_*^2\right) V_{f,s}^2 + \left[V_{f,s}^4 - \left(V_A^2 + a_*^2\right) V_{f,s}^2 + a_*^2 V_x^2\right]$$
$$= 4V_{f,s}^4 - 2\left(V_A^2 + a_*^2\right) V_{f,s}^2. \tag{4.35}$$

Using $V_{f,s}^4 = \left(V_A^2 + a_*^2\right) V_{f,s}^2 - a_*^2 V_x^2$ (see Eq. (4.31) above) we can rewrite the expression as

$$D = 4\left(V_A^2 + a_*^2\right) V_{f,s}^2 - 4a_*^2 V_x^2 - 2\left(V_A^2 + a_*^2\right) V_{f,s}^2$$
$$= 2\left(V_A^2 + a_*^2\right) V_{f,s}^2 - 4a_*^2 V_x^2. \tag{4.36}$$

Substituting both results for N and D back into Eq. (4.33) we obtain the expression for β in Eq. (4.29).

Problem 4.8 Show that in the opposite limit, *short wavelength* modes decouple from the cosmic rays in that they propagate at the thermal magnetosonic speed, but are nonetheless damped by cosmic rays since

$$V_p = V_{f,s} + i\frac{\mu}{2\kappa k}, \tag{4.37}$$

where $V_{f,s}$ is the fast/slow magnetosonic speed (see previous problem) for the thermal plasma (i.e., the dispersion relation contains only the thermal pressure P_{g0} with no contribution from P_{c0}), and

$$\mu = \frac{a_{c0}^2\left(V_p^2 - V_x^2\right)}{2\left[\left(V_A^2 + a_{g0}^2\right) V_{f,s}^2 - 2a_{g0}^2 V_x^2\right]}. \tag{4.38}$$

Solution First, since $V_{f,s}$ is the fast/slow magnetosonic speed for the *thermal* plasma, it follows that $a_{c0}^2 = 0$ and, therefore, $a_*^2 = a_{g0}^2 + a_{c0}^2 = a_{g0}^2$. Therefore, Eq. (4.31) can be written as

$$V_{f,s}^4 - \left(V_A^2 + a_{g0}^2\right) V_{f,s}^2 + a_{g0}^2 V_x^2 = 0. \tag{4.39}$$

As before we begin with Eq. (4.30) and assume that the phase speed V_p can be approximated by the fast/slow modes plus a small perturbation, $V_p = V_{f,s} + \varepsilon\mu$, where the perturbation is given by $\varepsilon = 1/i\kappa k$ and μ has to be determined. By substituting ε into Eq. (4.30) we find

$$V_p^5 - \frac{1}{\varepsilon}V_p^4 - \left(V_A^2 + a_*^2\right)V_p^3 + \frac{1}{\varepsilon}\left(V_A^2 + a_{g0}^2\right)V_p^2 + a_*^2 V_x^2 V_p - \frac{1}{\varepsilon}a_{g0}^2 V_x^2 = 0. \tag{4.40}$$

All appearing orders of V_p (up to the fifth) can be written as

$$\begin{aligned} V_p &= V_{f,s} + \varepsilon\mu & V_p^4 &= V_{f,s}^4 + 4\varepsilon V_{f,s}^3\mu + O\left(\varepsilon^2\right) \\ V_p^2 &= V_{f,s}^2 + 2\varepsilon V_{f,s}\mu + O\left(\varepsilon^2\right) & V_p^5 &= V_{f,s}^5 + 5\varepsilon V_{f,s}^4\mu + O\left(\varepsilon^2\right) \\ V_p^3 &= V_{f,s}^3 + 3\varepsilon V_{f,s}^2\mu + O\left(\varepsilon^2\right). \end{aligned} \tag{4.41}$$

Substituting these approximations into Eq. (4.40) and multiplying the equation with ε we obtain

$$\begin{aligned} &\varepsilon\left[V_{f,s}^5 + \cancel{5\varepsilon V_{f,s}^4\mu}\right] - \left[V_{f,s}^4 + 4\varepsilon V_{f,s}^3\mu\right] - \varepsilon\left(V_A^2 + a_*^2\right)\left[V_{f,s}^3 + \cancel{3\varepsilon V_{f,s}^2\mu}\right] \\ &\quad + \left(V_A^2 + a_{g0}^2\right)\left[V_{f,s}^2 + 2\varepsilon V_{f,s}\mu\right] + \varepsilon a_*^2 V_x^2\left[V_{f,s} + \cancel{\varepsilon\mu}\right] - a_{g0}^2 V_x^2 = 0. \end{aligned} \tag{4.42}$$

The three terms of order $O(\varepsilon^2)$ will be neglected for further calculations. In the following we rewrite the equation in terms of orders of ε, leading to

$$\begin{aligned} &-V_{f,s}^4 + \left(V_A^2 + a_{g0}^2\right)V_{f,s}^2 - a_{g0}^2 V_x^2 + \varepsilon V_{f,s}^5 - \varepsilon\left(V_A^2 + a_*^2\right)V_{f,s}^3 + \varepsilon a_*^2 V_x^2 V_{f,s} \\ &\quad + 2\varepsilon\mu\left(V_A^2 + a_{g0}^2\right)V_{f,s} - 4\mu\varepsilon V_{f,s}^3 = 0. \end{aligned} \tag{4.43}$$

Obviously, the first three terms are zero because of relation (4.39). All remaining terms include an ε, so we may divide the equation by $\varepsilon \neq 0$. We may also divide by $V_{f,s}$ for simplification. In the last line we also pull μ out,

$$V_{f,s}^4 - \left(V_A^2 + a_*^2\right)V_{f,s}^2 + a_*^2 V_x^2 - 2\mu\left[2V_{f,s}^2 - \left(V_A^2 + a_{g0}^2\right)\right] = 0. \tag{4.44}$$

By pulling the term proportional to μ to the right-hand side and dividing by the expression in square brackets we obtain

$$\mu = \frac{N}{D} = \frac{1}{2}\frac{V_{f,s}^4 - \left(V_A^2 + a_*^2\right)V_{f,s}^2 + a_*^2 V_x^2}{2V_{f,s}^2 - \left(V_A^2 + a_{g0}^2\right)}. \tag{4.45}$$

The numerator can be simplified by using the relation $a_{g0}^2 + a_{c0}^2 = a_*^2$, leading to

$$N = V_{f,s}^4 - \left(V_A^2 + a_*^2\right)V_{f,s}^2 + a_*^2 V_x^2$$

$$= \left[V_{f,s}^4 - \left(V_A^2 + a_{g0}^2\right) V_{f,s}^2 + a_{g0}^2 V_x^2\right] - a_{c0}^2 \left(V_{f,s}^2 - V_x^2\right)$$
$$= -a_{c0}^2 \left(V_{f,s}^2 - V_x^2\right), \qquad (4.46)$$

where we used the relation (4.39) for the term in square brackets. We find

$$\mu = -\frac{1}{2}\frac{a_{c0}^2\left(V_{f,s}^2 - V_x^2\right)}{2V_{f,s}^2 - \left(V_A^2 + a_{g0}^2\right)}. \qquad (4.47)$$

However, since $\varepsilon = 1/i\kappa k$, we have to multiply by i/i, leading to

$$V_p = V_{f,s} - i\frac{\mu}{\kappa k}. \qquad (4.48)$$

Problem 4.9 Derive Burgers' equation

$$\frac{\partial u_x^1}{\partial \tau} + \alpha u_x^1 \frac{\partial u_x^1}{\partial \xi} = \lambda \frac{\partial^2 u_x^1}{\partial \xi^2} \qquad (4.49)$$

from the $O(\varepsilon^2)$ expansion of the magnetized fluid equations, where

$$\alpha = \frac{\left[(\gamma_g + 1)\, a_{g0}^2 + (\gamma_c + 1)\, a_{c0}^2\right]\left(V_p^2 - V_x^2\right) + 3\left(V_A^2 V_p^2 - a_*^2 V_x^2\right)}{2\left[\left(a_*^2 + V_A^2\right) V_p^2 - 2a_*^2 V_x^2\right]} \qquad (4.50)$$

and

$$\lambda = \frac{\kappa a_{c0}^2 \left(V_p^2 - V_x^2\right)}{2\left[\left(a_*^2 + V_A^2\right) V_p^2 - 2a_*^2 V_x^2\right]}. \qquad (4.51)$$

Solution We begin with the 1D MHD transport equations (T-1)–(T-9) and normalize the set of equations to obtain a dimensionless description. Using the method of multiple scales (see Problem 2.16) we introduce a time scale T such that the relationship

$$\frac{V_p T}{L} = 1 \qquad (4.52)$$

holds. We also introduce the following normalizations,

$$x = L\bar{x} \qquad t = T\bar{t} \qquad \boldsymbol{B} = B_0\bar{\boldsymbol{B}}$$
$$P_{g,c} = P_{g0,c0}\bar{P}_{g,c} \qquad \rho = \rho_0\bar{\rho} \qquad \boldsymbol{u} = V_p\bar{\boldsymbol{u}}. \qquad (4.53)$$

We obtain

$$\frac{\partial \bar{\rho}}{\partial \bar{t}} + \frac{\partial}{\partial \bar{x}}(\bar{\rho}\bar{u}_x) = 0 \tag{4.54a}$$

$$\bar{\rho}\frac{d\bar{u}_x}{d\bar{t}} = -\frac{\bar{a}_{g0}^2}{\gamma_g}\frac{\partial \bar{P}_g}{\partial \bar{x}} - \frac{\bar{a}_{c0}^2}{\gamma_c}\frac{\partial \bar{P}_c}{\partial \bar{x}} - \frac{\bar{V}_A^2}{2}\frac{\partial}{\partial \bar{x}}\left(\bar{B}_x^2 + \bar{B}_z^2\right) \tag{4.54b}$$

$$\bar{\rho}\frac{d\bar{u}_y}{d\bar{t}} = \bar{V}_A^2\bar{B}_x\frac{\partial \bar{B}_y}{\partial \bar{x}} \tag{4.54c}$$

$$\bar{\rho}\frac{d\bar{u}_z}{d\bar{t}} = \bar{V}_A^2\boldsymbol{B}_x\frac{\partial \bar{B}_z}{\partial \bar{x}} \tag{4.54d}$$

$$\bar{B}_x = const. \tag{4.54e}$$

$$\frac{\partial \bar{B}_y}{\partial \bar{t}} = \frac{\partial}{\partial \bar{x}}\left[\bar{u}_y\bar{B}_x - \bar{u}_x\bar{B}_y\right] \tag{4.54f}$$

$$\frac{\partial \bar{B}_z}{\partial \bar{t}} = \frac{\partial}{\partial \bar{x}}\left[\bar{u}_z\bar{B}_x - \bar{u}_x\bar{B}_z\right] \tag{4.54g}$$

$$\frac{d\bar{P}_g}{d\bar{t}} + \gamma_g\bar{P}_g\frac{\partial \bar{u}_x}{\partial \bar{x}} = 0 \tag{4.54h}$$

$$\frac{d\bar{P}_c}{d\bar{t}} + \gamma_c\bar{P}_c\frac{\partial \bar{u}_x}{\partial \bar{x}} - \nu\frac{\partial^2 \bar{P}_c}{\partial \bar{x}^2} = 0, \tag{4.54i}$$

where the long wavelength parameter is defined as $\nu = \kappa/(V_p L)$, and where we used the definitions (4.26). The convective derivative is given by

$$\frac{d}{d\bar{t}} = \frac{\partial}{\partial \bar{t}} + \frac{\partial}{\partial \bar{x}}. \tag{4.55}$$

Similar to Problem 2.16 we introduce fast and slow variables $\xi = \bar{x} - \bar{t}$ and $\tau = \varepsilon\bar{t}$, with

$$\frac{\partial}{\partial \bar{x}} = \frac{\partial}{\partial \xi} \qquad\qquad \frac{\partial}{\partial \bar{t}} = \varepsilon\frac{\partial}{\partial \tau} - \frac{\partial}{\partial \xi},$$

together with the expansions

$$\begin{aligned} &\bar{\rho} = 1 + \varepsilon\bar{\rho}^1 + \ldots && \bar{u}_x = \varepsilon\bar{u}_x^1 + \ldots && \bar{u}_z = \varepsilon\bar{u}_z^1 + \ldots \\ &\bar{B}_z = \bar{B}_z^0 + \varepsilon\bar{B}_z^1 + \ldots && \bar{P}_g = 1 + \varepsilon\bar{P}_g^1 + \ldots && \bar{P}_c = 1 + \varepsilon\bar{P}_c^1 + \ldots, \end{aligned}$$

where it is convenient to assume $\bar{u}_y = 0$ and $\bar{B}_y = 0$.

Note that for all further calculations in this Problem we will omit the bars over the various quantities with the exception of the sound speeds and the Alfvén speed,

$$\bar{a}_{g0,c0} = \frac{a_{g0,c0}}{V_p} \qquad \bar{V}_A = \frac{V_A}{V_p} \tag{4.56}$$

Now we derive a general form of the set of equations with expansions to the second order:

Equation (T-1) is

$$\varepsilon^2 \frac{\partial \rho^1}{\partial \tau} - \varepsilon \frac{\partial \rho^1}{\partial \xi} - \varepsilon^2 \frac{\partial \rho^2}{\partial \xi} + \varepsilon \frac{\partial u_x^1}{\partial \xi} + \varepsilon^2 \frac{\partial u_x^2}{\partial \xi} + \varepsilon^2 \rho^1 \frac{\partial u_x^1}{\partial \xi} + \varepsilon^2 u_x^1 \frac{\partial \rho^1}{\partial \xi} = 0. \tag{4.57}$$

Equation (T-2) is

$$\begin{aligned} &\varepsilon^2 \frac{\partial u_x^1}{\partial \tau} - \varepsilon \frac{\partial u_x^1}{\partial \xi} - \varepsilon^2 \frac{\partial u_x^2}{\partial \xi} - \varepsilon^2 \rho^1 \frac{\partial u_x^1}{\partial \xi} + \varepsilon^2 u_x^1 \frac{\partial u_x^1}{\partial \xi} \\ &= -\varepsilon \frac{\bar{a}_{g0}^2}{\gamma_g} \frac{\partial P_g^1}{\partial \xi} - \varepsilon^2 \frac{\bar{a}_{g0}^2}{\gamma_g} \frac{\partial P_g^2}{\partial \xi} - \varepsilon \frac{\bar{a}_{c0}^2}{\gamma_c} \frac{\partial P_c^1}{\partial \xi} - \varepsilon^2 \frac{\bar{a}_{c0}^2}{\gamma_c} \frac{\partial P_c^2}{\partial \xi} \\ &\quad - \varepsilon \bar{V}_A^2 B_z^0 \frac{\partial B_z^1}{\partial \xi} - \varepsilon^2 \bar{V}_A^2 B_z^0 \frac{\partial B_z^2}{\partial \xi} - \varepsilon^2 \bar{V}_A^2 B_z^1 \frac{\partial B_z^1}{\partial \xi}. \end{aligned} \tag{4.58}$$

Equation (T-4) is

$$\begin{aligned} &\varepsilon^2 \frac{\partial u_z^1}{\partial \tau} - \varepsilon \frac{\partial u_z^1}{\partial \xi} - \varepsilon^2 \frac{\partial u_z^2}{\partial \xi} - \varepsilon^2 \rho^1 \frac{\partial u_z^1}{\partial \xi} + \varepsilon^2 u_x^1 \frac{\partial u_z^1}{\partial \xi} \\ &= \varepsilon \bar{V}_A^2 B_x \frac{\partial B_z^1}{\partial \xi} + \varepsilon^2 \bar{V}_A^2 B_x \frac{\partial B_z^2}{\partial \xi}. \end{aligned} \tag{4.59}$$

Note that B_x is normalized.

Equation (T-7) is

$$\begin{aligned} &\varepsilon^2 \frac{\partial B_z^1}{\partial \tau} - \varepsilon \frac{\partial B_z^1}{\partial \xi} - \varepsilon^2 \frac{\partial B_z^2}{\partial \xi} \\ &= \varepsilon B_x \frac{\partial u_z^1}{\partial \xi} + \varepsilon^2 B_x \frac{\partial u_z^2}{\partial \xi} - \varepsilon B_z^0 \frac{\partial u_x^1}{\partial \xi} - \varepsilon^2 u_x^1 \frac{\partial B_z^1}{\partial \xi} - \varepsilon^2 B_z^1 \frac{\partial u_x^1}{\partial \xi} - \varepsilon^2 B_z^0 \frac{\partial u_x^2}{\partial \xi}. \end{aligned} \tag{4.60}$$

Equation (T-8) is

$$\varepsilon^2 \frac{\partial P_g^1}{\partial \tau} - \varepsilon \frac{\partial P_g^1}{\partial \xi} - \varepsilon^2 \frac{\partial P_g^2}{\partial \xi} + \varepsilon^2 u_x^1 \frac{\partial P_g^1}{\partial \xi} + \gamma_g \varepsilon \frac{\partial u_x^1}{\partial \xi}$$

$$+ \gamma_g \varepsilon^2 \frac{\partial u_x^2}{\partial \xi} + \gamma_g \varepsilon^2 P_g^1 \frac{\partial u_x^1}{\partial \xi} = 0. \tag{4.61}$$

Equation (T-9) is

$$\varepsilon^2 \frac{\partial P_c^1}{\partial \tau} - \varepsilon \frac{\partial P_c^1}{\partial \xi} - \varepsilon^2 \frac{\partial P_c^2}{\partial \xi} + \varepsilon^2 u_x^1 \frac{\partial P_c^1}{\partial \xi} + \gamma_c \varepsilon \frac{\partial u_x^1}{\partial \xi} + \gamma_c \varepsilon^2 \frac{\partial u_x^2}{\partial \xi}$$
$$+ \gamma_c \varepsilon^2 P_g^1 \frac{\partial u_x^1}{\partial \xi} - \nu \varepsilon \frac{\partial^2 P_c^1}{\partial \xi^2} - \nu \varepsilon^2 \frac{\partial^2 P_c^2}{\partial \xi^2} = 0. \tag{4.62}$$

- The lowest order system of equations is then given by

$$\frac{\partial \rho^1}{\partial \xi} = \frac{\partial u_x^1}{\partial \xi} \qquad -\frac{\partial B_z^1}{\partial \xi} = B_x \frac{\partial u_z^1}{\partial \xi} - B_z^0 \frac{\partial u_x^1}{\partial \xi}$$
$$\frac{\partial u_x^1}{\partial \xi} = \frac{\bar{a}_{g0}^2}{\gamma_g} \frac{\partial P_g^1}{\partial \xi} + \frac{\bar{a}_{c0}^2}{\gamma_c} \frac{\partial P_c^1}{\partial \xi} + \bar{V}_A^2 B_z^0 \frac{\partial B_z^1}{\partial \xi} \qquad \frac{\partial P_g^1}{\partial \xi} = \gamma_g \frac{\partial u_x^1}{\partial \xi}$$
$$-\frac{\partial u_z^1}{\partial \xi} = \bar{V}_A^2 B_x \frac{\partial B_z^1}{\partial \xi} \qquad \frac{\partial P_c^1}{\partial \xi} = \gamma_c \frac{\partial u_x^1}{\partial \xi},$$

and we find immediately

$$\rho_1 = u_x^1 \qquad B_z^1 = -B_x u_z^1 + B_z^0 u_x^1$$
$$u_x^1 = \frac{\bar{a}_{g0}^2}{\gamma_g} P_g^1 + \frac{\bar{a}_{c0}^2}{\gamma_c} P_c^1 + \bar{V}_A^2 B_z^0 B_z^1 \qquad P_g^1 = \gamma_g u_x^1$$
$$u_z^1 = -\bar{V}_A^2 B_x B_z^1 \qquad P_c^1 = \gamma_c u_x^1.$$

Substituting u_z^1 in the expression for B_z^1 gives

$$B_z^1 = \frac{B_z^0 u_x^1}{1 - \bar{V}_A^2 B_x^2} = \frac{B_z^0 u_x^1}{1 - \bar{V}_x^2}, \tag{4.63}$$

where it can be shown from the normalizations that $\bar{V}_x^2 = \bar{V}_A^2 B_x^2$. Provided that the relation

$$1 = \bar{a}_{g0}^2 + \bar{a}_{c0}^2 + \frac{\bar{V}_A^2 B_z^{0\,2}}{1 - \bar{V}_x^2} \tag{4.64}$$

holds, we have the following eigenvector solutions

$$\left(\rho^1, u_x^1, u_z^1, B_z^1, P_g^1, P_c^1\right) = u_x^1 \left(1, 1, -\frac{\bar{V}_A^2 B_x B_z^0}{1 - \bar{V}_x^2}, \frac{B_z^0 u_x^1}{1 - \bar{V}_x^2}, \gamma_g, \gamma_c\right). \tag{4.65}$$

Note that the relation (4.64) is the normalized dispersion relation for the long wavelength limit Eq. (4.31),

$$V_p^4 - \left(V_A^2 + a_*^2\right) V_p^2 + a_*^2 V_x^2 = 0. \tag{4.66}$$

- The second order set of transport equations is given by (where we used the results from (4.65))

$$
\begin{aligned}
-\frac{\partial \rho^2}{\partial \xi} + \frac{\partial u_x^2}{\partial \xi} &= -\frac{\partial u_x^1}{\partial \tau} - 2u_x^1 \frac{\partial u_x^1}{\partial \xi} \\
&\quad - \frac{\partial u_x^2}{\partial \xi} + \frac{\bar{a}_{g0}^2}{\gamma_g}\frac{\partial P_g^2}{\partial \xi} + \frac{\bar{a}_{c0}^2}{\gamma_c}\frac{\partial P_c^2}{\partial \xi} \\
+\bar{V}_A^2 B_z^0 \frac{\partial B_z^2}{\partial \xi} &= -\frac{\partial u_x^1}{\partial \tau} - \bar{V}_A^2 \frac{B_z^{0^2} u_x^1}{\left(1-\bar{V}_x^2\right)^2}\frac{\partial u_x^1}{\partial \xi} & \text{(4.67a)} \\
\bar{V}_A^2 B_x \frac{\partial B_z^2}{\partial \xi} + \frac{\partial u_z^2}{\partial \xi} &= -\frac{\bar{V}_A^2 B_x B_z^0}{1-\bar{V}_x^2}\frac{\partial u_x^1}{\partial \tau} & \text{(4.67b)} \\
-\frac{\partial B_z^2}{\partial \xi} - B_x \frac{\partial u_z^2}{\partial \xi} + B_z^0 \frac{\partial u_x^2}{\partial \xi} &= -\frac{B_z^0}{1-\bar{V}_x^2}\frac{\partial u_x^1}{\partial \tau} - 2\frac{B_z^0}{1-\bar{V}_x^2} u_x^1 \frac{\partial u_x^1}{\partial \xi} & \text{(4.67c)} \\
\frac{\partial P_g^2}{\partial \xi} - \gamma_g \frac{\partial u_x^2}{\partial \xi} &= \gamma_g \frac{\partial u_x^1}{\partial \tau} + \gamma_g\left(\gamma_g + 1\right) u_x^1 \frac{\partial u_x^1}{\partial \xi} \\
\frac{\partial P_c^2}{\partial \xi} - \gamma_c \frac{\partial u_x^2}{\partial \xi} &= \gamma_c \frac{\partial u_x^1}{\partial \tau} + \gamma_c\left(\gamma_c + 1\right) u_x^1 \frac{\partial u_x^1}{\partial \xi} - \nu\gamma_c \frac{\partial^2 u_x^1}{\partial \xi^2}.
\end{aligned}
$$

To derive Burger's equation we transform equation (4.67b) to obtain an equation for u_z^2, and substitute the result back into equation (4.67c), so that

$$
\begin{aligned}
&-\frac{\partial B_z^2}{\partial \xi} + \bar{V}_A^2 B_x^2 \frac{\partial B_z^2}{\partial \xi} + \frac{\bar{V}_A^2 B_x^2 B_z^0}{1-\bar{V}_x^2}\frac{\partial u_x^1}{\partial \tau} + B_z^0 \frac{\partial u_x^2}{\partial \xi} \\
&= -\frac{B_z^0}{1-\bar{V}_x^2}\frac{\partial u_x^1}{\partial \tau} - 2\frac{B_z^0}{1-\bar{V}_x^2} u_x^1 \frac{\partial u_x^1}{\partial \xi}.
\end{aligned}
$$

After substituting the relation $\bar{V}_x^2 = \bar{V}_A^2 B_x^2$ in the second term on the left-hand side, we pull the third and fourth term to the right-hand side, and multiply with -1, so that

$$\left(1-\bar{V}_x^2\right)\frac{\partial B_z^2}{\partial \xi} = \frac{B_z^0}{1-\bar{V}_x^2}\frac{\partial u_x^1}{\partial \tau} + \frac{\bar{V}_A^2 B_x^2 B_z^0}{1-\bar{V}_x^2}\frac{\partial u_x^1}{\partial \tau} + 2\frac{B_z^0}{1-\bar{V}_x^2} u_x^1 \frac{\partial u_x^1}{\partial \xi} + B_z^0 \frac{\partial u_x^2}{\partial \xi}. \tag{4.68}$$

The first and second term on the right-hand side can be simplified by

$$\frac{B_z^0}{1-\bar{V}_x^2}\frac{\partial u_x^1}{\partial \tau}+\frac{\bar{V}_A^2 B_x^2 B_z^0}{1-\bar{V}_x^2}\frac{\partial u_x^1}{\partial \tau}=\left(\frac{1}{1-\bar{V}_x^2}+\frac{\bar{V}_A^2 B_x^2}{1-\bar{V}_x^2}\right)B_z^0\frac{\partial u_x^1}{\partial \tau}$$
$$=\frac{1+\bar{V}_x^2}{1-\bar{V}_x^2}B_z^0\frac{\partial u_x^1}{\partial \tau}.$$

Substituting this result into Eq. (4.68) and dividing by $1-\bar{V}_x^2$ we find

$$\frac{\partial B_z^2}{\partial \xi}=\frac{1+\bar{V}_x^2}{\left(1-\bar{V}_x^2\right)^2}B_z^0\frac{\partial u_x^1}{\partial \tau}+2\frac{B_z^0}{\left(1-\bar{V}_x^2\right)^2}u_x^1\frac{\partial u_x^1}{\partial \xi}+\frac{B_z^0}{1-\bar{V}_x^2}\frac{\partial u_x^2}{\partial \xi}. \tag{4.69}$$

Using this result to substitute $\partial B_z^2/\partial \xi$ in the momentum equation (4.67a) we find

$$-\frac{\partial u_x^2}{\partial \xi}+\bar{a}_{g0}^2\left[\frac{\partial u_x^2}{\partial \xi}+\frac{\partial u_x^1}{\partial \tau}+(\gamma_g+1)\,u_x^1\frac{\partial u_x^1}{\partial \xi}\right]$$
$$+\bar{a}_{c0}^2\left[\frac{\partial u_x^2}{\partial \xi}+\frac{\partial u_x^1}{\partial \tau}+(\gamma_c+1)\,u_x^1\frac{\partial u_x^1}{\partial \xi}-\nu\frac{\partial^2 u_x^1}{\partial \xi^2}\right]$$
$$+\bar{V}_A^2 B_z^0\left[\frac{1+\bar{V}_x^2}{\left(1-\bar{V}_x^2\right)^2}B_z^0\frac{\partial u_x^1}{\partial \tau}+2\frac{B_z^0}{\left(1-\bar{V}_x^2\right)^2}u_x^1\frac{\partial u_x^1}{\partial \xi}+\frac{B_z^0}{1-\bar{V}_x^2}\frac{\partial u_x^2}{\partial \xi}\right]$$
$$=-\frac{\partial u_x^1}{\partial \tau}-\bar{V}_A^2\frac{B_z^{0\,2}u_x^1}{\left(1-\bar{V}_x^2\right)^2}\frac{\partial u_x^1}{\partial \xi}.$$

We rearrange this equations to sort in terms of orders of u_x and obtain

$$-\frac{\partial u_x^2}{\partial \xi}+\bar{a}_{g0}^2\frac{\partial u_x^2}{\partial \xi}+\bar{a}_{c0}^2\frac{\partial u_x^2}{\partial \xi}+\frac{\bar{V}_A^2 B_z^{0\,2}}{1-\bar{V}_x^2}\frac{\partial u_x^2}{\partial \xi}$$
$$+\frac{\partial u_x^1}{\partial \tau}+\bar{a}_{g0}^2\frac{\partial u_x^1}{\partial \tau}+\bar{a}_{c0}^2\frac{\partial u_x^1}{\partial \tau}+\bar{V}_A^2 B_z^{0\,2}\frac{1+\bar{V}_x^2}{\left(1-\bar{V}_x^2\right)^2}\frac{\partial u_x^1}{\partial \tau}$$
$$+\bar{a}_{g0}^2(\gamma_g+1)\,u_x^1\frac{\partial u_x^1}{\partial \xi}+\bar{a}_{c0}^2(\gamma_c+1)\,u_x^1\frac{\partial u_x^1}{\partial \xi}+3\frac{\bar{V}_A^2 B_z^{0\,2}}{\left(1-\bar{V}_x^2\right)^2}u_x^1\frac{\partial u_x^1}{\partial \xi}$$
$$=\bar{a}_{c0}^2\nu\frac{\partial^2 u_x^1}{\partial \xi^2}.$$

The first line is zero because of the relation (4.64). We find

$$\left[1 + \bar{a}_{g0}^2 + \bar{a}_{c0}^2 + \bar{V}_A^2 B_z^{0^2} \frac{1 + \bar{V}_x^2}{\left(1 - \bar{V}_x^2\right)^2}\right] \frac{\partial u_x^1}{\partial \tau}$$

$$+ \left[\bar{a}_{g0}^2 (\gamma_g + 1) + \bar{a}_{c0}^2 (\gamma_c + 1) + 3 \frac{\bar{V}_A^2 B_z^{0^2}}{\left(1 - \bar{V}_x^2\right)^2}\right] u_x^1 \frac{\partial u_x^1}{\partial \xi} = \bar{a}_{c0}^2 \nu \frac{\partial^2 u_x^1}{\partial \xi^2}.$$

By using the relation (4.64) to substitute $\bar{V}_A^2 B_z^{0^2}/(1 - \bar{V}_x^2) = 1 - \bar{a}_*^2$ in the square brackets, we find

$$\frac{1}{1 - \bar{V}_x^2} \left[\left(1 + \bar{a}_*^2\right)\left(1 - \bar{V}_x^2\right) + \left(1 - \bar{a}_*^2\right)\left(1 + \bar{V}_x^2\right)\right] \frac{\partial u_x^1}{\partial \tau}$$

$$+ \frac{1}{1 - \bar{V}_x^2} \left\{\left[\bar{a}_{g0}^2 (\gamma_g + 1) + \bar{a}_{c0}^2 (\gamma_c + 1)\right]\left(1 - \bar{V}_x^2\right) + 3\left(1 - \bar{a}_*^2\right)\right\} u_x^1 \frac{\partial u_x^1}{\partial \xi}$$

$$= \bar{a}_{c0}^2 \nu \frac{\partial^2 u_x^1}{\partial \xi^2}.$$

In the first line, the expression in square brackets can be simplified to

$$\left(1 + \bar{a}_*^2\right)\left(1 - \bar{V}_x^2\right) + \left(1 - \bar{a}_*^2\right)\left(1 + \bar{V}_x^2\right) = 2\left(1 - \bar{a}_*^2 \bar{V}_x^2\right),$$

and by multiplying the entire equation with $1 - \bar{V}_x^2$ we obtain

$$2\left(1 - \bar{a}_*^2 \bar{V}_x^2\right) \frac{\partial u_x^1}{\partial \tau} + \left\{\left[\bar{a}_{g0}^2 (\gamma_g + 1) + \bar{a}_{c0}^2 (\gamma_c + 1)\right]\left(1 - \bar{V}_x^2\right)\right.$$

$$\left. + 3\left(1 - \bar{a}_*^2\right)\right\} u_x^1 \frac{\partial u_x^1}{\partial \xi} = \left(1 - \bar{V}_x^2\right) \bar{a}_{c0}^2 \nu \frac{\partial^2 u_x^1}{\partial \xi^2}.$$

Now, on dividing the equation by $2\left(1 - \bar{a}_*^2 \bar{V}_x^2\right)$ we obtain Burger's equation

$$\frac{\partial u_x^1}{\partial \tau} + \alpha u_x^1 \frac{\partial u_x^1}{\partial \xi} = \lambda \frac{\partial^2 u_x^1}{\partial \xi^2} \tag{4.70}$$

with

$$\alpha = \frac{\left[\bar{a}_{g0}^2 (\gamma_g + 1) + \bar{a}_{c0}^2 (\gamma_c + 1)\right]\left(1 - \bar{V}_x^2\right) + 3\left(1 - \bar{a}_*^2\right)}{2\left(1 - \bar{a}_*^2 \bar{V}_x^2\right)}$$

$$\lambda = \frac{\bar{a}_{c0}^2 \nu \left(1 - \bar{V}_x^2\right)}{2\left(1 - \bar{a}_*^2 \bar{V}_x^2\right)}.$$

Note that this is the normalized form of Burger's equation. Finally, we rewrite the solution in a non-normalized form, so that

$$\alpha = \frac{\left[a_{g0}^2\left(\gamma_g+1\right)+a_{c0}^2\left(\gamma_c+1\right)\right]\left(V_p^2-V_x^2\right)+3\left(V_p^4-a_*^2V_p^2\right)}{2\left(V_p^4-a_*^2V_x^2\right)}.$$

By using the dispersion relation (4.66) we can transform the last term in the numerator by $3\left(V_p^4-a_*^2V_p^2\right)=3\left(V_A^2V_p^2-a_*^2V_x^2\right)$, so that

$$\alpha = \frac{\left[a_{g0}^2\left(\gamma_g+1\right)+a_{c0}^2\left(\gamma_c+1\right)\right]\left(V_p^2-V_x^2\right)+3\left(V_A^2V_p^2-a_*^2V_x^2\right)}{2\left(V_p^4-a_*^2V_x^2\right)}.$$

The denominator can be transformed in the same way as we transformed the solution in Problem 4.7. For λ we find

$$\begin{aligned}\lambda = \frac{a_{c0}^2\kappa\left(V_p^2-V_x^2\right)}{2\left(V_p^4-a_*^2V_x^2\right)} &= \frac{a_{c0}^2\kappa\left(V_p^2-V_x^2\right)}{2V_p^2\left[2V_p^2-\left(V_A^2+a_*^2\right)\right]}\\ &= \frac{a_{c0}^2\kappa\left(V_p^2-V_x^2\right)}{2\left[\left(V_A^2+a_*^2\right)V_p^2-2a_*^2V_x^2\right]},\end{aligned}$$

where we used the dispersion relation (4.66) in the denominator. Note that λ resembles the long wavelength limit result (4.29). Note also that we transformed $\nu \to \kappa$, since we use the non-normalized form of the Burgers' equation.

4.4 Application 1: Diffusive Shock Acceleration

Problem 4.10 Suppose that an upstream energetic particle distribution proportional to p^{-a} is convected into a shock with compression ratio r from upstream. In the absence of particle injection at the shock itself, calculate the reaccelerated downstream energetic particle spectrum, and explain what happens if $a < q = 3r/(r1)$ or $a > q$.

Solution The general solution for diffusive shock acceleration theory is given by

$$f(0,p) = \frac{3}{u_1-u_2}p^{-q}\int_{p_{inj}}^{p}(p')^q\left[u_1 f(-\infty,p')+\frac{Q(p')}{4\pi p'^2}\right]\frac{dp'}{p'}, \tag{4.71}$$

where $q = 3r/(r-1)$ and $r = u_1/u_2$ is the shock compression ratio, and p_{inj} is the injection momentum (compare also with Eq. (5.59) in [5]). $Q(p)$ denotes the injection of particles at the shock.

If there is no particle injection at the shock, i.e. $Q(p) = 0$, and assuming that the particle background distribution (upstream) is given by $f(-\infty, p) = Ap^{-a}$, we find

$$f(0,p) = \frac{3u_1}{u_1 - u_2} A p^{-q} \int_{p_{inj}}^{p} (p')^{q-a-1}\, dp'$$
$$= \frac{3r}{r-1} A p^{-q} \int_{p_{inj}}^{p} (p')^{q-a-1}\, dp'. \tag{4.72}$$

The integral is easily solved and we obtain

$$f(0,p) = q\frac{A}{q-a} p^{-q} \left[(p')^{q-a}\right]_{p'=p_{inj}}^{p}$$
$$= q\frac{A}{q-a}\left[p^{-a} - p_{inj}^{q-a} p^{-q}\right], \tag{4.73}$$

where we used $q = 3r/(r-1)$.

- **Case:** $a < q = 3r/(r-1)$. In this case the upstream spectrum $f(-\infty, p)$ is harder (flatter) than the one that can be accelerated at the shock (see Fig. 4.1 as an example), then the transmitted spectrum (downstream) will be the harder upstream spectrum,

$$f(0,p) \propto p^{-a}. \tag{4.74}$$

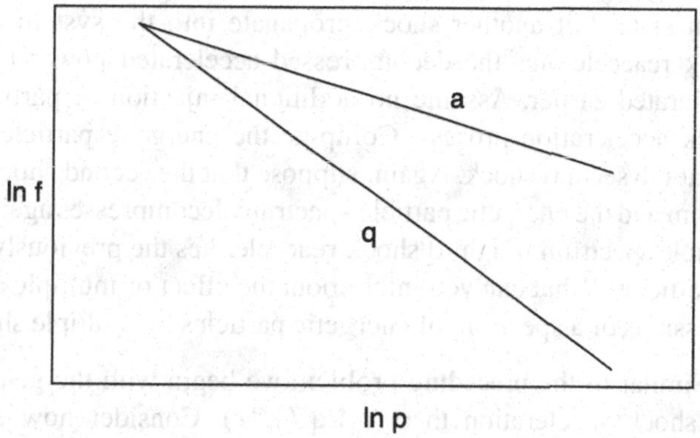

Fig. 4.1 Qualitatively, shown are the two spectra for the case $a < q$. The transmitted downstream spectrum is proportional to p^{-a}

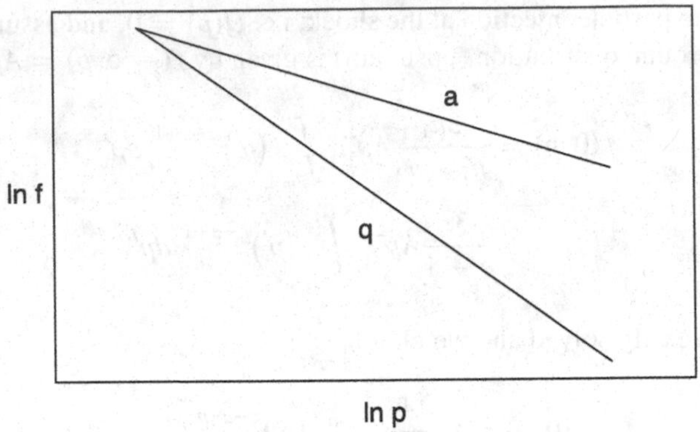

Fig. 4.2 Qualitatively, shown are the two spectra for the case $a > q$. The transmitted downstream spectrum is proportional to $p_{inj}^{q-a}p^{-q}$

- **Case:** $a > q$**.** In this case the upstream spectrum is softer (steeper) than the one that can be accelerated at the shock (see Fig. 4.2 as an example), thus, the shock accelerated spectrum will dominate,

$$f(0,p) \propto p_{inj}^{q-a} p^{-q}. \tag{4.75}$$

Problem 4.11 Suppose that a shock of compression ratio r accelerates n cm^{-3} particles injected as a monoenergetic source proportional to $\delta(p - p_0)$ at the shock, so producing a downstream energetic particle spectrum $\propto p^{-q}$. Now suppose the shock propagates out of the system and the compressed gas relaxes back to the ambient state. Let another shock propagate into the system and suppose that this shock reaccelerates the decompressed accelerated power law spectrum that was accelerated earlier. Assume no additional injection of particles into the diffusive shock acceleration process. Compute the energetic particle distribution reaccelerated at the second shock. Again, suppose that the second shock disappears out of the system and the energetic particle spectrum decompresses again. Derive the energetic particle spectrum if a third shock reaccelerates the previously accelerated spectrum of particles. What can you infer about the effect of multiple accelerations and decompressions of a spectrum of energetic particles by multiple shock waves?

Solution Similar to the preceding problem we begin with the general solution for diffusive shock acceleration theory, Eq. (4.71). Consider now a shock that propagates through a region with particle distribution $f(-\infty, p')$; in this case the distribution acts as an injection distribution $f(-\infty, p') = A\delta(p' - p_0)$, see, e.g., [3]. Substituting this distribution into Eq. (4.71) we find with $f(0,p) \equiv f(p)$

$$f_1(p) = \frac{3}{u_1 - u_2} p^{-q} \int_0^p (p')^q\, u_1 A \delta(p' - p_0) \frac{dp'}{p'}$$

$$= \frac{3u_1}{u_1 - u_2} A p^{-q} \int_0^p (p')^{q-1} \, \delta(p' - p_0) dp'. \tag{4.76}$$

Note that p_{inj} is set to zero here. This is somewhat arbitrarily; we just need to ensure that $0 < p_0 < p$. The integral is easily solved by

$$f_1(p) = qAp^{-q} (p_0)^{q-1} = \frac{Aq}{p_0} \left(\frac{p}{p_0} \right)^{-q}, \tag{4.77}$$

where we used $q = 3r/(r-1) = 3u_1/(u_1 - u_2)$.

Now the shock disappears from the system and the accelerated distribution $f_1(p)$ decompresses, so that the momentum of every particle decreases. Let us denote the new (decreased/decompressed) momentum with $\hat{p}$ and the decompressed distribution function with $\hat{f}_1(\hat{p})$, where $\hat{p}$ is a function of the accelerated momentum p. This distribution is taken as injection distribution into the second shock, so that the accelerated distribution after the second shock is given by

$$f_2(p) = qp^{-q} \int_{p_0}^p (p')^{q-1} \hat{f}_1(p') dp'. \tag{4.78}$$

The question now is, how to derive $\hat{f}_1(p')$ from Eq. (4.77), i.e., how to describe the decompression. To evaluate the decompression consider the collisionless transport equation (see [5], Sects. 5.2 and 5.7)

$$\frac{\partial f}{\partial t} + (\boldsymbol{u} + \boldsymbol{v}) \cdot \nabla f - \frac{p}{3} \nabla \cdot \boldsymbol{u} \frac{\partial f}{\partial p} = 0. \tag{4.79}$$

Obviously the force is described by

$$\frac{dp}{dt} = -\frac{p}{3} \nabla \cdot \boldsymbol{u}. \tag{4.80}$$

From Eq. (3.51), the continuity equation, we derive

$$\nabla \cdot \boldsymbol{u} = -\frac{1}{\rho} \frac{d\rho}{dt}, \tag{4.81}$$

where ρ is the background density. Substituting $\nabla \cdot \boldsymbol{u}$ we find

$$\frac{dp}{dt} = \frac{p}{3} \frac{1}{\rho} \frac{d\rho}{dt} \quad \rightarrow \quad \frac{dp}{p} = \frac{1}{3} \frac{d\rho}{\rho} \quad \rightarrow \quad \int_{p_1}^{p_2} \frac{dp}{p} = \frac{1}{3} \int_{\rho_1}^{\rho_2} \frac{d\rho}{\rho}. \tag{4.82}$$

The integrals are easily solved and we obtain

$$\frac{p_2}{p_1} = \left(\frac{\rho_2}{\rho_1}\right)^{1/3}, \tag{4.83}$$

where the indices 1 and 2 refer to two distinct scenarios. The idea is as follows: Assume an unshocked region in space with background density ρ_2. Suppose now that a shock passes through that region, compressing the background material to a higher density $\rho_1 > \rho_2$. After the shock has passed through that region the density decompresses back to ρ_2. Basically, index 1 can be interpreted as the shock (compression), where the particles have gained momentum p. Index 2 refers to the post-shock (decompression), when the momentum has decreased to $\hat{p}$. The shock compression ratio is then $r = \rho_1/\rho_2$, so that

$$\frac{\hat{p}}{p} = \left(\frac{1}{r}\right)^{1/3} = r^{-1/3}. \tag{4.84}$$

For convenience we introduce $r = R^{-3}$, so that $\hat{p} = Rp$.

Now, according to Liouville's theorem, the distribution f is a constant along the trajectory of a particle in phase space, implying that the distribution function after decompression $\hat{f}_1(\hat{p})$ is equal to $f_1(p)$ before decompression, therefore

$$\hat{f}_1(Rp) = f_1(p) \quad \rightarrow \quad \hat{f}_1(p') = f_1(p'/R), \tag{4.85}$$

where we used the coordinate transformation $p' = Rp$. We deduce that the momentum of every particle decreases according to the scaling $p \rightarrow Rp$, which means, that the shape of the distribution is retained but shifted down with a cutoff at Rp_0. In general, decompression means to change the momentum according to

$$p \rightarrow \frac{p}{R}. \tag{4.86}$$

- Substituting result (4.85) back into Eq. (4.78) we find

$$f_2(p) = qp^{-q} \int_{p_0 R}^{p} (p')^{q-1} f_1(p'/R)dp', \tag{4.87}$$

where, according to Eq. (4.77),

$$f_1(p'/R) = \frac{Aq}{p_0}\left(\frac{p'}{p_0 R}\right)^{-q}. \tag{4.88}$$

Note that the lower limit of the integral in Eq. (4.87) has been changed to $p_0 R$ to accommodate the shifted cutoff of the distribution function. Substituting $f_1(p'/R)$

in Eq. (4.87) we find

$$f_2(p) = \frac{Aq^2}{p_0}\left(\frac{p}{p_0R}\right)^{-q}\int_{p_0R}^{p}\frac{dp'}{p'} = \frac{Aq^2}{p_0}\left(\frac{p}{p_0R}\right)^{-q}\ln\left(\frac{p}{p_0R}\right). \tag{4.89}$$

Once the second shock has passed through, decompression leads to ($p \to p/R$)

$$f_2(p/R) = \frac{Aq^2}{p_0}\left(\frac{p}{p_0R^2}\right)^{-q}\ln\left(\frac{p}{p_0R^2}\right). \tag{4.90}$$

- For the third shock the accelerated spectrum is given by

$$\begin{aligned} f_3(p) &= qp^{-q}\int_{p_0R^2}^{p}(p')^{q-1}f_2(p'/R)dp' \\ &= qp^{-q}\int_{p_0R^2}^{p}(p')^{q-1}\frac{Aq^2}{p_0}\left(\frac{p'}{p_0R^2}\right)^{-q}\ln\left(\frac{p'}{p_0R^2}\right)dp' \\ &= \frac{Aq^3}{p_0}\left(\frac{p}{p_0R^2}\right)^{-q}\int_{p_0R^2}^{p}\ln\left(\frac{p'}{p_0R^2}\right)\frac{dp'}{p'} \\ &= \frac{Aq^3}{p_0}\left(\frac{p}{p_0R^2}\right)^{-q}\frac{1}{2}\left[\ln\left(\frac{p}{p_0R^2}\right)\right]^2. \end{aligned} \tag{4.91}$$

Note that the already shifted cutoff as the lower integration limit has shifted again by a factor of R to $p_0R \to p_0R^2$. After decompressing we find

$$f_3(p/R) = \frac{Aq^3}{p_0}\left(\frac{p}{p_0R^3}\right)^{-q}\frac{1}{2}\left[\ln\left(\frac{p}{p_0R^3}\right)\right]^2. \tag{4.92}$$

- For the fourth shock we find

$$\begin{aligned} f_4(p) &= qp^{-q}\int_{p_0R^3}^{p}(p')^{q-1}f_3(p'/R)dp' \\ &= qp^{-q}\frac{1}{2}\int_{p_0R^3}^{p}(p')^{q-1}\frac{Aq^3}{p_0}\left(\frac{p'}{p_0R^3}\right)^{-q}\left[\ln\left(\frac{p'}{p_0R^3}\right)\right]^2dp' \\ &= \frac{Aq^4}{p_0}\left(\frac{p}{p_0R^3}\right)^{-q}\frac{1}{2}\int_{p_0R^3}^{p}\left[\ln\left(\frac{p'}{p_0R^3}\right)\right]^2\frac{dp'}{p'} \\ &= \frac{Aq^4}{3\cdot 2p_0}\left(\frac{p}{p_0R^3}\right)^{-q}\left[\ln\left(\frac{p}{p_0R^3}\right)\right]^3. \end{aligned} \tag{4.93}$$

Note that the already shifted cutoff as the lower integration limit has shifted again by a factor of R to $p_0R^2 \rightarrow p_0R^3$. After decompressing we find

$$f_4(p/R) = \frac{Aq^4}{3!p_0}\left(\frac{p}{p_0R^4}\right)^{-q}\left[\ln\left(\frac{p}{p_0R^4}\right)\right]^3 . \tag{4.94}$$

In general we find for the i-th shock:

$$f_i(p/R) = \frac{Aq^i}{(i-1)!p_0}\left(\frac{p}{p_0R^i}\right)^{-q}\left[\ln\left(\frac{p}{p_0R^i}\right)\right]^{(i-1)} . \tag{4.95}$$

It can be shown that for an infinite number of shocks the distribution function approximates a power law,

$$f_\infty(p) \propto p^{-3}. \tag{4.96}$$

Chapter 5
The Transport of Low Frequency Turbulence

5.1 Mean Field Description of MHD Fluctuations

In this section we derive the transport equation for the small scale Elsässer variables starting from the evolution equations for the kinetic and magnetic energy. Note that the matrix $\nabla \boldsymbol{a} (= \nabla \otimes \boldsymbol{a})$ denotes a *dyadic product* and not a *vector gradient*. The reader is urged to caution, since this notation is not consistently used throughout the literature. Also, the symbol "·" denotes a dot product with the implication that $\nabla \cdot \boldsymbol{a}$ is the divergence of a vector $\boldsymbol{a}$. The multiplication of a vector $\boldsymbol{b}$ with a matrix is also indicated by the symbol "·", i.e., we write $\boldsymbol{b} \cdot \nabla \boldsymbol{a} = \boldsymbol{b} \cdot (\nabla \boldsymbol{a})$, meaning that the vector $\boldsymbol{b}$ is multiplied with the matrix $\nabla \boldsymbol{a}$. We will also make use of the identity

$$\nabla (fC) = f \nabla C + C \nabla f,$$

where the matrix $C\nabla f$ denotes again the dyadic product.

Problem 5.1 Complete the derivation of the transport equation for the Elsässer variables $\mathbf{z}^{\pm}$, starting from

$$\frac{\partial \mathbf{u}}{\partial t} + \mathbf{U} \cdot (\nabla \mathbf{u}) + \mathbf{u} \cdot (\nabla \mathbf{U}) - \frac{[\mathbf{B}_0 \cdot (\nabla \mathbf{b}) + \mathbf{b} \cdot (\nabla \mathbf{B}_0)]}{4\pi \rho_0} = -\frac{1}{\rho_0} \nabla \delta p^T + \mathbf{N}^{\mathrm{u}} \tag{5.1}$$

A. Dosch, G.P. Zank, *Transport Processes in Space Physics and Astrophysics*, Lecture Notes in Physics 918, DOI 10.1007/978-3-319-24880-6_5

and

$$\frac{\partial \mathbf{b}}{\partial t} + \mathbf{U} \cdot (\nabla \mathbf{b}) + \mathbf{u} \cdot (\nabla \mathbf{B}_0) - \mathbf{B}_0 \cdot (\nabla \mathbf{u}) - \mathbf{b} \cdot (\nabla \mathbf{U})$$
$$= -(\nabla \cdot \mathbf{U})\, \mathbf{b} - (\nabla \cdot \mathbf{u})\, \mathbf{B}_0 + \mathbf{N}^{\mathrm{b}}, \tag{5.2}$$

using the definitions for the Elsässer variables $\mathbf{z}^{\pm}$ and for the Alfvén speed $\mathbf{V}_A$,

$$\mathbf{z}^{\pm} = \mathbf{u} \pm \frac{\mathbf{b}}{\sqrt{4\pi\rho_0}} \qquad \mathbf{V}_A = \frac{\mathbf{B}_0}{\sqrt{4\pi\rho_0}}, \tag{5.3}$$

and the following definitions

$$\frac{\mathbf{U}}{\rho_0} \cdot \nabla \rho_0 = -\nabla \cdot \mathbf{U} \tag{5.4a}$$

$$\mathbf{U} \cdot \nabla \frac{1}{\sqrt{4\pi\rho_0}} = \frac{1}{\sqrt{4\pi\rho_0}} \nabla \cdot \frac{\mathbf{U}}{2} \tag{5.4b}$$

$$\mathbf{B}_0 \cdot \nabla \frac{1}{\sqrt{4\pi\rho_0}} = \nabla \cdot \mathbf{V}_A. \tag{5.4c}$$

Solution First we simplify the evolution equations for the turbulent kinetic (A) and magnetic (B) energy. Then we combine both equations to obtain transport equations for the forward and backward propagating Elsässer variables.

A. We begin with Eq. (5.1) and consider the last term on the left-hand side. A closer inspection reveals that

$$\frac{\mathbf{B}_0}{\sqrt{4\pi\rho_0}} \cdot \frac{\nabla \mathbf{b}}{\sqrt{4\pi\rho_0}} = \frac{\mathbf{B}_0}{\sqrt{4\pi\rho_0}} \cdot \left[\nabla \left(\frac{\mathbf{b}}{\sqrt{4\pi\rho_0}} \right) - \nabla \left(\frac{1}{\sqrt{4\pi\rho_0}} \right) \mathbf{b} \right]$$
$$= \frac{\mathbf{B}_0}{\sqrt{4\pi\rho_0}} \cdot \nabla \left(\frac{\mathbf{b}}{\sqrt{4\pi\rho_0}} \right) - \left[\mathbf{B}_0 \cdot \nabla \left(\frac{1}{\sqrt{4\pi\rho_0}} \right) \right] \frac{\mathbf{b}}{\sqrt{4\pi\rho_0}}$$
$$= \mathbf{V}_A \cdot \nabla \left(\frac{\mathbf{b}}{\sqrt{4\pi\rho_0}} \right) - \frac{\mathbf{b}}{\sqrt{4\pi\rho_0}} (\nabla \cdot \mathbf{V}_A), \tag{5.5}$$

where we used the chain rule in the first line. In the second line we expanded the square brackets and used the identity for dyadic products, $\boldsymbol{B}_0 \left[(\nabla f) \boldsymbol{b} \right] = \left[\boldsymbol{B}_0 \cdot (\nabla f) \right] \boldsymbol{b}$, in the second term. In the third line we used the relation (5.3) to substitute $\mathbf{B}_0 / \sqrt{4\pi\rho_0}$ in the first term and the relation (5.4c) to rewrite the second term. With this simplification the evolution equation for $\mathbf{u}$ can be rewritten as

$$\frac{\partial \mathbf{u}}{\partial t} + \mathbf{U} \cdot (\nabla \mathbf{u}) + \mathbf{u} \cdot (\nabla \mathbf{U}) - \mathbf{V}_A \cdot \nabla \left(\frac{\mathbf{b}}{\sqrt{4\pi\rho_0}} \right)$$
$$+ \frac{\mathbf{b}}{\sqrt{4\pi\rho_0}} (\nabla \cdot \mathbf{V}_A) - \frac{\mathbf{b}}{\sqrt{4\pi\rho_0}} \cdot \frac{\nabla \mathbf{B}_0}{\sqrt{4\pi\rho_0}} = -\frac{1}{\rho_0} \nabla \delta p^T + \mathbf{N}^u. \tag{5.6}$$

B. Starting with the transport equation for the turbulent magnetic fields (5.2) we may rewrite the equation as

$$\frac{\partial}{\partial t}\left(\frac{\mathbf{b}}{\sqrt{4\pi\rho_0}} \right) + \mathbf{U} \cdot \frac{\nabla \mathbf{b}}{\sqrt{4\pi\rho_0}} + \mathbf{u} \cdot \frac{\nabla \mathbf{B}_0}{\sqrt{4\pi\rho_0}} - \frac{\mathbf{B}_0}{\sqrt{4\pi\rho_0}} \cdot (\nabla \mathbf{u})$$
$$- \frac{\mathbf{b}}{\sqrt{4\pi\rho_0}} \cdot (\nabla \mathbf{U}) + (\nabla \cdot \mathbf{U}) \frac{\mathbf{b}}{\sqrt{4\pi\rho_0}} + (\nabla \cdot \mathbf{u}) \frac{\mathbf{B}_0}{\sqrt{4\pi\rho_0}} = \frac{\mathbf{N}^b}{\sqrt{4\pi\rho_0}},$$

where we pulled the terms $-(\nabla \cdot \mathbf{U})\,\mathbf{b} - (\nabla \cdot \mathbf{u})\,\mathbf{B}_0$ from the right to the left-hand side and multiplied the entire equation by $1/\sqrt{4\pi\rho_0}$, using the fact that ρ_0 is time-independent. A closer inspection of the second term on the left-hand side shows that

$$\mathbf{U} \cdot \frac{\nabla \mathbf{b}}{\sqrt{4\pi\rho_0}} = \mathbf{U} \cdot \nabla \left(\frac{\mathbf{b}}{\sqrt{4\pi\rho_0}} \right) - \left[\mathbf{U} \cdot \nabla \left(\frac{1}{\sqrt{4\pi\rho_0}} \right) \right] \mathbf{b}$$
$$= \mathbf{U} \cdot \nabla \left(\frac{\mathbf{b}}{\sqrt{4\pi\rho_0}} \right) - \frac{\mathbf{b}}{\sqrt{4\pi\rho_0}} \nabla \cdot \frac{\mathbf{U}}{2},$$

where we essentially used the results from Eq. (5.5), i.e., we used the chain rule and the identity for dyadic products. In the second line we used relation (5.4b) in the second term. Substituting this result back into the evolution equation for **b** and using the definition for the Alfvén speed (5.3) we find

$$\frac{\partial}{\partial t}\left(\frac{\mathbf{b}}{\sqrt{4\pi\rho_0}} \right) + \mathbf{U} \cdot \nabla \left(\frac{\mathbf{b}}{\sqrt{4\pi\rho_0}} \right) - \frac{\mathbf{b}}{\sqrt{4\pi\rho_0}} \left(\nabla \cdot \frac{\mathbf{U}}{2} \right) + \mathbf{u} \cdot \frac{\nabla \mathbf{B}_0}{\sqrt{4\pi\rho_0}}$$
$$- \mathbf{V}_A \cdot \nabla \mathbf{u} - \frac{\mathbf{b}}{\sqrt{4\pi\rho_0}} \cdot \nabla \mathbf{U} + (\nabla \cdot \mathbf{U}) \frac{\mathbf{b}}{\sqrt{4\pi\rho_0}} + (\nabla \cdot \mathbf{u})\, \mathbf{V}_A = \frac{\mathbf{N}^b}{\sqrt{4\pi\rho_0}}.$$

The third and seventh term can be combined, so that

$$\frac{\partial}{\partial t}\left(\frac{\mathbf{b}}{\sqrt{4\pi\rho_0}} \right) + \mathbf{U} \cdot \nabla \left(\frac{\mathbf{b}}{\sqrt{4\pi\rho_0}} \right) + \frac{\mathbf{b}}{\sqrt{4\pi\rho_0}} \left(\nabla \cdot \frac{\mathbf{U}}{2} \right) + \mathbf{u} \cdot \frac{\nabla \mathbf{B}_0}{\sqrt{4\pi\rho_0}}$$
$$- \mathbf{V}_A \cdot \nabla \mathbf{u} - \frac{\mathbf{b}}{\sqrt{4\pi\rho_0}} \cdot \nabla \mathbf{U} + (\nabla \cdot \mathbf{u})\, \mathbf{V}_A = \frac{\mathbf{N}^b}{\sqrt{4\pi\rho_0}}. \tag{5.7}$$

Elsässer Variables Now we combine the transport equations for **u** and **b** in such a way that $\mathbf{z}^+ = (5.6) + (5.7)$ and $\mathbf{z}^- = (5.6) - (5.7)$. We obtain

$$\frac{\partial \mathbf{z}^\pm}{\partial t} + \mathbf{U} \cdot (\nabla \mathbf{z}^\pm) \mp \mathbf{V}_A \cdot (\nabla \mathbf{z}^\pm) \pm \mathbf{z}^\mp \cdot \frac{\nabla \mathbf{B}_0}{\sqrt{4\pi\rho_0}} + \mathbf{z}^\mp \cdot \nabla \mathbf{U}$$
$$+ \frac{\mathbf{b}}{\sqrt{4\pi\rho_0}} \left(\nabla \cdot \mathbf{V}_A \pm \nabla \cdot \frac{\mathbf{U}}{2} \right) \pm (\nabla \cdot \mathbf{u}) \, \mathbf{V}_A = \mathbf{NL}_\pm - \frac{1}{\rho_0} \nabla \delta p^T,$$

with the nonlinear terms $\mathbf{NL}_\pm = \mathbf{N}^u \pm \mathbf{N}^b / \sqrt{4\pi\rho_0}$ on the right-hand side. A common assumption adopted in turbulence modeling is to assume that small scale fluctuations are incompressible, so that the mean density ρ_0 varies slowly and $\delta\rho = 0$ (see [5]). From the continuity equation we find that the fluctuating velocity field becomes solenoidal,

$$\frac{\partial \rho}{\partial t} + \nabla \cdot \rho \mathbf{u} = 0 \quad \rightarrow \quad \nabla \cdot \mathbf{u} = 0.$$

Thus, by neglecting the term proportional to $(\nabla \cdot \mathbf{u}) \, \mathbf{V}_A$ and by using

$$\frac{\mathbf{b}}{\sqrt{4\pi\rho_0}} = \pm \frac{\mathbf{z}^\pm - \mathbf{z}^\mp}{2},$$

we can simplify the equation to

$$\frac{\partial \mathbf{z}^\pm}{\partial t} + (\mathbf{U} \mp \mathbf{V}_A) \cdot \nabla \mathbf{z}^\pm + \frac{\mathbf{z}^\pm - \mathbf{z}^\mp}{2} \nabla \cdot \left[\frac{\mathbf{U}}{2} \pm \mathbf{V}_A \right] + \mathbf{z}^\mp \cdot \left[\nabla \mathbf{U} \pm \frac{\nabla \mathbf{B}_0}{\sqrt{4\pi\rho_0}} \right]$$
$$= \mathbf{NL}_\pm + S^\pm,$$

where we used $S^\pm = -\nabla \delta p^T / \rho_0$. Thus, the transport equation for the small scale Elsässer variables can be written as

$$\frac{\partial \mathbf{z}^\pm}{\partial t} + (\mathbf{U} \mp \mathbf{V}_A) \cdot \nabla \mathbf{z}^\pm + \frac{1}{2} \nabla \cdot \left[\frac{\mathbf{U}}{2} \pm \mathbf{V}_A \right] \mathbf{z}^\pm$$
$$+ \mathbf{z}^\mp \cdot \left[\nabla \mathbf{U} \pm \frac{\nabla \mathbf{B}_0}{\sqrt{4\pi\rho_0}} - \frac{1}{2} \mathbb{1} \nabla \cdot \left[\frac{\mathbf{U}}{2} \pm \mathbf{V}_A \right] \right] = \mathbf{NL}_\pm + S^\pm, \tag{5.8}$$

where $\mathbb{1}$ is the unity matrix.

5.2 The Transport Equation for the Magnetic Energy Density

One can introduce the following moments of the Elsässer variables

$$E_T = \frac{\langle \mathbf{z}^+ \cdot \mathbf{z}^+ \rangle + \langle \mathbf{z}^- \cdot \mathbf{z}^- \rangle}{2} = \langle u^2 \rangle + \langle b^2/4\pi\rho_0 \rangle \tag{5.9a}$$

$$E_C = \frac{\langle \mathbf{z}^+ \cdot \mathbf{z}^+ \rangle - \langle \mathbf{z}^- \cdot \mathbf{z}^- \rangle}{2} = \left\langle \mathbf{u} \cdot \mathbf{b}/\sqrt{4\pi\rho_0} \right\rangle \tag{5.9b}$$

$$E_D = \langle \mathbf{z}^+ \cdot \mathbf{z}^- \rangle = \langle u^2 \rangle - \langle b^2/4\pi\rho_0 \rangle, \tag{5.9c}$$

where E_T is twice the total energy in the fluctuations (the sum of kinetic and magnetic energy), E_C is the cross helicity, the difference in energy between forward and backward propagating modes, and E_D is the energy difference, i.e., the difference between twice the fluctuation kinetic energy and magnetic energy density (measured in Alfvén units), sometimes called the *residual energy*. By using the kinetic and magnetic energy

$$E_u = \frac{1}{2}\langle u^2 \rangle \quad \text{and} \quad E_b = \frac{1}{2}\langle b^2/4\pi\rho_0 \rangle \tag{5.10}$$

and by combining Eqs. (5.9a)–(5.9c) we also find the following useful relations,

$$r_A = \frac{E_u}{E_b} = \frac{E_T + E_D}{E_T - E_D} = \frac{1 + H_D}{1 - H_D}$$

$$H_D = \frac{E_D}{E_T} = \frac{r_A - 1}{r_A + 1}, \tag{5.11a}$$

where r_A is the Alfvén ratio and H_D is the normalized energy difference or residual energy. The total and residual energy can then be written in the form

$$E_T = 2E_u + 2E_b = 2E_b(r_A + 1) \tag{5.12a}$$

$$E_D = H_D E_T. \tag{5.12b}$$

A very general set of transport equations can be derived from Eq. (5.8) in terms of the above moments together with correlation length equations. The physical content is sometimes difficult to extract, so we make the following assumptions that are quite reasonable beyond some 1–2 AU.

(continued)

A. The Alfvén ratio is assumed to be constant.
B. The cross helicity E_C is assumed to be zero, i.e., the energy in inward and outward propagating modes is equal.
C. Structural similarity hypothesis.

Problem 5.2 Complete the derivation of the transport equation for E_T, Eq. (5.14), and hence derive the final form of the transport equation.

Solution Starting with Eq. (5.8) we can rewrite the second term in the last line as

$$\frac{\mathbf{z}^{\mp} \cdot \nabla \mathbf{B}_0}{\sqrt{4\pi\rho_0}} = \mathbf{z}^{\mp} \cdot \nabla \mathbf{V}_A + \frac{1}{2\rho_0} \mathbf{V}_A \left[\mathbf{z}^{\mp} \cdot \nabla \rho \right],$$

so that the evolution equation for the small scale Elsässer variables becomes

$$\frac{\partial \mathbf{z}^{\pm}}{\partial t} + (\mathbf{U} \mp \mathbf{V}_A) \cdot \nabla \mathbf{z}^{\pm} + \frac{1}{2} \nabla \cdot \left[\frac{\mathbf{U}}{2} \pm \mathbf{V}_A \right] \mathbf{z}^{\pm} + \mathbf{z}^{\mp} \cdot \nabla \mathbf{U} \pm \mathbf{z}^{\mp} \cdot \nabla \mathbf{V}_A$$
$$\pm \frac{1}{2\rho_0} \mathbf{V}_A \left[\mathbf{z}^{\mp} \cdot \nabla \rho \right] - \frac{1}{2} \mathbf{z}^{\mp} \cdot \mathbb{1} \nabla \cdot \left[\frac{\mathbf{U}}{2} \pm \mathbf{V}_A \right] = \mathbf{NL}_{\pm} + S^{\pm}. \tag{5.13}$$

To derive the evolution equation for the total energy E_T we multiply the evolution equation for $\mathbf{z}^+$ with $\mathbf{z}^+$ and the evolution equation for $\mathbf{z}^-$ with $\mathbf{z}^-$, i.e., $\mathbf{z}^+ \cdot \partial \mathbf{z}^+ / \partial t +$ ... and $\mathbf{z}^- \cdot \partial \mathbf{z}^- / \partial t + \ldots$. Then both equations are added up. By considering each term separately we find by using the moments of the Elsässer variables, Eqs. (5.9a)–(5.9c):

- First term: With definition (5.9a) the first term can easily be calculated as

$$\left\langle \mathbf{z}^+ \cdot \frac{\partial \mathbf{z}^+}{\partial t} + \mathbf{z}^- \cdot \frac{\partial \mathbf{z}^-}{\partial t} \right\rangle = \frac{1}{2} \frac{\partial}{\partial t} \left\langle \mathbf{z}^+ \cdot \mathbf{z}^+ + \mathbf{z}^- \cdot \mathbf{z}^- \right\rangle = \frac{\partial E_T}{\partial t}.$$

- Second term: Here we find

$$\left\langle \left(\mathbf{U} \cdot \nabla \mathbf{z}^+ \right) \cdot \mathbf{z}^+ + \left(\mathbf{U} \cdot \nabla \mathbf{z}^- \right) \cdot \mathbf{z}^- \right\rangle = \mathbf{U} \cdot \nabla E_T,$$

where we used

$$\left(\mathbf{U} \cdot \nabla \mathbf{z}^{\pm} \right) \cdot \mathbf{z}^{\pm} = U_i z_j^{\pm} \frac{\partial z_j^{\pm}}{\partial x_i} = \frac{U_i}{2} \frac{\partial z_j^{\pm 2}}{\partial x_i} = \frac{\mathbf{U}}{2} \cdot \nabla \mathbf{z}^{\pm 2}$$

together with Einstein's summation convention and the dyadic product. Note that $\nabla \mathbf{z}^{\pm 2}$ is the gradient, so that

$$\frac{\mathbf{U}}{2} \cdot \nabla \mathbf{z}^{+2} + \frac{\mathbf{U}}{2} \cdot \nabla \mathbf{z}^{-2} = \mathbf{U} \cdot \nabla \left(\frac{\mathbf{z}^{+2} + \mathbf{z}^{-2}}{2} \right) = \mathbf{U} \cdot \nabla E_T,$$

where we used definition (5.9a).

- Third term: Here we find

$$\langle -(\mathbf{V}_A \cdot \nabla \mathbf{z}^+) \cdot \mathbf{z}^+ + (\mathbf{V}_A \cdot \nabla \mathbf{z}^-) \cdot \mathbf{z}^- \rangle = -\mathbf{V}_A \cdot \nabla E_C,$$

where we basically used the results from the second term.
- Fourth term: Here, $(\nabla \cdot \mathbf{U})/2$ can be pulled out and we find

$$\left\langle \frac{1}{4} (\nabla \cdot \mathbf{U})\, \mathbf{z}^+ \cdot \mathbf{z}^+ + \frac{1}{4} (\nabla \cdot \mathbf{U})\, \mathbf{z}^- \cdot \mathbf{z}^- \right\rangle = \frac{1}{2} \nabla \cdot \mathbf{U} E_T,$$

where we used definition (5.9a).
- Fifth term: Here, we pull out $(\nabla \cdot \mathbf{V}_A)$ and find

$$\left\langle \frac{1}{2} (\nabla \cdot \mathbf{V}_A)\, \mathbf{z}^+ \cdot \mathbf{z}^+ - \frac{1}{2} (\nabla \cdot \mathbf{V}_A)\, \mathbf{z}^- \cdot \mathbf{z}^- \right\rangle = \nabla \cdot \mathbf{V}_A E_C,$$

where we used definition (5.9b).

For the remaining terms we will adopt the *structural similarity hypothesis*, where we approximate $z_i^+ z_j^- = a\mathbf{z}^+ \cdot \mathbf{z}^-$ for some constant a, i.e., the off diagonal elements are approximated by the trace of the matrix.

- Sixth term: We find

$$\langle (\mathbf{z}^- \cdot \nabla \mathbf{U}) \cdot \mathbf{z}^+ + (\mathbf{z}^+ \cdot \nabla \mathbf{U}) \cdot \mathbf{z}^- \rangle \approx 2a \nabla \cdot \mathbf{U} E_D + 2a E_D S_x^u,$$

where we used

$$\langle (\mathbf{z}^\pm \cdot \nabla \mathbf{U}) \cdot \mathbf{z}^\mp \rangle = \left\langle z_i^\pm z_j^\mp \right\rangle \frac{\partial U_j}{\partial x_i} \approx a \langle \mathbf{z}^\pm \cdot \mathbf{z}^\mp \rangle \frac{\partial U_j}{\partial x_i}$$
$$= a E_D \frac{\partial U_j}{\partial x_i} = a E_D \nabla \cdot \mathbf{U} + a E_D S_x^u,$$

and where $S_x^u = \sum_{i,j;i \neq j} \partial U_i / \partial x_j$ is the sum of the shear velocity gradient terms.
- Seventh term: By using the results from the previous term, we find immediately

$$\langle (\mathbf{z}^- \cdot \nabla \mathbf{V}_A) \cdot \mathbf{z}^+ - (\mathbf{z}^+ \cdot \nabla \mathbf{V}_A) \cdot \mathbf{z}^- \rangle = 0.$$

- Eighth term: This term vanishes identically,

$$\left\langle \frac{1}{2\rho}\mathbf{V}_A\left(\mathbf{z}^-\cdot\nabla\rho\right)\cdot\mathbf{z}^+ - \frac{1}{2\rho}\mathbf{V}_A\left(\mathbf{z}^+\cdot\nabla\rho\right)\cdot\mathbf{z}^-\right\rangle = 0,$$

since

$$\mathbf{V}_A\left(\mathbf{z}^\pm\cdot\nabla\rho\right)\cdot\mathbf{z}^\mp = V_{Ai}z_i^\mp z_j^\pm\frac{\partial\rho}{\partial x_j} \approx V_{Ai}aE_D\frac{\partial\rho}{\partial x_j}.$$

- Ninth term: By using definition (5.9c) we find

$$\left\langle -\frac{1}{4}\left[\mathbf{z}^-\cdot\mathbb{1}\left(\nabla\cdot\mathbf{U}\right)\right]\cdot\mathbf{z}^+ - \frac{1}{4}\left[\mathbf{z}^+\cdot\mathbb{1}\left(\nabla\cdot\mathbf{U}\right)\right]\cdot\mathbf{z}^-\right\rangle = -\frac{1}{2}\nabla\cdot\mathbf{U}E_D,$$

since

$$\mathbf{z}^\pm\cdot\mathbb{1}\left(\nabla\cdot\mathbf{U}\right)\cdot\mathbf{z}^\mp = z_j^\pm z_j^\mp\frac{\partial U_i}{\partial x_i} = (\mathbf{z}^\pm\cdot\mathbf{z}^\mp)(\nabla\cdot\mathbf{U}) = E_D(\nabla\cdot\mathbf{U}).$$

- Tenth term: By using the results from the previous term, we find immediately

$$\left\langle -\frac{1}{2}\left[\mathbf{z}^-\cdot\mathbb{1}\left(\nabla\cdot\mathbf{V}_A\right)\right]\cdot\mathbf{z}^+ + \frac{1}{2}\left[\mathbf{z}^+\cdot\mathbb{1}\left(\nabla\cdot\mathbf{V}_A\right)\right]\cdot\mathbf{z}^-\right\rangle = 0.$$

Combining all results we find the transport equation for E_T,

$$\frac{\partial E_T}{\partial t} + \mathbf{U}\cdot\nabla E_T - \mathbf{V}_A\cdot\nabla E_C + \frac{1}{2}\nabla\cdot\mathbf{U}E_T + \nabla\cdot\mathbf{V}_A E_C + \left(2a - \frac{1}{2}\right)\nabla\cdot\mathbf{U}E_D$$
$$+ 2aE_D S_x^u = \left\langle\left[\mathbf{NL}_+ + S^+\right]\cdot\mathbf{z}^+\right\rangle + \left\langle\left[\mathbf{NL}_- + S^-\right]\cdot\mathbf{z}^-\right\rangle. \tag{5.14}$$

Under the assumption that the cross helicity is zero and by using the relations (5.11a) and (5.12a) we find with $E_D = H_D E_T = H_D 2E_b(r_A+1)$, the following expression

$$2(r_A+1)\frac{\partial E_b}{\partial t} + 2(r_A+1)\mathbf{U}\cdot\nabla E_b + 2(r_A+1)\frac{1}{2}\nabla\cdot\mathbf{U}E_b$$
$$+ 2(r_A+1)\left(2a - \frac{1}{2}\right)\nabla\cdot\mathbf{U}H_D E_b + 2(r_A+1)2aH_D E_b S_x^u$$
$$= \left\langle\left[\mathbf{NL}_+ + S^+\right]\cdot\mathbf{z}^+\right\rangle + \left\langle\left[\mathbf{NL}_- + S^-\right]\cdot\mathbf{z}^-\right\rangle. \tag{5.15}$$

On dividing the entire equation by $2(r_A + 1)$ we obtain

$$\frac{\partial E_b}{\partial t} + \mathbf{U} \cdot \nabla E_b + \frac{1}{2}\nabla \cdot \mathbf{U} E_b + \left(2a - \frac{1}{2}\right) \nabla \cdot \mathbf{U} H_D E_b + 2a H_D E_b S_x^u$$
$$= \frac{\langle [\mathbf{NL}_+ + S^+] \cdot \mathbf{z}^+ \rangle}{2(r_A + 1)} + \frac{\langle [\mathbf{NL}_- + S^-] \cdot \mathbf{z}^- \rangle}{2(r_A + 1)}. \quad (5.16)$$

By neglecting the right-hand side and the *mixing terms* proportional to H_D we obtain the WKB equation (5.17).

Problem 5.3 Solve the steady-state WKB equation for the energy density of magnetic field fluctuations

$$\frac{\partial E_b}{\partial t} + \boldsymbol{U} \cdot \nabla E_b + \frac{1}{2}\nabla \cdot \boldsymbol{U} E_b = 0, \quad (5.17)$$

in a spherically symmetric steady flow for which $\boldsymbol{U} = U_0 \hat{\boldsymbol{r}}$, $U_0 = const.$, $\rho_0 = \rho_{00}(R_0/r)^2$, and where ρ_{00} is the density at a heliocentric distance R_0. Hence, show that $b^2/b_0^2 = (R_0/r)^3$.

Solution For a steady flow the energy density of magnetic field fluctuations is constant in time, thus $\partial E_b/\partial t = 0$. Considering a spherically symmetric flow, we transform the gradient and divergence into spherical coordinates,

$$\nabla_s E_b = \frac{\partial E_b}{\partial r} + \frac{1}{r}\frac{\partial E_b}{\partial \theta} + \frac{1}{r \sin\theta}\frac{\partial E_b}{\partial \phi}$$

$$\nabla_s \cdot \boldsymbol{U} = \frac{1}{r^2}\frac{\partial}{\partial r}\left(r^2 U_r\right) + \frac{1}{r\sin\theta}\frac{\partial}{\partial \theta}\left(\sin\theta\, U_\theta\right) + \frac{1}{r\sin\theta}\frac{\partial U_\phi}{\partial \phi}.$$

In a spherically symmetric flow all quantities are independent of θ and ϕ. Also, since $\boldsymbol{U} = U_0\hat{\boldsymbol{r}}$, we find $U_r = U_0$, so that $\boldsymbol{U} = (U_0, 0, 0)$ in spherical coordinates, and Eq. (5.17) becomes

$$U_0 \frac{\partial E_b}{\partial r} + \frac{1}{2}\frac{1}{r^2}\frac{\partial}{\partial r}\left(r^2 U_0\right) E_b = 0.$$

After dividing by $U_0 \neq 0$ and taking the derivative of the second term, we obtain

$$\frac{\partial E_b}{\partial r} + \frac{E_b}{r} = 0 \quad \Longrightarrow \quad \int \frac{dE_b}{E_b} = -\int \frac{dr}{r},$$

which can readily be solved by $\ln E_b = -\ln r + C_0$, where C_0 is an integration constant. We obtain the general solution of the WKB equation (5.17)

$$E_b = \frac{C}{r},$$

where $C = \exp C_0$ is another integration constant that has to be determined by the initial conditions. In order to calculate the constant C we recall from Eq. (5.10) that

$$E_b = \left\langle \frac{b^2}{4\pi\rho_0} \right\rangle = \left\langle \frac{b^2}{4\pi\rho_{00}\,(R_0/r)^2} \right\rangle = \frac{C}{r}, \tag{5.18}$$

where we used the density $\rho_0 = \rho_{00}(R_0/r)^2$ (see above). With the initial conditions $b = b_0$ at the radial distance $r = R_0$, we find for the constant

$$C = R_0 \left\langle \frac{b_0^2}{4\pi\rho_{00}} \right\rangle.$$

Substituting this result back into Eq. (5.18) we obtain the solution of the WKB equation (5.17)

$$E_b = \frac{b^2}{4\pi\rho_{00}\,(R_0/r)^2} = \frac{R_0}{r}\frac{b_0^2}{4\pi\rho_{00}} \quad \rightarrow \quad \frac{b^2}{b_0^2} = \left(\frac{R_0}{r}\right)^3.$$

5.3 Modelling the Dissipation Terms

Problem 5.4 Complete the derivation of the correlation length equation (5.22).

Solution We start with the covariance equation for L^T,

$$\frac{\partial L^T}{\partial t} + \mathbf{U}\cdot\nabla L^T + \frac{1}{2}\nabla\cdot\mathbf{U}L^T + 2\left(a - \frac{1}{4}\right)\nabla\cdot\mathbf{U}L^D + 2aL^D S_x^u = 0. \tag{5.19}$$

Zank et al. [6] argued that the velocity and magnetic field fluctuations posses equal areas under their respective correlation functions, from which one can infer that $\lambda^D = 0$. This is a somewhat severe restriction but maintains some tractability in the turbulence model. Since $L^D = \lambda^D E_D$ it follows that $L^D = 0$ so that the transport equation (5.19) can be simplified to

$$\frac{\partial L^T}{\partial t} + \mathbf{U}\cdot\nabla L^T + \frac{1}{2}\nabla\cdot\mathbf{U}L^T = 0.$$

With $L^T = 2E_T\lambda^T$ and recalling from Eq. (5.12a) that $E^T = 2E_b(r_A + 1)$, where r_A is the *constant* Alfvén ratio, we find $L^T = 4E_b(r_A + 1)\lambda^T$ and, therefore,

$$4(r_A+1)\frac{\partial E_b\lambda^T}{\partial t} + 4(r_A+1)\mathbf{U}\cdot\nabla(E_b\lambda^T) + 4(r_A+1)\frac{1}{2}\nabla:\mathbf{U}E_b\lambda^T = 0.$$

On dividing the equation by $2(r_A + 1) \neq 0$ we find

$$\lambda^T \frac{\partial E_b}{\partial t} + E_b \frac{\partial \lambda^T}{\partial t} + \lambda^T \mathbf{U} \cdot \nabla E_b + E_b \mathbf{U} \cdot \nabla \lambda^T + \frac{1}{2} \nabla \cdot \mathbf{U} E_b \lambda^T = 0. \tag{5.20}$$

Now we multiply the evolution equation for the magnetic field fluctuations, Eq. (5.16), by λ^T and obtain

$$\lambda^T \frac{\partial E_b}{\partial t} + \lambda^T \mathbf{U} \cdot \nabla E_b + \frac{1}{2} \lambda^T \nabla \cdot \mathbf{U} E_b + \lambda^T \left(2a - \frac{1}{2}\right) \nabla \cdot \mathbf{U} H_D E_b$$
$$+ \lambda^T 2a S_x^u H_D E_b = \lambda^T \left(\left\langle \mathbf{NL}_+ \cdot \mathbf{z}^+ \right\rangle + \left\langle \mathbf{NL}_- \cdot \mathbf{z}^- \right\rangle \right) + \lambda^T S,$$

where we used $S = \left\langle S^+ \cdot \mathbf{z}^+ \right\rangle + \left\langle S^- \cdot \mathbf{z}^- \right\rangle$. For the nonlinear dissipation term we use

$$\left\langle \mathbf{NL}_+ \cdot \mathbf{z}^+ \right\rangle + \left\langle \mathbf{NL}_- \cdot \mathbf{z}^- \right\rangle = -\frac{E_b^{3/2}}{\lambda},$$

see [5], Chap. 6.4 for a detailed description. Hence, we find

$$\lambda^T \frac{\partial E_b}{\partial t} = -\lambda^T \frac{E_b^{3/2}}{\lambda} + \lambda^T S - \lambda^T \mathbf{U} \cdot \nabla E_b - \frac{1}{2} \lambda^T \nabla \cdot \mathbf{U} E_b$$
$$- \lambda^T \left(2a - \frac{1}{2}\right) \nabla \cdot \mathbf{U} H_D E_b - \lambda^T 2a S_x^u H_D E_b. \tag{5.21}$$

Substituting this equation back into Eq. (5.20) we find

$$- \lambda^T \frac{E_b^{3/2}}{\lambda} + \lambda^T S - \lambda^T \mathbf{U} \cdot \nabla E_b - \frac{1}{2} \lambda^T \nabla \cdot \mathbf{U} E_b - \lambda^T \left(2a - \frac{1}{2}\right) \nabla \cdot \mathbf{U} H_D E_b$$
$$- \lambda^T 2a S_x^u H_D E_b + E_b \frac{\partial \lambda^T}{\partial t} + \lambda^T \mathbf{U} \cdot \nabla E_b + E_b \mathbf{U} \cdot \nabla \lambda^T + \frac{1}{2} \nabla \cdot \mathbf{U} E_b \lambda^T = 0.$$

The third and the eighth term cancel as well as the fourth and the tenth term. Then we move the first two terms to the right-hand side and divide the entire equation by E_b,

$$\frac{\partial \lambda^T}{\partial t} + \mathbf{U} \cdot \nabla \lambda^T - \lambda^T \left(2a - \frac{1}{2}\right) \nabla \cdot \mathbf{U} H_D - \lambda^T 2a S_x^u H_D = \lambda^T \frac{E_b^{1/2}}{\lambda} - \frac{\lambda^T S}{E_b}.$$

Identifying now λ^T with 2λ we find the transport equation for the correlation length

$$\frac{\partial \lambda}{\partial t} + \mathbf{U} \cdot \nabla \lambda - \lambda \left(2a - \frac{1}{2} \right) \nabla \cdot \mathbf{U} H_D - \lambda 2 a S_x^u H_D = E_b^{1/2} - \frac{\lambda S}{E_b}. \tag{5.22}$$

Problem 5.5 Integrate the steady-state spherically symmetric form of the transport equations,

$$U \frac{\partial E_b}{\partial r} + \frac{U}{r} E_b - \Gamma \frac{U}{r} E_b = -\frac{E_b^{3/2}}{\lambda} \tag{5.23a}$$

$$U \frac{\partial \lambda}{\partial r} + \Gamma \frac{U}{r} \lambda = \frac{E_b^{1/2}}{2} \tag{5.23b}$$

analytically, if $E_b(r = R_0) = E_{b0}$ and $\lambda(r = R_0) = \lambda_0$. Hence, show that asymptotically, in the limit of no mixing $\Gamma = 0$ (which is appropriate for either 2D or slab turbulence), one obtains the estimates

$$b^2/b_0^2 \sim (R_0/r)^{3.5} \qquad \lambda/\lambda_0 \sim (r/R_0)^{1/4}.$$

This model corresponds to Kolmogorov/von Karman turbulence in an expanding medium. Show that in the opposite limit of strong turbulence ($\Gamma = 1$), the solutions reduce asymptotically to

$$b^2/b_0^2 \sim (R_0/r)^4, \qquad \lambda/\lambda_0 \sim \textit{constant}.$$

This solution describes Taylor turbulence in a non-expansive medium.

Solution We begin by dividing Eq. (5.23a) by E_b and multiplying Eq. (5.23b) by a factor of 2. This yields

$$\frac{U}{E_b} \frac{\partial E_b}{\partial r} + \frac{U}{r} - \Gamma \frac{U}{r} = -\frac{E_b^{1/2}}{\lambda} \tag{5.24a}$$

$$E_b^{1/2} = 2U \frac{\partial \lambda}{\partial r} + 2\Gamma \frac{U}{r} \lambda. \tag{5.24b}$$

Substituting now the term proportional to $E_b^{1/2}$ in the first equation by the second, we obtain after some simplification the single differential equation

$$\frac{1}{E_b} \frac{\partial E_b}{\partial r} + (1 + \Gamma) \frac{1}{r} = -\frac{2}{\lambda} \frac{\partial \lambda}{\partial r}, \tag{5.25}$$

where we divided by U. Multiplying now by ∂r and integrating yields

$$\int_{E_{b0}}^{E_b} \frac{\partial E_b'}{E_b'} + (1+\Gamma)\int_{R_0}^{r} \frac{dr'}{r'} = -2\int_{\lambda_0}^{\lambda} \frac{d\lambda'}{\lambda'},$$

where we used the initial conditions E_{b0}, R_0, and λ_0. This equation can readily be solved and we obtain

$$\ln\left(\frac{E_b}{E_{b0}}\right) + (1+\Gamma)\ln\left(\frac{r}{R_0}\right) = -2\ln\left(\frac{\lambda}{\lambda_0}\right),$$

leading to the solution

$$\frac{E_b}{E_{b0}} = \left(\frac{R_0}{r}\right)^{(1+\Gamma)}\left(\frac{\lambda_0}{\lambda}\right)^2. \tag{5.26}$$

This equation can be solved for

$$E_b^{1/2} = \left(\frac{R_0}{r}\right)^{\frac{1+\Gamma}{2}} \frac{\lambda_0}{\lambda} E_{b0}^{1/2}. \tag{5.27}$$

This result is substituted back into Eq. (5.24b),

$$\left(\frac{R_0}{r}\right)^{\frac{1+\Gamma}{2}} \frac{\lambda_0}{\lambda} E_{b0}^{1/2} = 2U\frac{\partial \lambda}{\partial r} + 2\Gamma\frac{U}{r}\lambda.$$

Now, we multiply this equation by λ and divide by U, leading to

$$\left(\frac{R_0}{r}\right)^{\frac{1+\Gamma}{2}} \frac{\lambda_0}{U} E_{b0}^{1/2} = 2\lambda\frac{\partial \lambda}{\partial r} + 2\frac{\Gamma}{r}\lambda^2.$$

Note that $2\lambda\partial\lambda/\partial r = \partial\lambda^2/\partial r$, and by using the substitution $x = \lambda^2$ we can rewrite the equation as

$$\left(\frac{R_0}{r}\right)^{\frac{1+\Gamma}{2}} \frac{\lambda_0}{U} E_{b0}^{1/2} = \frac{\partial x}{\partial r} + 2\frac{\Gamma}{r}x.$$

Multiplying both sides with $r^{2\Gamma}$ (integrating factor) yields

$$r^{2\Gamma}\left(\frac{R_0}{r}\right)^{\frac{1+\Gamma}{2}} \frac{\lambda_0}{U} E_{b0}^{1/2} = \frac{\partial x}{\partial r} r^{2\Gamma} + 2r^{2\Gamma-1}\Gamma x \equiv \frac{\partial}{\partial r}\left(x r^{2\Gamma}\right).$$

Integrating both sides with respect to r and using the above initial conditions yields

$$R_0^{\frac{1+\Gamma}{2}} \frac{\lambda_0}{U} E_{b0}^{1/2} \int_{R_0}^{r} r'^{\frac{3\Gamma-1}{2}} dr' = \int_{R_0}^{r} \frac{\partial}{\partial r'} (x r'^{2\Gamma})$$

$$R_0^{\frac{1+\Gamma}{2}} \frac{\lambda_0}{U} E_{b0}^{1/2} \frac{2}{3\Gamma+1} \left[r^{\frac{3\Gamma+1}{2}} - R_0^{\frac{3\Gamma+1}{2}} \right] = \lambda^2 r^{2\Gamma} - \lambda_0^2 R_0^{2\Gamma},$$

where we substituted $x = \lambda^2$ in the second line and used λ_0 for the lower limit on the right-hand side, since $\lambda(r = R_0) = \lambda_0$. Rewriting this equation leads to

$$\lambda^2 r^{2\Gamma} = \lambda_0^2 R_0^{2\Gamma} + R_0^{\frac{1+\Gamma}{2}} \frac{\lambda_0}{U} E_{b0}^{1/2} \frac{2}{3\Gamma+1} \left[r^{\frac{3\Gamma+1}{2}} - R_0^{\frac{3\Gamma+1}{2}} \right]. \tag{5.28}$$

By using the definition

$$C = \frac{U}{E_b^{1/2}} \frac{\lambda_0}{R_0},$$

Eq. (5.28) can be solved after some straightforward algebra to

$$\frac{\lambda}{\lambda_0} = \left(\frac{R_0}{r}\right)^{\Gamma} \sqrt{1 + \frac{1}{C} \frac{2}{3\Gamma+1} \left[\left(\frac{r}{R_0}\right)^{\frac{3\Gamma+1}{2}} - 1 \right]} \tag{5.29}$$

$$\frac{E_b}{E_{b0}} = \left(\frac{R_0}{r}\right)^{(1-\Gamma)} \left[1 + \frac{1}{C} \frac{2}{3\Gamma+1} \left[\left(\frac{r}{R_0}\right)^{\frac{3\Gamma+1}{2}} - 1 \right] \right]^{-1}, \tag{5.30}$$

where we used the first expression (5.29) to substitute λ/λ_0 in Eq. (5.26) to derive the expression for E_b/E_{b0}. These two equations are the general solutions to the set of differential equations, (5.23a) and (5.23b), see also Zank et al. [6], JGR 101, A8.

Limits In the asymptotic limit ($r \gg R_0$) we obtain

$$\frac{\lambda}{\lambda_0} \sim \left(\frac{r}{R_0}\right)^{1/4} \qquad \text{for} \quad \Gamma = 0, \tag{5.31}$$

and

$$\frac{\lambda}{\lambda_0} \sim \textit{constant} \qquad \text{for} \quad \Gamma = 1. \tag{5.32}$$

Recall from Problem 5.3 that

$$\frac{b^2}{b_0^2} = \frac{E_b \rho_0}{E_{b0} \rho_{00}} = \frac{E_b}{E_{b0}} \left(\frac{R_0}{r} \right)^2 = \left(\frac{\lambda_0}{\lambda} \right)^2 \left(\frac{R_0}{r} \right)^{(3+\Gamma)},$$

where we substituted E_b/E_{b0} by Eq. (5.26) in the last step and where we used the relation $\rho_0 = \rho_{00}(R_0/r)^2$. Substituting λ/λ_0 by relation (5.31) and (5.32) we find

$$\frac{b^2}{b_0^2} = \left(\frac{R_0}{r} \right)^{3.5} \qquad \text{for} \quad \Gamma = 0,$$

and

$$\frac{b^2}{b_0^2} = \left(\frac{R_0}{r} \right)^{4} \qquad \text{for} \quad \Gamma = 1.$$

Problem 5.6 Determine the general solution to the stream-driven steady-state spherically symmetric form of the transport equations,

$$U \frac{\partial E_b}{\partial r} + \frac{U}{r} E_b - \Gamma \frac{U}{r} E_b = -\frac{E_b^{3/2}}{\lambda} + C_{sh} \frac{U}{r} E_b \tag{5.33}$$

$$U \frac{\partial \lambda}{\partial r} + \Gamma \frac{U}{r} \lambda = \frac{E_b^{1/2}}{2} - C_{sh} \frac{U}{2r} \lambda \tag{5.34}$$

analytically, if $E_b(r = R_0) = E_{b0}$ and $\lambda(r = R_0) = \lambda_0$. Hence, find asymptotic solutions for weak and strong mixing.

Solution Similar to the preceding problem we begin with dividing Eq. (5.33) by E_b and multiplying Eq. (5.34) by 2. This yields

$$\frac{U}{E_b} \frac{\partial E_b}{\partial r} + \frac{U}{r} - \Gamma \frac{U}{r} = -\frac{E_b^{1/2}}{\lambda} + C_{sh} \frac{U}{r} \tag{5.35a}$$

$$E_b^{1/2} = 2U \frac{\partial \lambda}{\partial r} + 2\Gamma \frac{U}{r} \lambda + C_{sh} \frac{U}{r} \lambda. \tag{5.35b}$$

Substituting the term proportional to $E_b^{1/2}$ in the first equation by the second, we obtain after some simplification

$$\frac{1}{E_b} \frac{\partial E_b}{\partial r} + (1 + \Gamma) \frac{1}{r} = -\frac{2}{\lambda} \frac{\partial \lambda}{\partial r}.$$

Note that this result is identical to Eq. (5.25). The solution is given by Eq. (5.27),

$$E_b^{1/2} = \left(\frac{R_0}{r}\right)^{\frac{1+\Gamma}{2}} \frac{\lambda_0}{\lambda} E_{b0}^{1/2}.$$

This result is substituted back into Eq. (5.35b),

$$\left(\frac{R_0}{r}\right)^{\frac{1+\Gamma}{2}} \frac{\lambda_0}{\lambda} E_{b0}^{1/2} = 2U\frac{\partial \lambda}{\partial r} + (2\Gamma + C_{sh})\frac{U}{r}\lambda. \tag{5.36}$$

Now, we multiply this equation by λ and divide by U, leading to

$$\left(\frac{R_0}{r}\right)^{\frac{1+\Gamma}{2}} \frac{\lambda_0}{U} E_{b0}^{1/2} = 2\lambda\frac{\partial \lambda}{\partial r} + \frac{(2\Gamma + C_{sh})}{r}\lambda^2. \tag{5.37}$$

Note that $2\lambda\partial\lambda/\partial r = \partial\lambda^2/\partial r$, and by using the transformation $x = \lambda^2$ we can rewrite this equation as

$$\left(\frac{R_0}{r}\right)^{\frac{1+\Gamma}{2}} \frac{\lambda_0}{U} E_{b0}^{1/2} = \frac{\partial x}{\partial r} + \frac{(2\Gamma + C_{sh})}{r}x.$$

Multiplying both sides with $r^{2\Gamma + C_{sh}}$ (integrating factor) yields

$$\begin{aligned} r^{2\Gamma + C_{sh}}\left(\frac{R_0}{r}\right)^{\frac{1+\Gamma}{2}} \frac{\lambda_0}{U} E_{b0}^{1/2} &= \frac{\partial x}{\partial r} r^{2\Gamma + C_{sh}} + (2\Gamma + C_{sh})\, r^{2\Gamma + C_{sh} - 1} x \\ &\equiv \frac{\partial}{\partial r}\left(x r^{2\Gamma + C_{sh}}\right). \end{aligned}$$

Integrating both sides with respect to r and using the above initial conditions yields

$$R_0^{\frac{1+\Gamma}{2}} \frac{\lambda_0}{U} E_{b0}^{1/2} \int_{R_0}^{r} r'^{\frac{3\Gamma-1}{2} + C_{sh}} dr' = \int_{R_0}^{r} \frac{\partial}{\partial r'}\left(x r'^{2\Gamma + C_{sh}}\right),$$

which can be solved by

$$\begin{aligned} R_0^{\frac{1+\Gamma}{2}} \frac{\lambda_0}{U} E_{b0}^{1/2} \frac{2}{3\Gamma + 1 + 2C_{sh}} \left[r^{\frac{3\Gamma+1}{2} + C_{sh}} - R_0^{\frac{3\Gamma+1}{2} + C_{sh}} \right] \\ = \lambda^2 r^{2\Gamma + C_{sh}} - \lambda_0^2 R_0^{2\Gamma + C_{sh}}, \end{aligned}$$

where we substituted $x = \lambda^2$ and used λ_0 for the lower limit on the right-hand side. Rewriting this equation gives

$$\lambda^2 r^{2\Gamma+C_{sh}}$$

$$= \lambda_0^2 R_0^{2\Gamma+C_{sh}} + R_0^{\frac{1+\Gamma}{2}} \frac{\lambda_0}{U} E_{b0}^{1/2} \frac{2}{3\Gamma+1+2C_{sh}} \left[r^{\frac{3\Gamma+1}{2}+C_{sh}} - R_0^{\frac{3\Gamma+1}{2}+C_{sh}} \right],$$

This equation can be solved after some lengthy but straightforward algebra to

$$\frac{\lambda}{\lambda_0} = \left(\frac{R_0}{r}\right)^{\Gamma+\frac{C_{sh}}{2}} \sqrt{1 + \frac{1}{C} \frac{2}{3\Gamma+1+2C_{sh}} \left[\left(\frac{r}{R_0}\right)^{\frac{3\Gamma+1}{2}+C_{sh}} - 1 \right]}$$

$$\frac{E_b}{E_{b0}} = \left(\frac{R_0}{r}\right)^{(1-\Gamma-C_{sh})} \left[1 + \frac{1}{C} \frac{2}{3\Gamma+1+2C_{sh}} \left[\left(\frac{r}{R_0}\right)^{\frac{3\Gamma+1}{2}+C_{sh}} - 1 \right] \right]^{-1},$$

where we used the constant C from the preceding problem. In the asymptotic limit ($r \ll R_0$) we obtain the same results as for the preceding problem. Compare also with Zank et al. [6], JGR 101, A8.

References

1. Gombosi, T.I.: Physics of the Space Environment. Cambridge Atmospheric and Space Science Series. Cambridge University Press, Cambridge (1998)
2. Gradshteyn, I.S., Ryzhik, I.M.: Table of Integrals, Series, and Products, 7th edn. Academic Press, New York (2000)
3. Melrose, D.B., Pope, M.H.: Proc. Astron. Soc. Aust. **10**, 222 (1993)
4. Schlickeiser, R.: Cosmic Ray Astrophysics. Springer, Berlin (2002)
5. Zank, G.P.: Transport Processes in Space Physics and Astrophysics. Lecture Notes in Physics, vol. 877, 1st edn. Springer, New York (2014)
6. Zank, G.P., Matthaeus, W.H., Smith, C.W.: J. Geophys. Res. **101**, 17093 (1996). doi:10.1029/96JA01275

A. Dosch, G.P. Zank, *Transport Processes in Space Physics and Astrophysics*,
Lecture Notes in Physics 918, DOI 10.1007/978-3-319-24880-6

Index

A. Dosch, G.P. Zank, *Transport Processes in Space Physics and Astrophysics*, Lecture Notes in Physics 918, DOI 10.1007/978-3-319-24880-6

Printed in the United States
By Bookmasters